Demolition Means Progress

HISTORICAL STUDIES OF URBAN AMERICA

Edited by Timothy J. Gilfoyle, James R. Grossman, and Becky M. Nicolaides

Demolition Means Progress

Flint, Michigan, and the Fate of the American Metropolis

ANDREW R. HIGHSMITH

The University of Chicago Press
Chicago and London

The University of Chicago Press, Chicago 60637
The University of Chicago Press, Ltd., London
© 2015 by The University of Chicago
All rights reserved. Published 2015.
Paperback edition 2016
Printed in the United States of America

25 24 23 22 21 20 19 18 17 16   3 4 5 6 7

ISBN-13: 978-0-226-05005-8 (cloth)
ISBN-13: 978-0-226-41955-8 (paper)
ISBN-13: 978-0-226-25108-0 (e-book)
DOI: 10.7208/chicago/9780226251080.001.0001

Library of Congress Cataloging-in-Publication Data

Highsmith, Andrew R., author.
    Demolition means progress : Flint, Michigan, and the fate of the American
metropolis / Andrew R. Highsmith.
        pages cm — (Historical studies of urban America)
    Includes bibliographical references and index.
    ISBN 978-0-226-05005-8 (cloth : alkaline paper) — ISBN 978-0-226-25108-0
(e-book) 1. Flint (Mich.)—History. 2. Flint (Mich.)—Economic conditions—
History. 3. Flint (Mich.)—Social conditions—History. 4. City planning—
Social aspects—Michigan—Flint. I. Title. II. Series: Historical studies of urban
America.
    F574.F62H54 2015
    977.4'37—dc23

                                                        2014045147

♾ This paper meets the requirements of ANSI/NISO z39.48-1992 (Permanence of
Paper).

For Bobby, Asha, Mira, and Aneel

Contents

Illustrations

Figures

Maps

Tables

Abbreviations

ABC: American Bowling Congress
BTU: Better Tomorrow for the Urban Child
CIO: Congress of Industrial Organizations
CRCF: Civic Research Council of Flint
CRO: Central Relocation Office
CSOH: Committee to Save Our Homes
DCD: Department of Community Development
FACI: Flint Area Conference, Inc.
FCIC: Flint Citizens Information Committee
FCSG: Flint Citizens' Study Group
FEPC: Fair Employment Practices Commission
FFEPC: Flint Fair Employment Practices Commission
FHA: Federal Housing Administration
FHC: Flint Housing Commission
GCDC: Genesee Community Development Conference
GCMPC: Genesee County Metropolitan Planning Commission
GM: General Motors Corporation
GMI: General Motors Institute
GNRP: General Neighborhood Renewal Plan
HEW: Department of Health, Education, and Welfare
HOLC: Home Owners' Loan Corporation
HOME: Housing Opportunities Made Equal
HUAC: House Committee on Un-American Activities
HUD: Department of Housing and Urban Development
IMA: Industrial Mutual Association
MAF: Manufacturers Association of Flint
MCRC: Michigan Civil Rights Commission
MSHD: Michigan State Highway Department

NAACP: National Association for the Advancement of Colored People
NAG: National Action Group
NAREB: National Association of Real Estate Boards
NFRC: New Flint Resistance Committee
OPEC: Organization of Petroleum Exporting Countries
OPM: Office of Price Management
PCP: Personalized Curriculum Program
SJCIA: St. John Citizens for Improvement Association
SOS: Save Our Schools
SSRP: Social Science Research Project
UAW: United Automobile Workers
ULF: Urban League of Flint
ULI: Urban Land Institute
UM-FLINT: University of Michigan–Flint

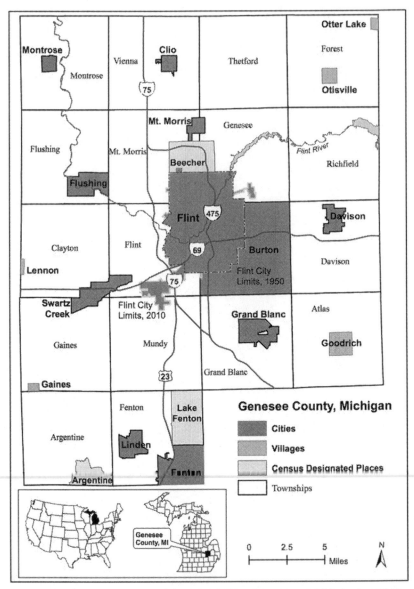

Genesee County, Michigan

- **Cities**
- **Villages**
- Census Designated Places
- Townships

MAP I.1. Genesee County, Michigan. *Source*: Michigan Center for Geographic Information, *Michigan Geographic Framework: Genesee County* (Lansing: Michigan Center for Geographic Information, 2009).

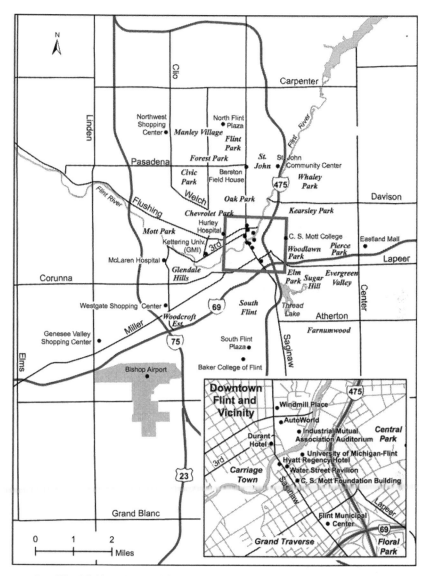

MAP I.2. Flint, Michigan. *Source:* Michigan Center for Geographic Information, *Michigan Geographic Framework: Genesee County* (Lansing: Michigan Center for Geographic Information, 2009).

Introduction

The city of Flint salutes you
To General Motors, we raise our voice in song
The city of Flint salutes you
To General Motors, the hand of friendship strong

The city of Flint salutes you
For the goal you've reached today
We are proud to be your neighbor
And we're glad that you came to stay
Lyrics from a song honoring the fiftieth anniversary
of the incorporation of General Motors (1958)

Demolition Means Progress
Phrase from a sign that once hung in front of a vacant
General Motors plant in Flint (1999–2008)

On November 23, 1954, an epic celebration erupted in Flint, Michigan, a mid-sized industrial city located seventy miles northwest of Detroit. At precisely 10:10 on that fall morning, workers cheered, factory whistles screeched, and "aerial bombs" exploded in the sky as a sparkling new sports coupe rolled off the assembly line at a General Motors Corporation (GM) plant just south of the city. The gold-plated, gold-trimmed Chevrolet Bel Aire was GM's fifty-millionth car produced in the United States, and hundreds of journalists, executives, and civic leaders had gathered at the company's Van Slyke Road assembly plant to bear witness to the historic accomplishment. As workers put the final touches on the coupe, plant supervisors shut down the assembly line for a speech by GM President Harlow Curtice. In an address broadcast across the country, Curtice identified the golden vehicle as a symbol of America's industrial supremacy: "Fifty million cars are more cars than any other country or any combination of other countries has ever produced. They represent a production feat that surpasses anything ever achieved by any other industrial organization."[1]

Like many of the revelers who had gathered to observe the event, Curtice viewed November 23 as a day to rejoice in GM's progress, the triumphs of American capitalism and democracy, and the seemingly boundless prosperity generated by the post–World War II economic boom.[2] The so-called Ve-

hicle City of Flint and its nearly two hundred thousand residents were ready
for a party.

Flint was a fitting place to celebrate GM's milestone. In 1908 local indus-
trialist William "Billy" Durant had founded the company. By 1954 metropolitan Flint—with approximately eighty thousand workers on the corporation's
payroll—was home to the largest agglomeration of GM factories and employees anywhere in the world.[3] To mark the occasion, over 150,000 people flocked
to Saginaw Street in downtown Flint for a "Golden Carnival" celebration that
included a parade, music, and speeches.[4] After Curtice and other executives
took their seats in front of the luxurious Durant Hotel, a "three-bomb salute"
announced the beginning of the parade. The showpiece of the parade was,
of course, the new Bel Aire, which sat atop the final float, but the hour-long
event featured a variety of entertaining performances. Among the main attractions were nine marching bands; a preview of GM's 1955 line of cars; a
display sponsored by the United Automobile Workers (UAW), Flint's largest
labor union; and a float dedicated to "Mr. and Mrs. USA," which carried a
young white family representing GM's customers. Heralded by journalists for
its extraordinary grandeur, the Golden Carnival was one of the most impressive celebrations of mass production and consumption in American history.[5]

The corporate and municipal officials who planned the Golden Carnival
hoped to showcase the bonds that connected Flint and General Motors, autoworkers and their employers, and loyal consumers with their trusted GM
products. Company managers sought to strengthen those ties by hosting
open houses in all of their plants nationwide on November 23. Following the
parade downtown, celebrants dispersed to visit the twelve major GM facilities scattered throughout Flint and suburban Genesee County. For Curtice
and other dignitaries, the celebration continued at the Industrial Mutual Association (IMA) Auditorium, where GM hosted a Golden Carnival luncheon.
There, Curtice, a longtime Flint resident, announced a three-million-dollar
donation from GM to fund the city's new cultural center.[6] Reflecting upon
GM's generosity and the momentousness of the occasion, F. A. Bower, a retired engineer from the company's Buick division, predicted, "With Mr. Curtice at the head, both the corporation and the city of Flint are assured of great
prosperity." Not to be outdone, Mayor George Algoe bragged that Flint was
the "envy of all other industrial cities."[7] With its high wages, low unemployment, internationally renowned public schools, and impressive rates of home
ownership, Flint was truly an amazing sight on that special November day.

The Golden Carnival, however, was a festival of both truth and fiction.
Although the splendor of the day made it difficult to detect, the 1954 extrava-
ganza concealed just as much as it revealed about the texture of life in Flint,

FIGURE I.1. General Motors' Golden Carnival, 1954. On November 23, 1954, Flint-area workers assembled GM's fifty-millionth American-made automobile. Corporate executives honored the occasion by organizing a series of celebrations, including this large parade in downtown Flint. Dubbed the Golden Carnival, the festivities symbolized GM's dominance within the automobile industry and Flint's emergence as a preeminent manufacturing center. Courtesy of the Richard P. Scharchburg Archives, Kettering University, Flint, MI.

the relationship between the city and GM, and the corporation's commitment to prosperity and opportunity in its hometown. In their stories about the tribute, reporters from the *Flint Journal*, the city's major daily newspaper, saluted the community's progress and GM's corporate triumphs. On that same day, however, the paper ran Jim Crow advertisements from local citizens seeking "Colored" housekeepers and white homebuyers. Furthermore, when

Harlow Curtice and his colleagues made their trek from the Van Slyke plant to the parade in downtown Flint, they traveled along Saginaw Street, the Vehicle City's most persistent racial fault line. As the float carrying the Bel Aire rolled north toward downtown, it passed by several all-white neighborhoods just west of Saginaw where African Americans were unwelcome. On the east side of the street sat Floral Park, one of only a few areas in the city where black people could obtain housing. Upon arriving downtown, Curtice and other executives viewed the parade from in front of the Durant Hotel, a whites-only establishment. Then, after the parade, Curtice attended a segregated luncheon, where he announced GM's gift to the city. The donation funded a cultural center near the all-white Woodlawn Park district, a neighborhood that black people could visit, but only during the daytime. Beneath the shine of the Golden Carnival, Flint was a profoundly segregated and unequal city, and the municipal and corporate officials who presided over the day's amusements were largely to blame.

Flint was also a city teetering on the brink of economic disaster. During the decade preceding the Golden Carnival, tens of thousands of white taxpayers had moved away from the city in search of newer and better housing, schools, and jobs in the segregated suburbs of Genesee County. These suburban migrants were part of one of the largest and most significant population shifts in American history.[8] Over the same period, GM and numerous other American employers implemented new investment strategies that redirected jobs, taxes, and capital from cities to suburbs. Significantly, although the organizers of the Golden Carnival described it as a celebration of Flint's place in the automobile industry, the state-of-the-art factory that birthed GM's fifty-millionth car was located not in the Vehicle City but in the nearby suburb of Flint Township. Because Flint's leaders consistently failed in their attempts to establish a tax- and resource-sharing metropolitan government, the suburban migrations of the postwar era ultimately caused economic devastation in the increasingly poor and black central city.[9]

In the end, F. A. Bower's predictions of stability and prosperity for the Vehicle City turned out to be false. During the economic crises of the 1970s, American businesses saw their sales decline precipitously. With its lineup consisting almost entirely of large, energy-inefficient cars and trucks, GM suffered devastating losses during the oil shocks of 1973–74 and 1979. American employers, GM included, responded to the slowdowns by launching an austerity campaign that resulted in millions of job losses. Between 1973 and 1987, the company's directors eliminated twenty-six thousand local jobs, pushing Flint's unemployment rate into the double digits.[10] The economic downturns of the 1970s and 1980s marked the end of the postwar boom and the onset of a new era of mass deindustrialization.[11]

The lean years, however, had only just begun. The high-tech, subprime real estate, and Wall Street booms of the 1990s and early 2000s did little to stem the loss of jobs from Flint and other manufacturing centers. During the 1990s GM's share of the domestic automobile market fell from 35 to 28 percent. Following the 1999 closure of Buick City, Flint's oldest and largest GM facility, the automaker's workforce in Genesee County dipped to just fifteen thousand. The outward flow of taxpayers and jobs triggered multimillion-dollar deficits for the city that ultimately led to two separate state takeovers of the municipal government. By 2009, a year that witnessed GM's once unimaginable descent into bankruptcy, the company employed fewer than eight thousand local workers, and the city's unemployment rate stood at 27.3 percent. Although a controversial federal loan helped return GM to profitability shortly after the bankruptcy filing, few of the jobs that left Flint during the late twentieth and early twenty-first centuries ever returned. A decade and a half into the new millennium, Flint—a city with more unemployed people than autoworkers—was the Vehicle City only in name.[12]

Demolition Means Progress is a book about the interlocking histories of racial and economic inequality, mass suburbanization, and deindustrialization in modern America. Although it is set in the Flint region between the 1930s and the present, it is in many respects a story about metropolitan America as a whole during those years. The focus throughout is on the political, economic, social, legal, and policy forces that transformed Flint from a segregated company town into a "hypersegregated" postindustrial metropolis. At first glance, Flint's story appears to confirm much of the conventional wisdom about the decline of America's cities—namely, that the nation's "urban crisis" began in the decades following World War II due to a combination of corporate abandonment, political neglect, and white racism.[13] For much of the twentieth century, observers from around the world looked to the partnership between Flint, the UAW, and GM as a microcosm of the American Dream. Buoyed by the postwar expansion of the automobile industry and the unprecedented collective bargaining agreements won by local trade unionists, the Flint of the 1940s and early 1950s was, for many people, a beacon of economic security, consumer abundance, and social opportunity. Today, by contrast, Flint is internationally renowned for its shuttered automobile factories, high crime and unemployment rates, racial segregation, and shrinking population. There can be little doubt that dramatic reversals, particularly in the economic sphere, have bedeviled the people of Flint, Detroit, and other urban centers over the past half-century.[14]

In truth, however, the driving forces in Flint's past—and, more broadly, in the history of metropolitan America as a whole—have always been renewal and reinvention more than decline and abandonment. Although it is undeniable that millions of urban dwellers experienced difficult structural challenges in the second half of the twentieth century, the residents of Flint and other cities, leaders and ordinary citizens alike, have proved to be unrelenting in their quests for revitalization, even in the face of enormous obstacles. Their attempts at reinvention and the often pernicious results of their efforts are central themes of this book.[15]

The account that follows borrows both its title and its overall conceptual orientation from the signs that once hung in front of GM's shuttered industrial facilities in the Flint area. In 1997 company executives announced plans to close a large complex of factories on Flint's predominantly black and poor north side. Shortly after the plants closed, workers placed placards in front of the facility that read "Demolition Means Progress" (see figure 10.4). The signs suggested that the Vehicle City and its people could not move forward to civic greatness until the old plants met the wrecking ball.

Much more than simply a corporate slogan, this suggestive phrase encapsulates the operating ethos of the nation's metropolitan leadership from at least the Great Depression of the 1930s through the early twenty-first century. Throughout that long period, the Flint area's leaders—like their counterparts in New York City, Los Angeles, Camden, New Jersey, and other metropolitan centers, large and small—repeatedly tried to revitalize their community by demolishing outdated and inefficient structures and institutions and building them anew.[16] As was the case elsewhere, the people of the Flint area had a variety of different visions and plans for revitalization, and they often disagreed with one another over how best to improve the metropolis. All of them, however, believed that the Vehicle City's best days were ahead. During the Depression, education officials hoped to renew Flint by remaking public schools into racially segregated community centers. In the 1940s and 1950s, federal housing administrators and local builders and lenders sought to renew the area's ramshackle suburbs by creating modern segregated subdivisions outside the city limits. Meanwhile, GM's directors worked to revolutionize automobile production by demolishing old urban factories and rebuilding them in nearby suburbs. When those efforts failed to create a renaissance, city leaders launched a plan to replace black neighborhoods with a freeway, modern factories, and new downtown attractions. In the end, though, each of these renewal efforts yielded a more impoverished city and a more divided metropolis. By the dawn of the twenty-first century, this long chain of revitalization attempts had helped to make metropolitan Flint one of the most racially segregated, economically polarized, and politically fragmented regions in the nation.

In Flint and elsewhere, a complex web of private actions and government policies connected racially segregated schools and neighborhoods with Jim Crow workplaces and development practices. Between the 1940s and the 1960s, just as millions of black southerners were setting out for Detroit, Chicago, and other large cities in what became known as the Second Great Migration, racial segregation increased markedly in Genesee County and many other metropolitan areas. Ironically, this occurred during an era in which northern cities such as Flint enjoyed strong reputations for racial progressivism.[17] Although well-paying industrial jobs were often plentiful in the Vehicle City, Flint and other northern metropolises were not promised lands. Upon arriving in Flint, black migrants encountered rampant segregation and discrimination. Neighborhood violence, "racial steering" by real estate professionals, job segregation, and other racist acts committed in the ostensibly private, nongovernmental sphere played important roles in maintaining Jim Crow in Flint and its suburbs.[18] Yet deliberate government policies also sustained segregation in the region. Working together, public officials and powerful corporate oligarchs from GM—who often exerted decisive control over the local policy-making process—maintained the color line by imposing housing, urban development, employment, and public education programs that were at times openly segregationist. More often, however, segregation arose from an amalgam of growth-minded renewal and development policies that were seemingly colorblind in nature.[19]

The persistent racial divide at the heart of Flint's story—a product of both popular forces and deliberate government actions—exposes the mythology of "de facto segregation." Although Americans first began using the phrase *de facto segregation* during the early twentieth-century debates over the rise of Jim Crow in the North, the term did not find a place in the national vernacular until the 1950s.[20] In the aftermath of the US Supreme Court's 1954 *Brown v. Board of Education* school desegregation decision, millions of people from a variety of ideological persuasions began using the terms *de jure* and *de facto* to denote supposedly fundamental distinctions between government-mandated segregation in the South and the private, extragovernmental, constitutionally permissible forms of "discrimination" that reputedly prevailed in the rest of the nation.[21] As scholars such as Matthew Lassiter have shown, however, the de jure–de facto distinction served only to mask the government interventions that maintained the color line throughout the United States. On the ground, the framework of de facto segregation allowed zoning commission-ers, school board members, and other government agents in places as diverse as Flint, Dallas, and Washington, DC, to deny responsibility for the public policies that created ghettos. More broadly, though, the language of de facto segregation was the key pillar in a powerful discourse on northern, western, and, ultimately, suburban racial innocence that allowed both ordinary whites and elected officials, regardless of region, to deny all legal and moral respon-sibility for Jim Crow.[22]

The fog of de facto segregation has also made it much harder to grasp the full scope of segregationist practice in the modern United States. This dif-ficulty stems in part from the overemphasis on housing within the de facto narrative. Specifically, the concept of de facto segregation turns on the notion that residential real estate markets are the foundations for the racial divide—that is, the roots from which most other forms of Jim Crow, especially in the educational arena, grow. Although most scholars of metropolitan his-tory now reject many aspects of the de facto framework, its emphasis on the overarching, constitutive power of housing seems to have lingered within the literature on racial segregation. One of the principal manifestations of this housing-centered scholarship has been the marginalization of schools and education policies within the field of metropolitan history.[23]

A growing body of literature suggests, however, that housing and edu-cation policies have historically operated in mutually constitutive fashion, working in tandem as part of larger, more complex networks of Jim Crow. Such was the case in a host of metropolitan areas during the twentieth cen-tury. In Nashville, Tennessee; Hartford, Connecticut; and Raleigh, North Carolina, for instance, education authorities, at times working alongside housing officials, played key roles both in maintaining pupil segregation and in defining the boundaries of segregated residential districts.[24] This was also true of Flint, where segregated school-neighborhood units were the focal points for the city's broader public sphere. By incorporating schools into the broader history of metropolitan racial inequality, this study aims to provide a more comprehensive assessment of the roots and structure of the color line.

As part of that effort, this book offers a new three-part typology to replace the outdated language of de facto and de jure segregation. Regardless of the era or region, segregation in the modern United States has virtually always proceeded from some combination of statutory and legal requirements, the discriminatory administration or implementation of public policies and pro-grams, and popular forces. Therefore, the pages that follow introduce the terms *legal segregation, administrative segregation,* and *popular segregation* to denote the shifting structures of the color line. Hereinafter, the term *legal segregation* signifies the types of racial apartheid arising from statutory, constitutional, or judicial mandates. State laws requiring Jim Crow schools and judicial rulings enforcing racially restrictive housing covenants would fall under this heading. The phrase *administrative segregation*, by contrast, refers to the forms of racial separation resulting from the administration of government policies and programs or the exercise of the state's bureaucratic powers. In contrast to legal segregation, wherein the law itself or its constitutionally designated interpreters specifically mandate racial partitioning, this second mode of separation largely results from the implementation of public policies that do not expressly require Jim Crow. Federal mortgage insurance policies favoring segregation, school district gerrymandering by race, and racial separation achieved through public housing site locations would all be examples of this variety of Jim Crow.

Historically, the innumerable government actions included within the categories of legal and administrative segregation have been indispensable to upholding the color line.[25] In reality, though, Jim Crow has always been much more than a top-down system of laws and policies. As the Flint case confirms, private citizens—acting either in adherence to established laws, judicial rulings, and administrative policies or, as has often been the case since the civil rights legislative revolution of the 1960s, in violation of them—have also been important agents in the maintenance of the color line. The final category, *popular segregation*, thus encompasses patterns of racial partition driven to at least some significant degree by nonstatist forces and actors, though often with the endorsement of statutes, judicial decrees, or administrative procedures. Because government officials and state actions often play a role in creating or reinforcing patterns of popular segregation, the term is in no way synonymous with the concept of de facto segregation. On the contrary, whereas the de facto term refers narrowly (and problematically) only to forms of racial separation that are purely extragovernmental, the more capacious category of popular segregation includes many types of Jim Crow rooted in government policies, so long as they are also driven to some significant extent by nonstatist forces. Notwithstanding the state's role in enabling certain forms of popular segregation, this type of racial separation usually manifests itself most visibly at the level of the individual, household, neighborhood, or private organization. Acts of popular segregation might include realtors who violate fair housing laws by hiding property listings from nonwhite buyers, white parents who prohibit their children from associating with people of color, business owners who refuse to serve black customers, or racists who harass a Latino person visiting a white neighborhood.

Significantly, these three forms of segregation are rarely mutually exclusive. Indeed, hybrid forms of segregation are quite common, and overlaps between the three categories are often pronounced—particularly in the case of popular segregation, which in many instances derives from a complex combination of privately held animosities, legally protected rights to discriminate, and market-based forms of Jim Crow rooted in public policies.[26] Stated another way, these three categories tend to be intersectional and permeable rather than unique and pure. As such, they constitute a nuanced, flexible, and accurate framework for understanding the very real and meaningful, if often subtle, variations among different modes of Jim Crow. In addition, because each of these three types of segregation has existed throughout the United States at one point or another, this typology, unlike the de facto–de jure binary, makes no presumptions about region or temporality.

Without ignoring popular and legal forms of Jim Crow, the following account focuses carefully on the key role that administrative segregation has played in shaping the modern American metropolis. Of all the forms of administrative segregation that emerged during the twentieth century, few were as consequential as redlining, the practice of denying or curtailing mortgage insurance, loans, and other goods and services based upon geographic, socioeconomic, and often racial considerations. Over the past generation, scholars such as Kenneth Jackson and David Freund have persuasively documented the causes and legacies of this phenomenon. During the 1930s representatives of a federal New Deal agency called the Home Owners' Loan Corporation (HOLC), in the course of making so-called residential security maps for the Flint area and 238 other metropolitan regions, helped to enshrine and codify the already popular notion that racial integration was a threat to property values and thus a significant mortgage risk factor. In so doing, HOLC officers played an important role in shaping the practice of redlining. Across the United States, local lenders and officials from the Federal Housing Administration (FHA) adopted the HOLC's racist standards for measuring mortgage risk and systematically redlined neighborhoods occupied by African Americans and other people of color. In metropolitan regions across the nation, including the Flint area, redlining helped to widen economic gaps between cities and suburbs while hardening the color lines surrounding minority neighborhoods.[27]

However, redlining was not simply a racial policy. It also had a powerful class dimension. This lesser-known facet of redlining was especially visible outside of Flint's city limits for a time between the mid-1930s and the early 1950s, when federal underwriters used mortgage disinvestment—or, as was often the case, the threat of it—to stimulate the modernization of poor and working-class suburbs, including those that were all white.[28] During this period, FHA officials, often with the support of local builders and lenders, routinely redlined white working-class areas outside the city limits while favoring all-white middle- and upper-class neighborhoods in Flint and a small number of well-equipped suburban municipalities. By doing so, federal administrators hoped to motivate elected officeholders in Flint's hardscrabble suburbs to develop new building and zoning codes and improve utilities and services—all with an eye toward boosting real estate values and making these communities eligible for government-backed mortgage insurance. The Vehicle City was not unique, though, as redlining affected similarly situated suburbs in Chicago, Detroit, Denver, and other metropolitan areas during this time.[29]

By withholding investment in such a fashion, federal and local actors engaged in what might be called *suburban redlining*.[30] Largely confined to the period from the mid-1930s to the early 1950s, the practice of suburban redlining emerged out of class considerations that were integral to the federal government's early home-ownership programs. As the Flint case suggests, impoverished all-white suburbs were often as susceptible to federal redlining as were minority neighborhoods in cities. Nevertheless, not all suburbs suffered from redlining during this period. Rather, redlining affected certain types of impoverished and working-class suburbs, namely those with minimal services, small tax bases, poor schools, and low-quality housing. Things began to change between the 1930s and the 1950s, however, when suburban governments implemented new land use plans, built schools and roads, and funded major utility and service improvements. In order to fund such projects, suburban policy makers across the region raised taxes substantially, but they often did so with strong support from white homeowners, many of whom desperately wanted modern urban amenities. By the close of the 1950s, the FHA had acknowledged these renewal and development efforts by largely abandoning its suburban redlining policies, a decision that led to a prolonged building boom in Flint's segregated suburbs.

As large numbers of white homebuyers, business owners, and investors began flocking to the outlying areas of Flint and other cities, a powerful new ethos of *suburban capitalism* began to take shape throughout the United States. At base, this philosophy revolved around the pillars of suburban-centered economic development, racial segregation, and metropolitan political fragmentation. Unlike many of their forebears from the Gilded Age and Progressive era, who believed that cities were the nation's economic engines, proponents of suburban capitalism looked to segregated suburbs as the surest path to growth and prosperity. After making their initial investments—

usually in new homes, businesses, or community improvement projects—
suburban capitalists looked to insulate their economic interests from urban
officials by effectively seceding from big cities.[31] In practice, then, suburban
capitalism entailed the shifting of economic resources from cities to segregated suburbs along with an explicitly antiurban political program. Such hostility toward cities and urban dwellers was hardly new, having first surfaced
in the railroad and streetcar suburbs of Boston, New York City, and other
metropolises during the late nineteenth- and early twentieth-century battles
over annexation. Still, it is possible to distinguish between these two strands
of suburban separatism. Whereas the earlier variety of secessionism grew primarily out of local concerns, the suburban capitalism of the postwar era was
a national phenomenon with deep roots in federal New Deal policies.[32]

In Flint and elsewhere, the Federal Housing Administration played an in-
strumental role in the rise of suburban capitalism. Mindful of the sacrifices
they had made to improve living standards and secure FHA-backed investments, suburban capitalists and their elected representatives often supported
the incorporation of their communities and vociferously opposed annexation, metropolitan government, and other urban growth initiatives. Over
time, regional disputes over desegregation, land use, incorporation, and municipal boundaries—all exacerbated by the FHA's policies—left the city of
Flint isolated from its neighbors and without room to expand. Much the same
occurred across the United States, particularly in the Northeast and Midwest,
where the rise of suburban capitalism helped to isolate and devastate urban
economies from Milwaukee to Baltimore. Federal redlining polices thus did
much more than simply subsidize racial segregation. They also structuralized
racial and economic inequalities by motivating white suburbanites to erect
legal and political barriers around themselves and their wealth-generating
investments.[33]

The sheer depth of the Flint area's political fragmentation first became
recognizable during the postwar jurisdictional battles over GM's network
of suburban facilities. During the two decades following the onset of World
War II, GM executives, like their peers in many other economic sectors, devised an aggressive corporate growth strategy that centered on the spatial decentralization of capital. As part of this progrowth agenda, American business leaders implemented sweeping suburban investment campaigns that
shifted immense amounts of resources and millions of jobs away from the
nation's large urban centers.[34] Nowhere was this plan more evident than in
the Vehicle City. Between 1940 and 1960, GM opened eight new industrial
facilities in Genesee County, all of them in the suburbs. By 1960 the twenty-
one thousand men and women who labored in GM's new suburban plants

accounted for over a third of the automaker's overall workforce in the Flint metropolitan area. That percentage would only grow over time.[35]

Not all of GM's suburban factories were examples of corporate abandonment, capital flight, or "spatial mismatch," however.[36] Neither, for that matter, were many of the other suburban enterprises that sprouted in postwar America. Rather, the investment decisions made by local GM executives and other likeminded business leaders during this period were part of a larger agenda of *metropolitan capitalism* in which cities such as Flint were indispensable. On the surface, GM's philosophy of metropolitan capitalism seemed to share a great deal in common with the suburban capitalist separatism embraced by many white property owners, lenders, and federal mortgage underwriters. After all, adherents of both systems supported racial segregation and implicitly favored suburbs over cities as the best sites for development. However, proponents of these two progrowth philosophies disagreed about how best to maximize the return on their investments. While suburban capitalists endorsed the fragmentation of local government, proponents of metropolitan capitalism hoped to maintain close ties to cities, municipal policy makers, and urban dwellers.

The contrast between the spatial and political approaches employed at Ford Motor Company in metropolitan Detroit and those utilized at GM in Genesee County helps to illustrate the differences between these two forms of capitalism. At Ford, executives pursued an aggressive suburban capitalist growth strategy that began in the early twentieth century with the shifting of industrial production from Detroit to the nearby white suburbs of Highland Park and Dearborn. After opening the new facilities, Ford's leaders waged successful campaigns to protect them from annexation by Detroit officials. The desire to escape big-city taxes and regulations, along with the quest to associate with smaller and presumably more manipulable local governments, seems to have played a pivotal role in driving Ford's version of suburban capitalism.[37]

Automobile executives charted a somewhat different course in the Flint area, however. There company leaders also wanted to control local government. Furthermore, like Henry Ford and his associates, GM's directors preferred new factory sites in suburbs very close to the municipal border that were easily accessible to most city dwellers, including African Americans. The opening of such facilities thus seldom resulted in meaningful spatial mismatches between urban residents and suburban jobs. Unlike the decision makers at Ford, however, GM's leaders in Flint harbored no animus toward the city or its elected officials. In fact, when GM chiefs ordered new plants built in suburban Genesee County, they did so with the full support of Flint's municipal leadership. Moreover, representatives of GM built facilities in the Genesee County suburbs with the explicit hope that the city would one day acquire them, through either annexation or the creation of a regional government. Accordingly, GM officials actively supported the annexation of all of their outlying factories in Genesee County as well as a 1958 plan known as "New Flint," which promised to consolidate the Vehicle City and its urbanized suburbs under a more efficient metropolitan "super government."

In the end, a powerful coalition of suburban capitalists, policy makers, and judges defeated New Flint, proving along the way that there were clear limits to GM's local political power. Even in defeat, however, GM's leaders evinced a genuine commitment to remaining in Flint—at least for a time. Their dedication to producing automobiles in postwar Flint suggests the need for an alternative periodization of urban deindustrialization. Recent scholarship on the history of capital migrations in modern America suggests that the deindustrialization of the nation's urban centers began during the postwar era as business owners sought out new investment opportunities in suburbs, rural areas, the "Sunbelt" states of the South and West, and abroad.[38] While this may well have been the case in places such as Detroit and Oakland, it was not true of Flint; Youngstown, Ohio; Gary, Indiana; or other cities where deindustrialization set in later.[39] Although GM and other local companies clearly shifted capital from the city to the suburbs during the postwar decades, Flint's industrial crisis did not originate in that era. On the contrary, it was not until the 1970s and 1980s when company officials, frustrated by their declining ability to control local politics and enticed by new opportunities overseas and in the states of the South and West, gave up on their metropolitan capitalist growth strategy and began disinvesting from the Flint area altogether. The decision to reject New Flint and metropolitan capitalism in favor of a more divisive and fragmentary suburban capitalist order thus had extraordinary implications for the city's fate—implications that, when viewed from the local vantage point, seem almost as significant as the rise of neoliberalism and economic globalization.[40]

The concepts of metropolitan capitalism and suburban capitalism have a much broader and enduring significance, however. With all their connotations of competition, profit seeking, and power, the two terms provide a useful lens for viewing the history of suburbanization in the postwar United States. Conflicts over race, space, capital, and power were almost constant features of the nation's metropolitan political landscape in the postwar decades. Furthermore, the pursuit of economic gain and capitalist expansion— whether by individual homeowners looking to increase their property values, mortgage lenders in search of profits, small business owners looking for new customers, or major corporations seeking to enrich stockholders—played a major role in driving suburban growth. Such structural imperatives were not the only forces responsible for mass suburbanization, of course. Concerns about space and privacy and a whole host of additional factors, many of them social and cultural, also led many Americans to favor suburbs over cities. Still, when taken together, the frameworks of suburban and metropolitan capitalism shed fresh new light on the political economy of suburbanization. Indeed, to the extent to which the quests for economic security and profit helped to power the rush to suburbia, metropolitan and suburban capitalism were two of the dominant philosophies of the twentieth century.

In detailing Flint's experiences with structural racism and metropolitan economic development, *Demolition Means Progress* builds upon a recent body of scholarship that questions the utility of regionally based historical frameworks.[41] Beyond illuminating the myths embedded in the de facto–de jure / North-South segregation binary, Flint's story begs for a reconsideration of the well-known Rust Belt–Sunbelt economic model. Over the past thirty years, historians of American business and politics have made a compelling case that the decades following World War II witnessed the simultaneous economic ascendance of the suburbs and the states of the South and West. Because of these local and regional shifts, northern industrial cities such as Flint, Cincinnati, and Philadelphia saw their economic fortunes decline sharply. Many Americans, in turn, began using the terms "Sunbelt" and "Rust Belt" to contrast the fast-growing states of the nation's southern and western rims with the fading industrial centers of the Midwest and Northeast.[42] In reality, though, neither the Rust Belt nor the Sunbelt has ever been a coherent geographic region. Rather, the Rust Belt and Sunbelt are narratives of regional development that mask important countervailing realities at the local level. While the discourse on the Sunbelt tends to camouflage social inequalities by emphasizing the economic dynamism of the South and West, the story of the Rust Belt often functions as an exclusionary declension narrative—one that revolves much too closely around the experiences of white working-class men.[43] Without question, the twenty-five years following World War II were a golden era of job security and consumer affluence for millions of white male workers in the United States. To many of these individuals, the industrial crises of the 1970s and 1980s seemed to mark the end of an exceptional epoch in American history. Yet for African Americans and others, the postwar decades were an era marked by widespread discrimination and segregation. With its emphasis on the loss of postwar abundance, the story of Rust Belt decline obscures this differential experience.[44]

Equally important, the Rust Belt and Sunbelt narratives conceal signifi-cant economic shifts that have occurred within metropolitan areas. By the 1970s urban economic crises were unfolding not only in northern industrial centers such as Flint and Cleveland but also in Atlanta, Los Angeles, and other ostensibly booming Sunbelt metropolises. Even during the worst of the 1970s and 1980s crises, however, many of the nation's suburbs, including a large number of Flint's, remained economically vibrant. Although pros-perous, overwhelmingly white Genesee County suburbs such as Fenton and Linden were (and are) technically part of the region known as the Rust Belt, they had (and continue to have) very little in common with the rapidly dein-dustrializing cities of Flint and Detroit. By the same token, economically de-pressed neighborhoods within cities such as Houston and Phoenix stand in stark contrast to many of the dynamic, fast-growing suburbs that surround them.[45] Ultimately, then, the patterns of uneven local development that reign in Flint and the nation as a whole point away from regional frameworks and toward the need for more precise metropolitan-oriented geographic models. Likewise, the saga of Flint's transformation confirms that it is wisest to inves-tigate the origins of America's urban-industrial crisis from a metropolitan vantage point.

As historians Robert Self, Matthew Lassiter, and others have demon-strated, the metropolitan lens also helps to clarify the full range of actors and forces undergirding racial and spatial inequality in modern America.[46] "White flight" is unquestionably one such force—and a powerful one at that. On its own, however, the white flight narrative, with its focus squarely on developments occurring inside cities, is insufficient to explain the racial and spatial transformations that took place in the nation's metropolitan centers after World War II. At first blush, the striking demographic shifts that cre-ated black majorities in Flint and other American cities seem to bolster the explanatory power of white flight.[47] Between 1950 and 2010, the number of whites living in Flint dropped precipitously, from 149,100 to 38,328, while the city's African American population increased from 13,906 to 57,939.[48] As was the case in New Orleans, Chicago, and other cities, the majority of whites who left Flint moved to segregated suburbs. In some instances the desire to flee integration—or its specter—was clearly an important factor driving suburbanization.[49] However, there were also numerous other forces at work. Large numbers of whites departed Flint and other urban centers long before African Americans and other people of color appeared in their neighbor-hoods, and many others migrated directly to the suburbs from other parts of the country without ever residing in the city. Additionally, beyond the con-ditions inside cities that drove some whites away, an array of "pull" forces helped to make suburbs appealing in their own right. The allure of federally insured homes, the opening of new commuter-friendly superhighways, suburban job growth , the quest for safe streets and better schools, and, crucially, local public policies favoring suburbanization all existed alongside—and at times reinforced—racial avoidance as factors driving the rush to the suburbs.

Whether in discussions of racial succession or elsewhere, the centrality of the local in metropolitan history is a key theme of this book.[50] Although a glimpse of the past reveals many instances in which local actors, conditions, and policies exerted a decisive influence over the Flint area's trajectory, the importance of such forces became especially apparent during two distinct moments in the Vehicle City's history: the campaigns waged between the 1930s and the 1950s to modernize the region's working-class suburbs and the 1960s and 1970s battles over "slum clearance" and open housing. During the decades surrounding World War II, residents of Flint's underdeveloped suburbs often complained bitterly about poor schools, bumpy roads, inadequate utilities, and other nuisances that detracted from property values and a high standard of living. In their concerted attempts to overcome federal redlining and alleviate such frustrations, suburban officeholders routinely called for tax increases. Far from being angered by such requests, suburbanites in the Flint area often supported local tax hikes, at least for a time during the postwar era. This widespread sentiment in favor of higher taxes stemmed not only from national-level pressures created by the FHA's redlining policies, but also from the severity of conditions in many suburban areas and a number of other local considerations—considerations that led many suburbanites in Genesee County to choose a different course from the tax-revolting path taken by their counterparts in other metropolitan areas.[51]

The significance of the local was also on display during the urban renewal and fair housing struggles of the 1960s and 1970s. As in other cities, Flint's open occupancy movement emerged out of a set of highly localized battles over urban renewal and freeway construction. In 1958 spokespersons for the city announced their intention to clear the overwhelmingly black Floral Park and St. John Street neighborhoods to make way for new freeways and an industrial park. Municipal and corporate leaders promised that the redevelopment of such "blighted" neighborhoods would produce new jobs and economic growth. Rather than opposing the clearance plan, as African Americans in Oakland, New York City, and other larger metropolises did, black activists in Flint, after considering their small numbers and relative lack of political power, opted to pursue an aggressive civil rights agenda from within the prorenewal coalition.[52] These organizers consistently failed to achieve their objectives, however. Between the late 1950s and the late 1970s, GM executives and Flint's political leadership demolished St. John and Floral Park, replacing them with a freeway, an industrial park, segregated public housing complexes, and vacant land. Ultimately those policy interventions destroyed thousands of local jobs while strengthening the region's color lines.

The durability of the racial divide in the Flint area and across the United States underscores the need for an expansive chronological framework for analyzing the history of administrative segregation. In 1968, a watershed year in the civil rights movement, lawmakers in Flint, the state of Michigan, and the nation's capital all passed fair housing legislation. Even after the enactment of such formal protections, however, both government policies and private actions continued to bolster segregation and inequality in Flint and other communities.[53] In the Flint area, a controversial federal housing assistance program for low-income home purchasers known as Section 235 proved to be an important vehicle for deepening such divisions. Theoretically, Section 235 was supposed to open up new residential options for African Americans and the poor in both cities and suburbs. In actuality, though, the Section 235 program hardened the color line and fanned racial tensions, particularly in the inner-ring suburb of Beecher, which received a disproportionate share of low-cost government housing units.[54] Equally important, the Section 235 program unleashed a wave of federalized predatory lending on people of color and the poor that grimly foreshadowed the subprime mortgage crisis of the early twenty-first century.[55] In the cases of Beecher and untold other communities, the government "subsidies" offered through the Section 235 program were really more like debts—financial shackles that exploited poor and working-class families and impeded the struggle for racial and economic justice.[56]

Demolition Means Progress is a two-part book consisting of ten chronologically and thematically arranged chapters focusing on the residential color line, school segregation, employment discrimination, suburban development, urban renewal, and deindustrialization. Part I, "Company Town," explores the structures of racial and economic inequality and metropolitan development in the Flint area from the early 1900s through the 1950s. As GM rose to prominence during the first half of the twentieth century, Flint's rapid growth resulted in a series of housing shortages, infrastructure crises, and public health disasters. The opening chapters analyze local and federal responses to Flint's growth in the areas of housing, education, and employment as well as the economic, geographic, and racial considerations that drove city leaders' policy choices. Chapter 1 addresses home construction efforts during the interwar era, the growth of racially restrictive housing covenants, the Vehicle City's deep-seated culture of Jim Crow, and the birth of federal redlining initiatives. Chapter 2 investigates the origins and early development of local GM executive Charles Stewart Mott's "community education" initiative as well as the board of education policies that helped to maintain school and neighborhood segregation in Flint. In chapter 3, the focus moves to racial and gender discrimination in the workplace and the battles over fair employment that broke out in the 1940s and 1950s.

In chapters 4 and 5, the setting shifts from the city to the suburbs. Chapter 4 explains how local residents and elected officials in the working-class areas of the out-county overcame suburban redlining by raising taxes, urbanizing their governments and services, luring middle-class homeowners, maintaining racial segregation, and becoming politically independent from surrounding communities. Chapter 5 addresses the decentralization of capital in postwar Flint and the ensuing conflicts that occurred between suburban capitalists and those who endorsed GM's philosophy of metropolitan capitalism. Part I closes with the pivotal defeat of the New Flint plan—a loss that shook GM's faith in local politics while imperiling the city's economy.

Part II, "Fractured Metropolis," begins with chapters 6 and 7, which explore the struggles over open housing, urban renewal, and neighborhood transition that broke out in Genesee County during the 1960s and 1970s. After years of relative calm, the fair housing movement first gained momentum in the late 1950s and early 1960s in response to the city's controversial plan for "slum clearance." Although activists eventually succeeded in their quest to win a municipal open occupancy ordinance, racial discrimination persisted. In the end, the city's urban renewal program hardened the color line while hastening the process of deindustrialization. Moreover, the clearance of black neighborhoods contributed to a destructive wave of blockbusting and racial succession within the city and several inner-ring suburbs.

By the early 1970s the battles over neighborhood succession had migrated across Flint's border into the suburb of Beecher. There, as chapter 8 demonstrates, the federal government's Section 235 housing program unleashed a rash of predatory lending and waves of racial succession that helped to "tip" Beecher toward a black majority. However, Beecher's experience proved to be somewhat exceptional, as suburban policy makers in other areas of the county successfully employed exclusionary zoning and other forms of administrative segregation to forestall the "Beecherization" of their communities. The defense of the color line outside the city was part of a wider campaign for suburban independence that accelerated in the 1960s and 1970s. The secession of Flint's suburbs—which marked the culmination of the FHA's original suburban capitalist development strategy—made it virtually impossible for Flint's economy to thrive.

In chapter 9, the campaign to desegregate Flint's public schools takes center stage. In 1975, after many years of delay, officials from the federal Department of Health, Education, and Welfare (HEW) ordered the Flint Board of Education to desegregate its schools. Shortly thereafter, the Charles Stewart Mott Foundation, which had helped to fund Flint's public schools for many decades, abruptly withdrew its support for community education and shifted its financial resources toward the city's ongoing efforts to revitalize the central business district. Nevertheless, the school board continued to embrace the neighborhood schools and community education policies that had kept pupils and their families segregated. After several years of negotiations, the conflict ultimately ended in 1980 when HEW functionaries endorsed a version of the school board's voluntary desegregation plan. As elsewhere, Flint's voluntary plan failed to deliver racially integrated schools.

Demolition Means Progress ends with the devastating economic crises that swept across Flint and other industrial cities in the late twentieth and early twenty-first centuries. During those decades, the combined effects of corporate austerity, mass suburbanization, and capital disinvestment from the inner city wreaked havoc on Flint's economy. Predictions that Flint would become a ghost town turned out to be untrue, however, as civic leaders and residents continued in their quest to revitalize and repopulate the city. In most cases, though, their efforts failed to deliver economic opportunities and racial equity to the Vehicle City and its growing black majority.

The story that follows is one of a city and a region transformed by an almost ceaseless quest for revitalization. The deeply entrenched patterns of racial, spatial, and economic inequality that grew out of such renewal efforts were neither innocent nor exceptional. On the contrary, the extreme social injustices that Flint activists confronted before and after World War II had their roots in a deep national well of private antipathies, growth-obsessed public policies, and discriminatory government actions that first converged during the early twentieth century. In their attempts to stimulate economic growth, foster community solidarity, and revitalize areas perceived to be in decline, civic leaders in Flint and other American cities ultimately created dense webs of legal, popular, and administratively driven inequality. Although the details of Flint's story are different, of course, from those of other places, the Vehicle City's experience with the unending process of urban renewal remains more representative than peculiar. Indeed, Flint's tangled history of social division, with all of its local variations, is every bit as American as the golden Chevrolet Bel Aire that rolled through the Vehicle City's segregated streets on November 23, 1954.

PART ONE

Company Town

City Building and Boundary Making

On the eve of the Allied victory in World War II, as Michigan's defense plants hummed and the nation's economy boomed, Carl Crow, the official historian of the Buick division of General Motors, penned a glowing tribute to his employer and its hometown. Published in 1945, Crow's encomium included the phrase "Buick is Flint and Flint is Buick," a simple but revealing statement that captured the close relationship between GM and the Vehicle City. Comparing the company-town bond to that of a "self-sacrificing father and a successful son," GM's chronicler saluted the people of Flint for enabling GM's rise to industrial supremacy. Because of GM's triumphs, Crow averred, Flint's citizens had obtained a degree of security and prosperity that made them the envy of the world.[1] Like many of the city's promoters, Crow believed that Flint's successes symbolized the fruition of the American Dream of progress, prosperity, and opportunity. "America is a thousand Flints," he concluded, because the city and its people exemplified the principles and aspirations that made the United States a beacon of hope for the wider world.[2]

Although Flint's standing as a leading industrial city would ultimately prove fleeting, Crow's optimism at war's end was understandable. By the time he had finished writing *The City of Flint Grows Up*, the United States sat on the cusp of one of the longest periods of economic expansion and consumer prosperity in human history. As one of the world's preeminent manufacturing centers, Flint played a major role in driving the nation's post–World War II economic boom. Equally important, the cars and trucks made in the Vehicle City helped to fuel the transportation revolution that transformed the United States into a predominantly suburban nation. In exchange for these accomplishments, Flint's autoworkers earned impressively high wages, particularly after they unionized in the 1930s. Hoping to claim a share of that bounty, tens of thousands of migrants flocked to the Vehicle City during the first half of the twentieth century. Their arrivals helped to make Flint the eleventh-fastest-growing city in the United States, a fact that seldom escaped the area's boosters. Nevertheless, conditions on the ground in this densely populated working-class town were much more complicated than Crow and others were willing to acknowledge. To wit, the Flint that Crow championed was also a harshly divided city.[3]

During the decades preceding World War II, a potent combination of private discrimination, federal housing and development initiatives, corporate practices, and municipal public policies converged to make Flint one of the most racially segregated cities in the United States. As was the case in most other urban communities nationwide, housing proved to be a key venue for the establishment and maintenance of Flint's color line. In the 1910s and 1920s, GM executives and local real estate developers worked to resolve the area's deep housing shortage by building new homes and neighborhoods for migrant workers. Because the deeds to these new properties contained racially restrictive covenants, however, they were available only to white buyers. By the 1930s federal and local policy makers had begun playing a more active role in shaping the Flint area's residential housing market. Still, though, segregation was the rule. Across the nation, in fact, New Deal housing programs hardened popular, legal, and administratively driven forms of residential segregation. During the 1930s officers from the federal Home Owners' Loan Corporation helped to codify racist standards for measuring mortgage risks, neighborhood stability, and market vitality. Later adopted by officials from the Federal Housing Administration, those standards led to the systematic practice of mortgage redlining. Working alongside local realtors, builders, and municipal officials, FHA representatives in Flint and other places established rules for metropolitan real estate development that all but required discrimination against African Americans.

For a time during the 1930s, 1940s, and early 1950s, many of these same federal policy makers also engaged in suburban redlining. This punitive practice forced officials from the area's white working-class suburbs to provide new services and establish political independence from Flint as a prerequisite to obtaining federal mortgage insurance. Over time, the redlining of Flint's dilapidated suburbs fostered the growth of a socially and politically divisive brand of suburban capitalism. As with whites-only schools, parks, and workplaces, the segregated and politically fragmented residential arrangements that proliferated during this period fit comfortably within Flint's civic culture of Jim Crow.

Company and Town

Flint's phoenixlike emergence on the national scene was a testament to copi-
ous amounts of hard work, government support, and imperial ambition. The
city's identity as a commercial and industrial center first began to take shape
in 1819, when an enterprising settler named Jacob Smith established a fur
trading post on the Flint River. Prior to then, several groups of Ojibwe Indians had shared the sparsely developed lands of the Saginaw Valley with an assortment of trappers, subsistence farmers, and territorial officials. By 1820 the
US government had acquired all of the Ojibwe lands in southeastern Michigan through a mix of treaties, purchases, and raw violence.[4] Flint's ascent
as an urban and commercial hub coincided with this era of Native American dispossession and displacement. Located between Detroit and Saginaw,
Smith's settlement on the banks of the Flint River was an excellent stopping
point for both traders and travelers. An influx of settlers during the antebel-
lum period helped to turn Flint into a bustling hamlet of nearly twenty thousand permanent residents. Encouraged by this growth, community leaders
launched a successful campaign for municipal incorporation in 1855. Soon
after that, the city's burgeoning lumber and carriage manufacturing industries began attracting thousands of workers to Genesee County. Nevertheless,
Flint remained a tiny, relatively unknown city until the turn of the twentieth
century, when the arrival of the "horseless carriage" inaugurated a new epoch
in the community's history.

The birth of the automobile industry helped to transform Flint from a
small town into a metropolitan center. The shift began shortly after engineers
unveiled the first generation of gasoline-powered cars in the 1880s, at which
point area investors launched a campaign to recruit automobile companies to
the Flint region. Civic boosters achieved a major victory in 1904 when one of
the area's leading industrialists, James H. Whiting, acquired the Buick Motor
Company and relocated its manufacturing operations to Flint's north side.
Soon afterward, Whiting hired a Flint-based carriage maker and speculator
named Billy Durant to run his new venture. Durant promptly moved Buick's
industrial operations ninety miles southeast to the city of Jackson, Michigan,
however, claiming that Buick's Flint facilities were too small and outdated. In
response, civic leaders, in what would become a recurring theme in the city's
history, launched a major fundraising campaign to bring Buick back to Flint.
Within weeks, local bankers and industrialists had pledged over five hundred
thousand dollars in stock subscriptions, which provided Durant with the capital necessary to build a suitable assembly plant on Flint's north side. By 1906
Durant and Buick had returned to Flint to establish long-term operations.[5]

Two years later, Durant founded the General Motors Corporation. Eager to build an industrial empire, Durant promptly acquired the Olds, Cadillac, and Oakland automobile companies. On the strength of those investments, GM quickly became one of the world's leading automobile producers, bringing windfall profits to Durant, Whiting, Charles Stewart Mott, and other local industrialists. By 1929 Durant's company had produced ten million cars, and GM was well on its way to becoming the world's largest industrial firm. As GM's birthplace and manufacturing headquarters, Flint grew rapidly during this freewheeling era of corporate expansion.[6] Between 1900 and 1930, Flint's population soared from just 13,103 to 156,492.[7]

As GM grew and opened new manufacturing and assembly facilities, the city began to resemble Pittsburgh, Cleveland, and other industrial metropolises. On the near north side—to the west of the gritty St. John Street neighborhood, the Flint River, and the Chesapeake and Ohio rail lines—sat GM's massive complex of Buick plants. With an employee roll that routinely exceeded twenty-five thousand, "The Buick" was Flint's largest employer and virtually a city unto itself. Nearby there were two major industrial facilities operated by GM's AC Spark Plug division. At its peak, AC employed nearly twenty thousand workers who manufactured spark plugs, oil filters, and other auto parts. On Flint's west side, in a valley surrounding the Flint River, visitors could find GM's Fisher Body 2 plant and a large complex of Chevrolet plants known as "Chevy in the Hole." There nearly twenty thousand workers produced GM's top-selling line of cars and trucks. Just to the northwest of Chevy in the Hole, on the corner of Third and Chevrolet Avenues, stood the General Motors Institute of Technology, an elite division of the company dedicated to training automobile engineers and corporate managers. Another of Flint's major industrial facilities was the Fisher Body 1 plant, located on the city's far south side. By 1955 eight thousand workers at Fisher Body 1 manufactured automobile bodies for the north side Buick plants.[8] An industrial marvel, Flint was home to more GM workers than any other city in the world and second only to Detroit in annual vehicle production. "It is to the automobile," claimed the *New York Times Magazine* in 1937, "what Pittsburgh is to steel, what Akron is to rubber."[9]

Although the Vehicle City's booming economy attracted migrants from virtually every part of the world, Flint was much less diverse than most urban centers. As late as 1930, over 80 percent of Flint's residents were native-born whites, most of them Protestants. Approximately three-quarters of Flint's people hailed from either Michigan or the nearby states of Ohio, Indiana, Illinois, and Wisconsin. Nearly 15 percent of Flint's residents in 1930 were born in the South, many from the states of Arkansas, Kentucky, Missouri, and Ten-

FIGURE 1.1. Aerial view of "The Buick," n.d., circa 1925. The Buick manufacturing and assembly complex on Flint's north side was one of GM's largest and most important industrial facilities in the United States. During the postwar era, approximately twenty-five thousand men and women worked within this sprawling compound. With numerous GM facilities scattered throughout Genesee County, metropolitan Flint was the company's international manufacturing headquarters for most of the twentieth century. Courtesy of the Richard P. Scharchburg Archives, Kettering University, Flint, MI.

nessee. In a striking departure from many other urban areas, foreign-born residents accounted for just 14 percent of the city's overall population on the eve of the Depression. Among these immigrants, the largest numbers were from England, Scotland, and Anglophone sections of Canada. Whereas Catholics hailing from Poland, Italy, and elsewhere constituted up to a third of the population in Detroit, Chicago, and other midwestern cities, southern and eastern Europeans accounted for only 19 percent of Flint's already small immigrant total. Similarly, only a tiny fraction of the city's total population— approximately 1 percent—were of either Hispanic or Asian descent. For their part, African Americans represented just 3.6 percent of the Vehicle City's relatively homogeneous Depression-era population.[10]

Upon arriving in the Flint area, most migrants knew that they wanted to make a living at General Motors. They knew much less, however, about where, with whom, and under what conditions they would live. For many newcomers, then, the quest to locate housing also involved a broader process of discovering the region, its built environment, and its people. Among many poor and working-class whites, that process centered on outlying areas, where land was least expensive, taxes were low, and building and zoning regulations were minimal.[11] By 1930 there were almost sixty thousand people living in suburban Flint, most of them residing in the townships of Burton, Flint, Genesee, and Mt. Morris, which formed a contiguous border around the city. These semirural areas just beyond the city limits contained a small number of professionally built subdivisions along with numerous informal working-class settlements, most of which lacked urban amenities such as parks, sewers, and paved roads.[12] Residents of Flint's underdeveloped suburbs tended to be poor white migrant workers, most of them relatively uneducated, and the housing stock included many one-room shacks and temporary homes. Because many poor and working-class migrants could not afford to hire professional developers, do-it-yourself building was an extremely common practice.[13] In nearby suburbs such as Genesee Township, for instance, owner-built homes made up well over half of the overall housing stock.[14] In describing the housing that ringed the city's borders, Flint's building inspector Peter J. Weidner noted that the Vehicle City was "fringed on all sides with shanty towns."[15]

Flint's overwhelmingly poor and working-class suburbs were unique in some ways and representative in others. Across North America, metropolitan regions featured widely differential class profiles that were ultimately refracted in the suburban landscape. In larger cities such as New York, Chicago, and Los Angeles, the suburban periphery was a variegated landscape of rich, middling, and poorer communities characterized by varying degrees of privilege and social exclusivity. In more heavily blue-collar metropolises such as Flint and Detroit, by contrast, roughly hewn working-class settlements dominated the suburban periphery. In short, the Flint of the early twentieth century lacked picturesque, high-end suburbs such as Riverside, outside of Chicago; Llewellyn Park, New Jersey, near New York City; or Pacific Palisades in Southern California. Instead, Flint's wealthiest citizens tended to cloister their affluence in residential settings fairly close to the city center. As a result, Flint's suburbs were more gritty and undeveloped than those in many other metropolitan areas.[16]

Housing and living conditions were far more variable inside the city. Because of Flint's rapid growth, standard affordable housing was extremely difficult to locate for many migrants. Tents, tarpaper shacks, and other "homes of the homeless" were thus quite common throughout the city. As one observer sarcastically noted about Flint's cramped, substandard living arrangements, "they keep them [residents] so thick that their feet hang out the windows."[17] Nevertheless, the city also claimed many middle-class and even wealthy neighborhoods, most of them on the west and northwest sides.[18] Unlike the informal working-class settlements that straddled the city limits,

FIGURE 1.2. Substandard working-class housing, n.d., circa 1910. Flint's rapid growth during the first half of the twentieth century resulted in severe housing shortages. Upon arriving in the Flint area, thousands of migrants erected their own homes, often in shantytowns near GM factories or in the largely undeveloped working-class suburbs that proliferated just beyond the city limits. From the mid-1930s through the early 1950s, officials from the Federal Housing Administration used the power of mortgage redlining to force suburbanites to improve government services and eradicate low-cost homes such as these. Courtesy of the Richard P. Scharchburg Archives, Kettering University, Flint, MI.

Glendale Hills and other premier west side neighborhoods had full access to municipal services and utilities, featured new homes of sound construction, and contained a highly educated, economically stable mix of skilled autoworkers and professionals. According to a 1934 survey sponsored by the federal Civil Works Administration, the neighborhoods of the west side were "almost entirely white" with "few machine operators or unskilled workers" and virtually no immigrants from outside northern Europe.[19]

To the southeast of these exclusive west side enclaves were a number of older neighborhoods ringing the city's downtown commercial core. During the 1920s and 1930s, many of these formerly wealthy areas experienced population loss, the conversion of large homes to rental units, and the spread of rooming houses and commercial structures. As a result, once privileged communities such as Grand Traverse and the area that came to be known as Carriage Town began to show signs of housing dilapidation and social instability. Although these areas featured many attractive Victorian homes and a relatively well-educated population, the arrival of small numbers of African Americans, southern European immigrants, and other "undesirable" social groups led to waves of white departures and disinvestment.[20]

Not all of Flint's close-in neighborhoods suffered such high rates of turn-over, however. Woodlawn Park, just to the east of downtown, was perennially one of Flint's most exclusive residential areas. Home to Charles Stewart Mott, GM's largest stockholder, and other company brass, this all-white community featured a mix of elegant mansions and attractive brick and masonry homes that were among the newest, priciest, and most carefully restricted in the city.[21] It also contained many of the city's most important civic and cultural institutions including Central High School and, by the 1950s, the Flint Public Library and the Flint Institute of Arts. Although Woodlawn Park bordered two poorer neighborhoods that housed numerous black and immigrant families, racially restrictive housing covenants and a host of popular and administratively driven segregationist devices kept the area all white and ethnically homogeneous.

Several miles north of Woodlawn Park was the St. John Street neighborhood, home to the city's main vice district, its largest concentration of poverty, and its second-highest proportion of African Americans. Surrounded on three sides by the Buick factories, the Flint River, and a labyrinth of rail lines, this teardrop-shaped community formed the historic heart of the so-called North End, a collection of gritty working-class neighborhoods between Saginaw Street and the riverfront stretching from Fifth Avenue on the edge of downtown all the way to the city's northern border. A writer for the *Flint Journal* once remarked on its near complete isolation, noting, "St. John is virtually an island in a city. On its west side the massive Buick complex is like a mountain range. On the east is the Flint River. There are few ways to get in and out."[22] By all accounts, St. John was a heavily polluted neighborhood, and it contained some of the oldest, most dilapidated housing in the city.[23] However, the area was also home to a large number of small business owners and professionals and was one of the most racially and ethnically diverse sections of the city. According to one survey, "Nativity is highly mixed with negroes largely from the south and a large number of white occupants from central, eastern and southern Europe. There is also a considerable number from Asia Minor and from Mexico."[24]

For those seeking to escape the soot and noise of St. John, the community of Floral Park presented many attractive options. Located on the southwestern border of Woodlawn Park in an area bounded by Fifth Street, the Grand Trunk Railroad tracks, South Saginaw Street, and Lapeer Road, Floral Park was home to a diverse array of immigrants, African Americans, and working-class whites. Prior to World War II, blacks could find housing only in a small section of the neighborhood near the Grand Trunk rail lines. Despite such rigid segregation, however, the area housed an eclectic mix of professionals, domestics, shopkeepers, and factory workers. Unlike the inhabitants of St. John—who suffered from industrial pollution and spatial isolation— Floral Park residents benefited from the neighborhood's location near downtown shops and the Thread Lake recreation area. The Clifford Street Community Center, Quinn Chapel AME Church, the Michigan Theater, and the Golden Leaf nightclub anchored the south side's highly segregated yet bustling public sphere. To be sure, poverty and substandard housing were common in Floral Park, just as they were in St. John. Nonetheless, black residents of the south side enjoyed a somewhat privileged social status that correlated with the neighborhood's distance from the North End.[25]

Due primarily to their affordability, Floral Park and St. John were popular destinations for poor and working-class migrants of all racial and ethnic backgrounds during the first half of the twentieth century. Over time, however, whites discovered well-worn paths out of the two neighborhoods, many of them leading to the segregated working-class districts of Flint's east side or the new frontiers of suburbia. For African Americans, by contrast, Flint offered precious few housing options. The relative immobility of African Americans stemmed in part from the peculiar urban and industrial geographies that isolated St. John from the rest of the city. However, the hydra of popular, administrative, and, occasionally, legal Jim Crow helped to shape the region's settlement patterns even more than the Flint River divide, rail lines, and other physical barriers. In the first half of the twentieth century, these three modes of segregation found a powerful synthesis in the spread of court-sanctioned racially restrictive housing covenants, which helped to ossify Flint's already stark color lines.

Building a Segregated Metropolis

General Motors executives played a quiet but important role in creating and maintaining residential Jim Crow in Flint. The company's first major foray into the housing sector occurred in 1919, when executives established a new company division called the Modern Housing Corporation. It was the first branch of the corporation devoted to residential construction and sales, and it grew out of the embarrassment that company chiefs suffered because of the shantytowns that had sprouted near GM facilities in Flint, Detroit, and other plant cities. By 1933, workers from Modern Housing had erected nearly three thousand new homes in three subdivisions on Flint's northwest side. Later named Civic Park, Chevrolet Park, and Mott Park, GM's new neighborhoods offered buyers a choice of twenty-eight different home designs and an array of amenities including sewer and water connections, gas and electric service, and newly constructed streets and sidewalks. In their attempts to lure young families with children, GM officials persuaded members of the Flint Board of Education to construct two new public schools to serve the neighborhoods. The company also offered generous financing terms that included loans at 6 percent interest and down payments of only 10 percent.[26]

Between 1919 and the early 1930s, thousands of working-class and professional families rushed to acquire GM's new houses. However, the homes were not available to all buyers. Hoping to protect home values and a high quality of life in the new subdivisions, officials from GM attached stringent, legally enforceable deed and building restrictions to all the properties. Specifically, GM mandated that only single-family homes could be built in its new subdivisions and that occupants could not keep livestock, sell liquor, or construct outdoor cesspools or privies on their properties. The covenants also required racial segregation, stipulating that homes "could not be leased to or occupied by any person or persons not wholly of the white or Caucasian race."[27] Because the US Supreme Court, through a 1926 ruling, and later the FHA endorsed the use of such agreements, they contributed to a hybrid form of segregation rooted simultaneously in popular attitudes, legal decisions, and the administrative practices of public officials.[28]

As was the case in other communities throughout the United States, much of the new housing constructed in the Flint area between 1900 and 1950 fell under similarly restrictive covenants.[29] In Woodlawn Park, for instance, the James A. Welch Company offered new home sites in "a district fully restricted and possessing the proper social atmosphere—where environs will always be pleasant and where values will steadily increase."[30] The covenants for Woodlawn Park stipulated, "No poultry, cattle, live stock, except watch dogs and family pets and driving horses, shall be kept." Recognizing the special needs of the neighborhood's wealthy residents, Welch also crafted a racial restriction that allowed visits and overnight stays from service workers while still prohibiting black residency: "No negroes or persons of negro extraction (except while employed thereon as servants) shall occupy any of the land."[31] The use of these restrictive covenants provided homeowners in Woodlawn Park and other areas with the privilege of preselecting their future neighbors in perpetuity.

Throughout the region, realtors routinely referenced restrictive covenants in their marketing campaigns for new housing developments. In the spring of 1930, George Kellar advertised lots for sale in the Brookside subdivision just east of Woodlawn Park. Like Welch, Kellar promised buyers "adequate restrictions to protect your home investment." That same year, realtor Don Waters marketed "beautiful homesites, adequately restricted—moderately

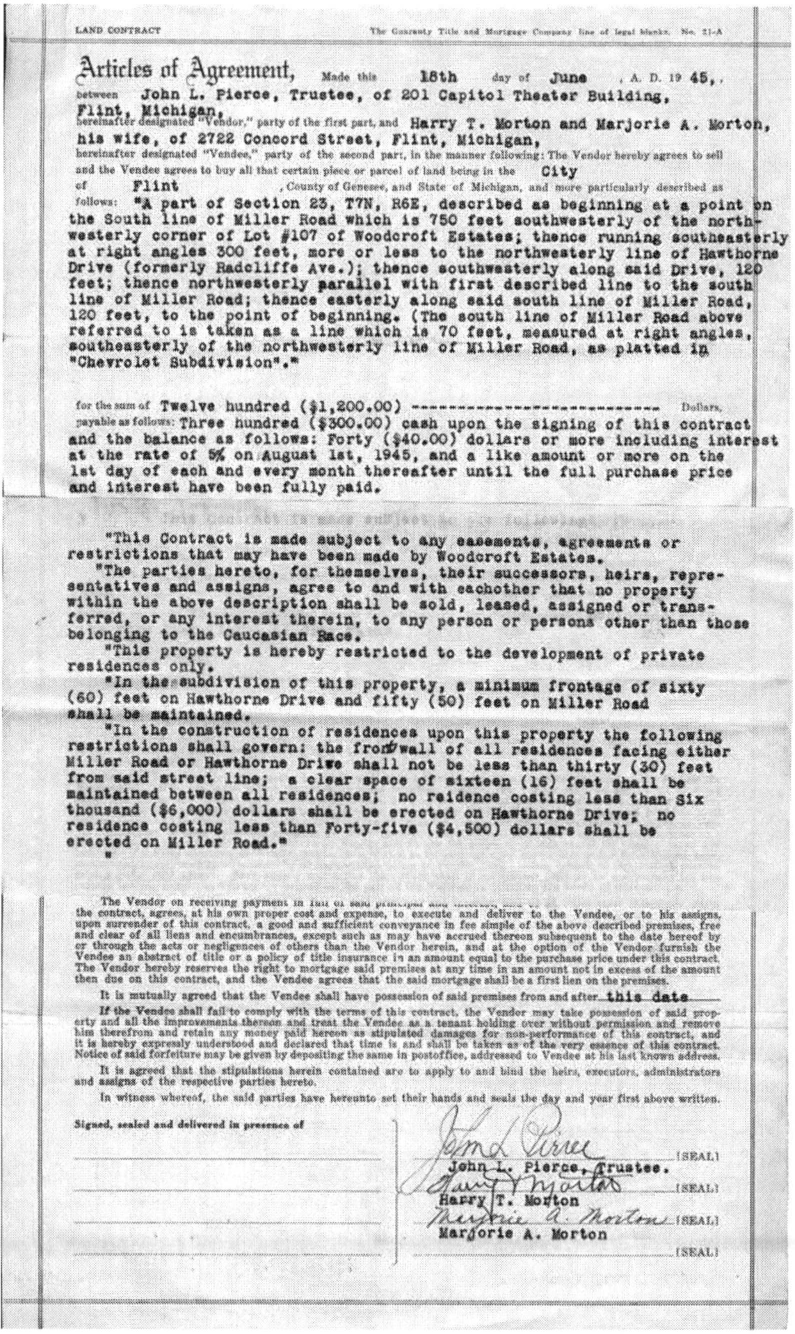

Articles of Agreement, Made this **18th** day of **June** , A. D. 19 **45**,.
between **John L. Pierce, Trustee, of 201 Capitol Theater Building, Flint, Michigan,**
hereinafter designated "Vendor," party of the first part, and **Harry T. Morton and Marjorie A. Morton, his wife, of 2722 Concord Street, Flint, Michigan,**
hereinafter designated "Vendee," party of the second part, in the manner following: The Vendor hereby agrees to sell and the Vendee agrees to buy all that certain piece or parcel of land being in the   **City**
of   **Flint** , County of Genesee, and State of Michigan, and more particularly described as
follows: **"A part of Section 23, T7N, R6E, described as beginning at a point on the South line of Miller Road which is 750 feet southwesterly of the northwesterly corner of Lot #107 of Woodcroft Estates; thence running southeasterly at right angles 300 feet, more or less to the northwesterly line of Hawthorne Drive (formerly Radcliffe Ave.); thence southwesterly along said Drive, 120 feet; thence northwesterly parallel with first described line to the south line of Miller Road; thence easterly along said south line of Miller Road, 120 feet, to the point of beginning. (The south line of Miller Road above referred to is taken as a line which is 70 feet, measured at right angles, southeasterly of the northwesterly line of Miller Road, as platted in "Chevrolet Subdivision"."**

for the sum of **Twelve hundred ($1,200.00)** — — — — — — — — — — — — — — — — — — Dollars,
payable as follows: **Three hundred ($300.00) cash upon the signing of this contract and the balance as follows: Forty ($40.00) dollars or more including interest at the rate of 5% on August 1st, 1945, and a like amount or more on the 1st day of each and every month thereafter until the full purchase price and interest have been fully paid.**

"This Contract is made subject to any easements, agreements or restrictions that may have been made by Woodcroft Estates.
"The parties hereto, for themselves, their successors, heirs, representatives and assigns, agree to and with eachother that no property within the above description shall be sold, leased, assigned or transferred, or any interest therein, to any person or persons other than those belonging to the Caucasian Race.
"This property is hereby restricted to the development of private residences only.
"In the subdivision of this property, a minimum frontage of sixty (60) feet on Hawthorne Drive and fifty (50) feet on Miller Road shall be maintained.
"In the construction of residences upon this property the following restrictions shall govern: the front wall of all residences facing either Miller Road or Hawthorne Drive shall not be less than thirty (30) feet from said street line; a clear space of sixteen (16) feet shall be maintained between all residences; no reidence costing less than Six thousand ($6,000) dollars shall be erected on Hawthorne Drive; no residence costing less than Forty-five ($4,500) dollars shall be erected on Miller Road."

The Vendor on receiving payment in full of said principal and interest, and at the option of said contract, agrees, at his own proper cost and expense, to execute and deliver to the Vendee, or to his assigns, upon surrender of this contract, a good and sufficient conveyance in fee simple of the above described premises, free and clear of all liens and encumbrances, except such as may have accrued thereon subsequent to the date hereof by or through the acts or negligences of others than the Vendor herein, and at the option of the Vendor furnish the Vendee an abstract of title or a policy of title insurance in an amount equal to the purchase price under this contract. The Vendor hereby reserves the right to mortgage said premises at any time in an amount not in excess of the amount then due on this contract, and the Vendee agrees that the said mortgage shall be a first lien on the premises.

It is mutually agreed that the Vendee shall have possession of said premises from and after **this date**.

If the Vendee shall fail to comply with the terms of this contract, the Vendor may take possession of said property and all the improvements thereon and treat the Vendee as a tenant holding over without permission and remove him therefrom and retain any money paid hereon as stipulated damages for non-performance of this contract, and it is hereby expressly understood and declared that time is and shall be taken as of the very essence of this contract. Notice of said forfeiture may be given by depositing the same in postoffice, addressed to Vendee at his last known address.

It is agreed that the stipulations herein contained are to apply to and bind the heirs, executors, administrators and assigns of the respective parties hereto.

In witness whereof, the said parties have hereunto set their hands and seals the day and year first above written.

Signed, sealed and delivered in presence of

—————————————   *John L. Pierce* [SEAL]
—————————————   **John L. Pierce, Trustee.**
—————————————   *Harry T. Morton* [SEAL]
—————————————   **Harry T. Morton**
—————————————   *Marjorie A. Morton* [SEAL]
       Marjorie A. Morton
—————————————   [SEAL]

FIGURE 1.3. Restrictive housing covenant, 1945. During the first half of the twentieth century, restrictive housing covenants were a popular tool for maintaining racial segregation in the United States. The widespread use of court-sanctioned agreements such as the one shown here, which covered a parcel of land in the elite Woodcroft Estates subdivision on Flint's far west side, contributed to the Vehicle City's perennial standing as one of the nation's most segregated urban centers. Courtesy of the Genesee Historical Collections Center, University of Michigan–Flint.

priced," in the Glendale Hills neighborhood. Waters also reminded prospective purchasers that all of his properties were subject to the terms established by the segregated National Association of Real Estate Boards (NAREB), "an organization governed by a code of ethics, conceived for the protection of property buyers, owners and sellers." As part of its ethics code, NAREB explicitly prohibited member realtors from selling property to nonwhites in segregated white neighborhoods.[32]

By the close of the 1930s, the widespread use of restrictive covenants by local residents had helped make Flint the third most segregated city in the nation, surpassed only by Miami, Florida, and Norfolk, Virginia.[33] Housing, however, was but one of the many domains in which segregation occurred. Like their counterparts in other cities, blacks in interwar Flint faced a rigid and deeply reticulated system of popular, legal, and administrative Jim Crow at work, at school, and in most places of public accommodation.[34] At GM, black workers found it next to impossible to gain employment except as janitors or foundry workers. Likewise, most downtown retailers refused to hire black salespersons and clerks. In the realm of education, housing segregation and the gerrymandering of attendance districts combined to keep black and white pupils apart. With few exceptions, this vast network of Jim Crow also extended into the consumer sphere, where white proprietors of barbershops, restaurants, movie theaters, hotels, bowling alleys, and taverns either denied service to black customers or established designated areas or times to accommodate them.[35] For instance, prior to the 1940s black moviegoers could attend shows at the Michigan Theater near Floral Park, but only if they sat separately in the balcony, or "crow's nest."[36] Managers of the IMA Auditorium north of downtown required visiting bands to perform two acts, one for white patrons from 9 p.m. to midnight, and the second for African Americans, from 1 to 5 a.m.[37] The structures of Jim Crow were so thoroughly embedded in the city's culture that the color line extended beyond death, with racially restrictive burial covenants covering nearly all privately owned cemeteries in the county until 1964.[38] "Jim Crow was the rule in Flint, that's plain as day," black resident Mary Helen Loving remembered.[39]

Along with realtors, homeowners, and other private citizens, municipal officials played an important role in maintaining segregation in interwar Flint. Outside of the well-defined borders of St. John and Floral Park, black pedestrians and motorists faced near-constant harassment by officers from the Flint Police Department. Referring to the heavily policed Flint River color line that divided St. John Street from the all-white east side, Eugene Simpson, who went on to become pastor of Mt. Tabor Baptist Church, remembered that during the 1930s "Blacks weren't allowed across the river unless

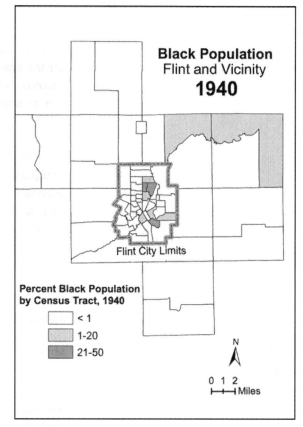

MAP 1.1. Black population by census tract, Flint and vicinity, 1940. *Source:* Minnesota Population Center, *National Historical Geographic Information System: Version 2.0* (Minneapolis: University of Minnesota, 2011).

they were going to school [or work]."[40] Local school and recreation officials also practiced administrative Jim Crow. At Central High School, members of the board of education prohibited African American children from using the swimming pool until 1944.[41] Likewise, park managers at Berston Field House and other public facilities maintained segregationist policies until the 1940s. Black children wishing to use the Berston Field House Library could access books only during official school-sponsored visits. For black swimmers the Berston pool was open only on Wednesdays. To accommodate black residents the rest of the week, city officials operated outdoor sprinklers across the street. "We would always have access to the pool after it had been used by the white community," north side resident Max Brandon remembered. "And when we finished using the pool that evening, all the water would be drained out."[42]

FIGURE 1.4. Flint Masonic Temple Association minstrel show, 1944. Blackface performances such as this one, held at Flint's Masonic Temple in April 1944, were common throughout the United States until well into the 1950s. In Flint and elsewhere, minstrelsy was an integral part of Jim Crow culture. Courtesy of the Richard P. Scharchburg Archives, Kettering University, Flint, MI.

Like segregated schools and parks, minstrelsy fit comfortably within Flint's civic culture. In 1940 George Oscar Bowen of the Flint Board of Commerce published his popular *Book of Songs*, which contained the lyrics to minstrel songs such as "Old Black Joe" and "Carry Me Back to Old Virginny."[43] Kiwanis Club members, who, until at least the 1950s, organized annual minstrel shows to fund charitable causes, may have used Bowen's book. During the 1933 Kiwanis show, Flint's mayor Raymond Brownell, along with city commissioners Harry Comins and George Barnes, school superintendent L. H. Lamb, city parks chairperson Arthur Sarvis, Flint Board of Education member Forest Boswell, and numerous GM executives donned blackface to raise funds for a children's health camp.[44] At around the same time, an unnamed teacher at Fairview Elementary School forced young Eugene Simpson to perform minstrel songs for his class. According to Simpson, "[T]he teacher would have me stand in front of the class and I would sing, 'I'm a manish [*sic*] pickaninny, blacker than a crow.'"[45] Regardless of the context, minstrel routines were spectacles of racial power designed, in the words of black resident Mary Helen Loving, "to remind black people of their position."[46]

In the decades preceding World War II, whites in Flint and many other American cities forged a broad cultural and political consensus on the need for strict racial segregation. By the mid-1930s, when federal officials began taking a more active interest in metropolitan real estate markets, Flint was already a segregated and unequal city. Clearly, then, Flint's deeply rooted patterns of Jim Crow predated the establishment of the Home Owners' Loan Corporation, the Federal Housing Administration, and other important federal agencies created as part of the New Deal. Nevertheless, the 1930s still marked a turning point in the history of the color line, for it was during that decade that government officials helped to give birth to a new federalized system of administrative segregation.

The Origins of Federal Redlining

The Great Depression dealt a vicious blow to the American automobile industry and its plant cities. In response to the stock market collapse and sagging consumer demand for cars and other durable goods, automobile manufacturers in the United States cut new vehicle production by as much as 75 percent.[47] The decline of manufacturing devastated Flint's economy, generating unemployment rates in the early 1930s that routinely approached 50 percent. By 1938 the number of Flint families on relief had reached 19,658, nearly half of the households in the city.[48] Banks in Flint and across the nation responded to the crisis by sharply curtailing lending, which helped to decimate the housing market. Between 1928 and 1933, new residential construction in the United States fell by 95 percent while spending on home repairs dropped by 90 percent.[49] In the Flint metropolitan area, new home starts declined from an average of 1,786 per year during the 1920s to just 126 by 1933.[50] The crash also caused a sharp decline in the quality of the Flint area's housing stock. According to a 1934 federal study, nearly 15 percent of Flint's housing units required major structural repairs, and at least 2 percent of the city's homes warranted immediate demolition. Of the inhabited homes within the city, more than 20 percent lacked a working bath, and 25 percent had no access to hot water.[51]

Grave housing shortages also plagued the region during this time. Citywide, there were approximately thirty-five thousand dwelling units during the 1930s, nearly all of which were detached single-family wood-frame structures. On average, just over four persons lived in each of the city's houses, but because the housing stock consisted primarily of small one- and two-bedroom "workman's cottages," the average Flint bedroom accommodated two people, with some sections of the city posting even higher occupancy rates. The fact that thousands of unemployed autoworkers had already left the city prior to the 1934 survey made such figures even more startling. While many left the state altogether, others followed an emerging North American trend by setting out for Flint's largely undeveloped suburbs, where they built their own homes, grew crops, and raised livestock to make ends meet.[52] In all, Flint's population declined by approximately five thousand during the 1930s. Meanwhile, its gritty, haphazardly developed suburbs gained over twenty thousand new residents, most of them poor and working-class whites.[53] Although the outmigrations of unemployed autoworkers significantly reduced overcrowding in the city, Flint's housing shortage persisted. According to building inspector Peter Weidner and other public officials, the problem stemmed primarily from the confluence of Flint's rapid, unplanned growth during the 1910s and 1920s—which had created severe housing shortages that lingered until well after World War II—and the almost complete collapse of the nation's residential building industry during the 1930s. Conservatively, Weidner and other experts estimated that the city needed at least five thousand new homes to meet the housing needs of its nearly 150,000 residents. To remedy this problem, local officials in Flint and other cities pinned their hopes on developments in Washington, DC.

President Franklin D. Roosevelt and federal lawmakers wasted little time in addressing the nation's burgeoning housing crisis. In June 1933 the president signed legislation creating the Home Owners' Loan Corporation to help reduce housing foreclosures and revive the nation's real estate industry. Seemingly overnight, the HOLC helped to stabilize local housing markets across the country by acquiring hundreds of thousands of delinquent home loans and refinancing them under more forgiving terms. By 1936, when the HOLC ceased its loan acquisition program, federal officials had purchased and refinanced the mortgages on more than 20 percent of the nation's nonagricultural, owner-occupied housing units.[54] In the process of doing so, the HOLC helped to reshape the ways in which Americans thought about home finance, neighborhood stability, and the real estate market.[55] Although the agency's loan acquisition program lasted for just three years, its policies introduced millions of Americans to the concept of low-interest, long-term, fully amortized home mortgages. Prior to the Depression, private lenders in Flint and elsewhere typically offered only short-term high-interest mortgages, often with large "balloon" payments due up front and at the end of the loan period.[56] By demonstrating that lenders could generate higher profits by replacing onerous mortgages with fully amortized long-term low-interest loans, the HOLC revolutionized the world of home finance.

Yet the HOLC's policies also hardened the color line by enshrining a racially and socioeconomically biased calculus for measuring risk, value, and stability in residential neighborhoods. As part of its operations, the HOLC created "residential security maps" for Flint and 238 other American cities.

Designed to illustrate mortgage risk factors, housing conditions, and the overall desirability of residential districts, the security maps ranked neighborhoods on a descending scale from A to D. The neighborhoods with the most desirable and valuable housing received a grade of A, color-coded with green on most security maps, while "second-grade" B neighborhoods were typically blue in color. The least desirable residential areas earned C and D grades, with C neighborhoods often colored yellow and "fourth-grade" D areas usually shaded in red. The term *redlining*, used in reference to the discriminatory mortgage insurance and lending practices later employed by bankers and FHA officials, derived from the HOLC's use of red in mapping D-grade neighborhoods.[57]

Officials from the HOLC often subcontracted with local realtors to create the residential security maps. In Flint prominent brokers such as H. H. Darby, Claude Perry, Robert Gerholz, and Mark Piper performed in that role.[58] Typically the HOLC's surveyors classified and ranked neighborhoods based on eight criteria: the intensity of sale and rental demand; the percentage of homeowners; the age and type of buildings; economic stability; the social status of residents; the sufficiency of public utilities; the accessibility of schools, churches, and businesses; and the building, deed, and zoning restrictions established to protect the area from "inharmonious" social groups and incompatible land uses. In order to receive an A rating, a neighborhood had to be only partially developed and situated within a new, well-planned area, and its housing stock had to be in high demand regardless of the economic climate. Surveyors from the HOLC relied heavily on the preexisting lending habits of local bankers to determine which neighborhoods met these criteria. Consequently, each of the HOLC's neighborhood descriptions included sections on the availability of local mortgage funds.[59] An A neighborhood, one HOLC report maintained, was usually "synonymous with the areas where good mortgage lenders with available funds are willing to make their maximum loans to be amortized over a 10–15 year period." Neighborhoods in the B category, by contrast, were usually fully developed and consisted of sturdily built homes that lacked some high-end features. "The second grade or B grade areas," the HOLC's 1937 guidelines for surveyors revealed, "are like a 1935 automobile—still good, but not what people are buying today who can afford a new one."[60]

Implicitly, HOLC officials assumed that first- and second-grade neighborhoods would be all white. Therefore the agency's formal instructions to surveyors included no direct references to the racial demography necessary for a neighborhood to obtain a rating of either A or B. In the descriptions of typical C and D neighborhoods, however, race and ethnicity appeared prominently. Besides containing more obsolete homes, offering poor access to transportation facilities, and providing inadequate utility services, third-grade areas, in the HOLC's calculus, often had no racially restrictive housing covenants and were experiencing "infiltration of a lower grade population." Third-grade areas also had a preponderance of "jerrybuilt" housing as well as "neighborhoods lacking homogeneity." At the bottom of the HOLC's continuum, fourth-grade, or D, areas often suffered from severe housing decay and the widespread influx of nonwhite residents. The least desirable neighborhoods, according to HOLC officials, "are characterized by detrimental influences in a pronounced degree, undesirable population or an infiltration of it, low percentage of home ownership, very poor maintenance, and often vandalism present."[61]

Between 1935 and 1940, representatives from the HOLC's city survey program visited and rated thousands of urban and suburban neighborhoods in 239 metropolitan areas. They conducted their survey of Flint and its suburbs during the summer of 1937. As HOLC agents and local realtors traveled around the area, they carried one-page instruction sheets, which they used to complete "area descriptions" for residential districts. The instruction sheets reminded surveyors to assess the quality of each neighborhood's parks and recreation services, "scenic features," transportation infrastructure, zoning, residential restrictions, schools, churches, business centers, and utilities. Additionally, HOLC leaders instructed mapping consultants to search for nuisances, "such things as obnoxious odors, noises, traffic conditions, fire hazards from certain types of plants such as cleaning plants, refineries, slaughterhouses, disposal and reclaiming establishments." According to the HOLC's instruction form, "infiltrations of lower grade population or different racial groups," along with the "encroachments of apartments commercial or industrial properties," were also nuisances that detracted from a neighborhood's quality.[62] To measure the favorability of an area's social characteristics, the HOLC asked its agents to list the percentage of Negroes, foreign-born residents, and families on relief in each neighborhood and to assess the risk of "infiltration" by these undesirable social groups. By the government's explicit standards, then, racial, ethnic, and class segregation were essential components of neighborhood stability.

Overall, the Flint region rated very poorly under the HOLC's criteria. Even though agents surveyed fifty residential areas scattered throughout Genesee County, only two neighborhoods received A ratings: an especially tony section of Woodlawn Park near downtown and the elite Woodcroft Estates subdivision on Flint's far west side. These two premier residential areas featured many expensive new homes protected by strict racial restrictions. The seven blue areas on Flint's residential security map included several neighborhoods adjoining Woodcroft Estates and Woodlawn Park, GM's Modern Housing Corporation subdivisions, and a few middle-class enclaves on the far west side of the city. Surveyors reported that these neighborhoods contained no Negroes, only a handful of foreign-born occupants, and "few—if any" families on relief.[63] Of the remaining forty-one residential areas surveyed, eighteen received a rating of C. These districts, sprinkled throughout the city and its contiguous suburbs, earned their ratings due to a variety of factors including the prevalence of substandard working-class housing, the encroachment of rental units and commercial establishments, the presence of foreign-born and poor families, high unemployment rates, and proximity to factories, slums, and other nuisances. Significantly, none of the region's C-grade districts contained any black residents. Nonetheless, HOLC survey teams viewed the presence of foreign-born individuals—especially Mexicans, Asians, and southern and eastern Europeans—as a reliable indication that an area was in decline.[64]

Just as individual realtors and bankers had done for decades, the HOLC's survey teams characterized black people, the poor, various classes of immigrants, and all varieties of racial integration as detrimental influences on real estate values, neighborhood stability, and quality of life. Consequently, the three small areas in the Flint region that contained African American occupants in 1937 each received a grade of D. Surveyors explained these ratings with brief annotations such as, "Undesirables—aliens and negroes." Factors such as proximity to black residential areas and the absence of either physical or social barriers between "inharmonious" racial and ethnic groups also played a role in determining which neighborhoods received the lowest ratings. In all but one case, the all-white neighborhoods that bordered areas with black residents received either C or D grades. The sole exception was neighborhood "B-6," which formed the western edge of Woodlawn Park. This all-white enclave housed an assortment of middle-class and wealthy property owners, many of whom inhabited newly built single-family homes. Nevertheless, the area bordered Lapeer Road, the dividing line between Woodlawn Park and the racially mixed neighborhood of Floral Park. As HOLC surveyors noted, neighborhood B-6 was "[t]oo close to 'C' and 'D' areas to the west." Rather than assign the neighborhood a C or D grade, however, HOLC consultants issued a B designation, concluding, "Will hold up. Pride of ownership."[65] West Woodlawn Park no doubt contained many proud homeowners, yet the neighborhood's blue grade stemmed also from its abundance of racially restrictive housing covenants and the impermeability of the Lapeer Road color line that separated it from Floral Park. During the 1930s and 1940s, many

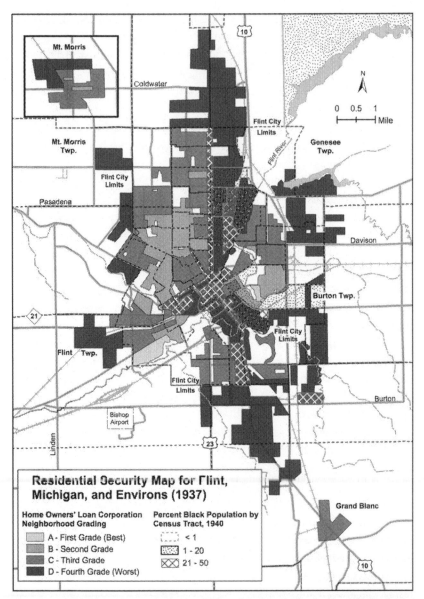

Residential Security Map for Flint,
Michigan, and Environs (1937)

Home Owners' Loan Corporation
Neighborhood Grading

- A - First Grade (Best)
- B - Second Grade
- C - Third Grade
- D - Fourth Grade (Worst)

Percent Black Population by
Census Tract, 1940

- < 1
- 1 - 20
- 21 - 50

MAP 1.2. Residential security map for the city of Flint and Genesee County (1937) and the distribution of the black population, Flint and vicinity, 1940. Between 1935 and 1940, representatives of the federal Home Owners' Loan Corporation created residential security maps for the Flint region and 238 other metropolitan areas across the United States. The HOLC's maps helped to codify and systematize the widely held notion that racial integration was a significant mortgage risk factor and, in the process, played a major role in shaping the practice of redlining in twentieth-century America. In Flint and elsewhere, lenders and officials from the Federal Housing Administration adhered to the HOLC's racist standards for measuring mortgage risk and systematically redlined neighborhoods occupied by African Americans and other people of color. By the late 1940s and early 1950s, federal redlining and other modes of administra-

black residents of Floral Park referred to the Lapeer Road boundary as the Mason-Dixon Line.[66]

Despite their discriminatory guidelines for rating neighborhoods, HOLC officers agreed to refinance numerous mortgages in C and D areas, including many held by blacks. Nationwide, in fact, African Americans accounted for approximately 5 percent of the HOLC's refinanced mortgages.[67] In Flint the HOLC exceeded the national average by refinancing over a quarter of the single-family home mortgages held by African Americans—though that accounted for just 237 properties in all.[68] However, HOLC policies clearly dictated that mortgages in "risky" C and D areas "should be made and serviced on a different basis than in the First and Second grade areas."[69] To account for the supposedly increased risk of operating in C and D areas, HOLC representatives in Flint and elsewhere encouraged local banks to charge higher fees and interest rates when lending in "low-grade" residential neighborhoods. In reality, then, the HOLC's policies did even more than promote segregation and provide a blueprint for the future redlining practices of the FHA. They also contributed in some measure to the economic exploitation of African Americans and other residents of so-called declining neighborhoods.[70]

While it is undeniable that federal surveyors paid careful attention to race and ethnicity when assessing mortgage risks, the Flint case suggests that social class, the provision of municipal services, and other factors were equally important considerations in the creation of the HOLC's residential security maps. Significantly, of the twenty-three Flint neighborhoods that received D ratings, only three contained black residents. Moreover, none of the all-white suburban districts surveyed in 1937 received a grade of A. Among the suburban areas included in the Flint security map, only two small slivers of land in the city of Mt. Morris—both of them well-developed middle-class residential districts—secured B grades. Remarkably, all other suburban neighborhoods in the Flint region earned either C or D ratings. For the entire city of Grand Blanc, located in south suburban Flint, HOLC appraisers issued a C grade while warning of a "trend toward 'D' rating." In addition, the HOLC categorically assigned D grades to all of the close-in working-class residential areas that ringed the city of Flint's borders.[71] Beyond weighing ra-

tive segregation had helped to ossify Flint's color lines while triggering long-term disinvestment from the inner city. Ironically, though, dozens of Flint's working-class suburbs, many of them all white, also suffered during this period due to the FHA's lesser-known suburban redlining policies. *Sources*: Records of the Federal Home Loan Bank Board, Records Relating to the City Survey File, 1935–1940, box 23, Record Group 195, National Archives II, College Park, MD; Minnesota Population Center, *National Historical Geographic Information System: Version 2.0* (Minneapolis: University of Minnesota, 2011). Map by Gordon Thompson.

cial and ethnic considerations, HOLC surveyors looked closely at the class composition of residential areas, unemployment figures, building and zoning codes, housing and school quality, utility services, tax rates, the development of urban infrastructure, and whether local governments were autonomous. The HOLC's agents listed no African American residents in their area descriptions of suburban Flint, but they did find dozens of poor and working-class neighborhoods with shabbily constructed homes that lacked municipal water, sanitary sewers, paved roads, and other amenities. Such areas fared very poorly in the surveys. For neighborhood D-22, a residential area in Burton Township just south of the city, surveyors explained their D rating by noting, "Comparatively new section of cheap construction. Laborer's homes. Outside the city." Similarly, regarding neighborhood D-20 just beyond Flint's southeastern border, a map consultant wrote, "Cheap laborers' cottages. Some as small as two rooms. Easternmost three streets are outside the city limits."[72]

Under the HOLC's complicated assessment system, Flint's heavily working-class suburbs actually rated far worse than many neighborhoods in the city; and the Flint case was not unique. In fact, much the same held true in many other metropolitan centers—especially those with large working-class populations and underdeveloped physical and political infrastructures. In the Detroit area, for example, HOLC officers surveyed eighty-eight suburban communities, all but a few of them restricted to whites, yet sixty-one of them received designations of red or yellow.[73] Likewise, HOLC's map of the Denver metropolitan region included five segregated white suburbs that all received C or D ratings.[74] The Chicago security map fit the same general pattern, with all but the most carefully restricted, politically independent, economically stable, and infrastructure-rich suburban areas garnering grades of red or yellow. In their comments on the working-class suburb of Stone Park, for instance, HOLC surveyors explained their D rating by noting, "There is no paving, no sidewalks, no sewers, and no trees. It is a community of indiscriminate owner-built shacks, the occupants of which are dependent on well water and the use of 'out houses,' which are everywhere in evidence."[75] The suburbs of New York City, Cleveland, and other cities fared somewhat better under the HOLC's stringent standards, with many of them obtaining first- or second-grade ratings, but the improved evaluations were due to the fact that a disproportionate number of these communities were incorporated, possessed strict zoning and building codes, offered high-quality municipal services, and included many attractive middle-class residential districts.[76] When it came to assessing real estate risk, HOLC's appraisers were as concerned with issues related to social class, urbanization, political autonomy, and capital investment as they were with racial segregation.

A thorough review of the HOLC's nationwide city survey program reveals several important truths about the rationale that informed neighborhood appraisals. Clearly, federal surveyors and underwriters looked askance at all-black, all-Latino, and racially integrated neighborhoods wherever they existed. Still, those same individuals also assigned C and D ratings to hundreds of working-class suburbs, including huge sections of Genesee County, primarily because these areas were unincorporated or lacked zoning and building regulations, municipal services, high-quality housing and schools, and other signs of affluence and urbanization. It is important to note that just as in the case of black neighborhoods, HOLC officials did not hesitate to refinance mortgages in poor and working-class white suburbs, even those that received C and D ratings. Nevertheless, by categorizing such communities as risky places for investment, appraisers from the HOLC helped to establish and codify the standards that lenders and FHA officials would later use to engage in suburban redlining.

The Federal Housing Administration

Just one year after President Roosevelt signed the Home Owners' Loan Act, Congress passed the landmark National Housing Act of 1934, which triggered the formation of the Federal Housing Administration. Like the HOLC, the FHA emerged out of a growing belief that the health of the private housing market depended upon the spread of long-term low-interest fully amortized loans. In support of that goal, the FHA offered insurance on home mortgages issued by local lenders. The goal of the FHA's insurance policy was to induce bankers to issue more mortgages by reducing the risks of lending. By almost all measures, the results of its efforts were astounding. Between 1934 and 1972, the FHA insured eleven million mortgages, helping to increase home-ownership rates nationwide from 44 to 63 percent.[77] In the Vehicle City, where single-family homes already predominated, home-ownership rates reached as high as 77 percent under the FHA's insurance program.[78]

Among their many achievements, FHA policy makers helped to standardize building regulations and zoning codes to protect residential neighborhoods from inharmonious land uses. Administrators from the FHA also established national benchmarks for appraising housing quality, neighborhood stability, and actuarial risks. Following a basic mathematical formula, FHA underwriters studied and rated neighborhoods and established strict guidelines for local lenders on which sections of metropolitan areas were eligible to receive federally insured mortgages. In order to guarantee uniformity, the FHA produced underwriting manuals and other widely circulated guidelines that outlined the agency's policies on insuring home mortgages. Like the HOLC, the FHA rated residential areas on a descending scale from A to D based on a wide variety of factors including a neighborhood's economic stability and tax structure, the proximity of "undesirable" groups and adverse environmental influences, the quality of schools and utility services, an area's geographic positioning vis-à-vis civic institutions and commercial facilities, and social appeal. Officials from the FHA did not always use the same color scheme as the HOLC, but their racial and spatial logic was essentially the same.

The FHA's approach to lending in third- and fourth-grade neighborhoods was much different from the HOLC's, however. Drawing from academic research in the field of real estate economics, FHA administrators concluded that most third- and fourth-grade neighborhoods were too risky from an investment perspective to be eligible for federal mortgage insurance.[79] Therefore FHA officials opted to deny mortgage insurance to most of the nation's third- and fourth-grade neighborhoods. Over time, this practice of disinvestment came to be known as *redlining.*

In creating standards for neighborhood appraisals, representatives of the FHA drew heavily on the logic embedded in the HOLC's residential security maps. According to real estate theorist Homer Hoyt, who worked at the FHA's Economics and Statistics Division, "reject" neighborhoods marked by low rental values, dilapidated housing, or the presence of nonwhites were "undesirable for loan purposes." "On the other hand," Hoyt suggested, "where rents are high, the percentage of owner occupancy is high, the condition of the buildings good and there is no race other than white, there will be found areas that rate high for loan purposes."[80] Federal appraisers also rated neighborhoods based upon their proximity to "low-grade" districts and their accessibility to poor people and nonwhites. "A neighborhood should be graded down," FHA records explained, "even though it is now a very good neighborhood if it is in the path of a low-grade neighborhood that is growing in its direction." Under FHA policy, therefore, appraisers needed to consider not only a neighborhood's current status, but also its future prospects for excluding "undesirable races" through zoning and deed restrictions.[81] Ultimately the federal government's policies on race and lending found expression in the FHA's official *Underwriting Manual,* which stated, "If a neighborhood is to retain stability, it is necessary that properties shall continue to be occupied by the same racial group."[82]

From the 1930s until well into the 1960s, most American neighborhoods with African American occupants simply could not obtain high enough ratings to qualify for either FHA-insured mortgages or conventional home loans. Even if a black neighborhood scored highly based on ostensibly nonracial factors such as utility services or access to public transportation, FHA officials instructed appraisers to crunch numbers to ensure a D grade. In his guidelines for appraisers, Hoyt went so far as to write, "All areas that have the lowest rents in the city, the greatest number of buildings needing major repairs, an intermixture of races, and which are generally regarded as vice or slum areas, should not be given a higher rating than one on stability of the neighborhood, protection from adverse influences, and appeal of the neighborhood." "Such ratings," he continued, "will take off 44 from the total rating and almost assure the rejection of the neighborhood unless the score on all the other factors is perfect. . . . Since slum areas may frequently be rated as perfect with respect to adequate transportation and sufficiency of utilities and conveniences which would give them 25 points, it is necessary to rate stability, protection from adverse influences, and appeal of neighborhood no higher than one to avoid giving passing grades to sections that are unquestionably slums."[83] According to the FHA's leaders, neighborhoods that contained black people were "unquestionably slums," or well on their way toward becoming them, and thus ineligible for federal mortgage insurance.

In most cases FHA officials exhibited a strong preference for insuring mortgages on new homes over existing structures. Because cities contained a disproportionate number of older homes, the FHA's policies thus reflected a strong antiurban bias. Nevertheless, in Flint and other metropolitan areas, lenders and government underwriters initially adhered to the HOLC's approach by favoring modern all-white neighborhoods in the city over poorly serviced, predominantly jerrybuilt suburbs.[84] They engaged in suburban redlining because most outlying areas, particularly in working-class districts, lacked the legal and physical infrastructure necessary to meet federal requirements for mortgage insurance. Federal appraisers paid special attention to the poor housing, inadequate utilities, and substandard services that abounded in most suburban areas. They considered race as well, but racial restrictions were insufficient to secure federally insured investment. Officials from the FHA also looked for evidence of urbanization, political autonomy, economic vitality, and other signs that suburban capitalism had taken hold.

Specifically, FHA underwriters wanted suburban governments to enact progrowth economic policies and develop better roads and schools, cleaner water, modern sewer systems, and more restrictive building and zoning codes. Prior to the 1950s, however, many of the nation's suburbs, particularly those in working-class strongholds such as Genesee County, remained largely undeveloped. In the county as a whole, for instance, only the city of Flint and portions of Mt. Morris and Burton Townships could claim fully operational public water and sewer systems. In addition, very few of the area's suburbs possessed vibrant and independent economies. The result of such under-development, according to one FHA report, was "a concentration" of feder-ally backed building in the city of Flint and a small number of incorporated, well-equipped suburban municipalities such as Davison and Mt. Morris.[85] Meanwhile, federal officials redlined most of the area's suburban villages and all but three of the eighteen townships in Genesee County—those being Bur-ton, Flint, and Mt. Morris. Although the FHA insured numerous mortgages in well-developed middle-class sections of suburban Flint, residents in the remainder of the metropolitan region simply could not secure government-backed loans—at least until their elected representatives had met federal standards. What residential development that did occur in these underdevel-oped areas during the 1930s, 1940s, and early 1950s was largely the result of either self-building or privately financed construction. After surveying Flint's "peripheral belt" in the 1940s, famed urban planner Edmund N. Bacon de-fended the FHA's suburban redlining policy, concluding, "By the principles of good business, expressed by the FHA, as well as sound common sense, it would be poor judgment to pour good money for new construction into the bottomless pit of a blighted [suburban] neighborhood."[86]

By engaging in suburban redlining, FHA officials created powerful incen-tives for those on the ground who wanted to revitalize suburbs and make them politically independent. Officials in Genesee County and throughout the nation responded to such inducements by rushing to expand suburban government and infrastructure during the decades bracketing World War II. In doing so, local actors in the Flint region were tapping into a longer tradi-tion of suburban city and community building that had first emerged dur-ing the late nineteenth and early twentieth centuries. Unlike their predeces-sors, however, who received little or no pressure from Washington officials to modernize and urbanize their communities, the suburban capitalists of the mid-twentieth century felt obligated to act in response to federal threats of disinvestment.[87] For suburban officeholders in Genesee County, the les-sons of suburban redlining were clear: in order to receive federal and local backing for new housing, Flint's underdeveloped suburbs would have to lure more middle-class and wealthy home buyers while maintaining political in-dependence, a well-developed physical infrastructure, high-quality munici-pal services, a growth-friendly system of taxation and regulation, a tightly restricted real estate market, and rigid racial segregation. Residents of the seg-

TABLE 1.1. Number of new dwellings constructed in selected units of the Flint metropolitan area by year, 1930–52

	City of Flint	Contiguous inner-ring suburbs (Burton, Flint, Genesee, and Mt. Morris Townships)	Balance of Genesee County (including rural areas)	TOTAL
1930	360	827	552	1,739
1931	122	119	79	320
1932	12	150	100	262
1933	13	68	45	126
1934	20	143	95	258
1935	68	225	150	443
1936	261	302	201	764
1937	257	585	390	1,232
1938	138	437	291	866
1939	439	739	493	1,671
1940	629	543	362	1,534
1941	720	753	502	1,975
1942	250	389	259	898
1943	356	115	77	548
1944	301	152	101	554
1945	243	181	121	545
1946	705	739	493	1,937
1947	875	876	584	2,335
1948	658	679	453	1,790
1949	978	538	359	1,875
1950	2,015	775	525	3,315
1951	1,035	667	450	2,152
1952	1,400	700	475	2,575
TOTAL	11,855	10,702	7,157	29,714

Source: Federal Housing Administration, "An Analysis of the Flint, Michigan SMA (Genesee County) as of January 1953," n.d., Reports of Housing Market Analysis, 1937–1963, box 10, Record Group 31, Records of the Federal Housing Administration, National Archives II, College Park, MD.

regated working-class communities surrounding the city would have to embrace the tenets of suburban capitalism or suffer redlining and other forms of disinvestment.

While government officeholders in suburban Flint rushed to dig wells, lay sewer lines, recruit new businesses, implement zoning codes, and incorporate, the FHA, at least in its first two decades, turned its attention toward premier white neighborhoods in the central city and its most urbanized middle-class suburbs.[88] The results were quite impressive. Indeed, the FHA's "selective credit" mortgage insurance program helped to resurrect metropolitan Flint's dormant real estate industry.[89] In 1933, just a year prior to the passage of the National Housing Act, the city of Flint issued a mere 13 permits for new homes. Seven years later, that number had increased to 629.[90] By the end of the 1940s, builders were constructing nearly one thousand new homes per year in the city, and the home-ownership rate in Flint had increased to nearly 80 percent.[91]

Nevertheless, the FHA's racial and spatial calculus guaranteed that the housing boom occurred only in exclusive and fully serviced residential neighborhoods with all-white populations. Local building activity in the city thus centered on the elite, carefully restricted neighborhoods of Woodlawn Park and Woodcroft Estates as well as the quasi-suburban all-white neighborhoods that formed the inside edges of Flint's municipal boundary. In Woodlawn Park and the neighboring Brookside subdivision, both of which rated highly on the HOLC's security maps, builders constructed 838 new homes in the 1940s. Woodcroft Estates and its adjoining neighborhood to the east gained nearly 750 new units over the same period. After the neighborhoods forming the city's outer edge acquired municipal services, they too received the FHA's support. Just during the 1940s the city issued over four thousand new building permits for homes located in the city's outermost census tracts.[92] Of these new permits, nearly half went to local developers Robert Gerholz and Gerald F. Healy, who together built sixteen hundred west side homes for white buyers on land previously owned by GM's Modern Housing Corporation.[93]

Administrators from the FHA and their local representatives provided builders and lenders with explicit instructions on which areas of the metropolis were safe for new housing and home mortgages. Just as important, however, federal guidelines designated sites that were too risky for investment. In the minds of FHA underwriters and members of the real estate industry, few areas were as risky as neighborhoods inhabited by African Americans. Consequently, most real estate developers adhered to the FHA's explicit recommendations (as well as their own preexisting guidelines) and abstained from building new housing for black buyers during the middle decades of the twentieth century.

This neglect of the African American housing market proved to be especially disastrous during the 1940s, when a series of labor shortages caused by World War II and the postwar economic boom helped to trigger a mass migration of black workers to Flint and other cities. Between 1940 and 1947, the city's black population nearly doubled to approximately twelve thousand. Yet over that time span, builders erected only twenty-five new homes for black purchasers, all of them within the St. John and Floral Park districts.[94] In a 1941 column published in the *Brownsville Weekly News*, an African American newspaper, an unnamed author described the FHA's policies toward black loan seekers, noting, "Flint colored people cannot secure an FHA loan to improve their property nor to build any."[95] Statistics compiled by federal census officials in 1940 lent credence to such complaints. That year, "nonwhites" held just 276 home mortgages in the city of Flint, compared to a figure of 11,977 for whites.[96]

On the rare occasions on which federal officials agreed to insure home mortgages for black buyers, it was only after careful negotiations with builders and bankers. In Flint those deliberations often hinged on finding developers willing to build for African Americans and securing segregated sites for new developments. In 1944 the Urban League of Flint (ULF) launched an ambitious effort to get fifty new homes built for black buyers in Floral Park. After finding no local builders willing to construct the homes, league officials obtained a commitment from Merrill and Company, a New York City building firm. Yet after ten months of site negotiations, the FHA agreed to insure the development only on the condition that the new single- and two-family homes be constructed in a segregated, deteriorated section of Floral Park.[97] Several years later, a similar situation occurred when federal representatives and local builder Ira MacArthur agreed to build two dozen homes for black buyers, but only if they were located in the St. John neighborhood.[98] Reacting against the federal and local policies that circumscribed choices even for well-qualified black buyers, Urban League officials Frank J. Corbett and Arthur J. Edmunds observed, "The main deterrent in the Negro's efforts to improve his housing conditions is not the lack of interest in or desire for adequate and decent homes. Instead, it is the controls that restrict his housing opportunities to the extent that he can only live in the worst neighborhood in the city."[99]

Beyond contributing to legal, popular, and administrative modes of residential segregation, the FHA's mortgage insurance program exacerbated the shortage of affordable rental housing in the city. Between 1942 and 1948, as white home-ownership rates in the area soared, five thousand rental properties—many of them in Floral Park and St. John—quietly disappeared from Flint's housing stock.[100] Hoping to capitalize on the vibrant real estate market that the FHA had helped to create, thousands of white landlords put their units up for sale during and after the war.[101] The result was a rash of evictions that left many migrant families on the brink of homelessness. According to Jake Waldo, an official with the local Congress of Industrial Organizations (CIO) chapter, property owners evicted between eight and twenty black renters per week during the mid- and late 1940s.[102] The combined weight of Flint's rising population, the mass evictions of area renters, and the rigidity of the residential color line caused housing shortages to multiply. In the fall of 1943, the *Flint Journal* noted that desperate home seekers in St. John had converted dozens of abandoned storefronts and automobile garages into temporary shelters, while hundreds of black migrants were "doubling up" with friends and relatives in apartments.[103] The situation was genuinely dire for thousands of new migrants.

Citywide, Flint's overall vacancy rate declined during the 1940s from 2.5 to less than 0.5 percent. In St. John and Floral Park, though, vacancy rates were negligible.[104] Over time, the demand for low-cost rental units provided enough of an incentive for many area homeowners to subdivide single-family homes into tiny apartments. Like their counterparts in booming industrial cities such as St. Louis, Baltimore, and Oakland, local landlords in Flint rented "kitchenettes" and other small, sparsely appointed apartments for exorbitant rates.[105] Because of the city's terrible housing shortages, though, the competition for—and, by extension, the cost of—even substandard units was very high. Between 1940 and 1955, the median monthly rent in the city jumped from twenty-five to seventy-five dollars, one of the largest increases recorded in the United States.[106] "It is next to impossible," Urban League official Charles Eason observed, "to find a Negro home in Flint today where there is not some doubling up."[107]

The 1940s brought a few noteworthy victories in the battle for fair housing. In 1947 FHA officers formally expunged all racial references from the federal *Underwriting Manual*.[108] Just a year later, the US Supreme Court issued its landmark ruling in *Shelley v. Kraemer*, which held that racially restrictive housing covenants were not legally enforceable. During a 1949 speech to the Detroit Mortgage Bankers Association, FHA commissioner Franklin D. Richards trumpeted the new policy changes and encouraged lenders to consider African American borrowers.[109] His successor, Walter L. Greene, went even further, arguing, "FHA policy is to insure projects for open occupancy. We have done so in the past and we shall continue to do so."[110] In reality, though, the FHA did not abandon racial integration as a mortgage risk factor until well into the 1960s. Moreover, despite the revisions to the *Underwriting Manual* and the *Shelley* ruling, which together undermined the legal foundations for residential Jim Crow, real estate professionals and policy makers in Flint and other cities continued to follow the FHA's lead by maintaining popular and administrative modes of segregation.[111]

When confronted by civil rights activists who demanded better housing options for African Americans, federal officials accepted little responsibility for the color line. Instead, FHA spokespersons blamed local real estate developers, "the market," and white homeowners for discrimination against black people.[112] To the FHA's Greene, the problem stemmed largely from bankers' ignorance of the purchasing power of African Americans.[113] According to Albert M. Cole, the chief administrator from the federal Housing and Home Finance Agency, "the real problem [of discrimination] lies with citizens, the businessmen, the builders, the lenders, the realtors and the civic leaders."[114] For their part, local builders, lenders, and property brokers charged the government with establishing strict guidelines and "market rules" for assaying mortgage risk and housing values. In truth, however, popular and administrative methods of segregation worked in tandem to maintain rigid racial separation and unequal housing in the nation's metropolitan centers.

Both before and long after World War II, nongovernmental actors such as bankers, realtors, builders, and homeowners played an important role in forging Jim Crow. Yet with the establishment of the HOLC and the FHA, the federal government helped to codify and nationalize segregationist practices.[115] Through its mortgage insurance and neighborhood appraisal programs, the FHA became one of the nation's primary sponsors of popular, administrative, and, for a time, legal modes of segregation. At the same time, the FHA used its power over home finance policies to create the foundation of a new suburban capitalist order predicated on the decentralization of investment and metropolitan political fragmentation. Nevertheless, housing was only one venue in which the broader struggle over race and development unfolded in the twentieth-century metropolis. Although local activists focused their efforts on battling residential segregation, educational policies were equally important to maintaining the color line.

From Community Education to
Neighborhood Schools

Between the 1930s and the 1970s, the Vehicle City gained international acclaim for both its stylish cars and its neighborhood-based approaches to public education. As with the home mortgage programs of the HOLC and the FHA, Flint's system of "community education" grew out of the turmoil of the Great Depression. In the mid-1930s, in response to the city's rapid growth and a series of polarizing labor organizing drives led by the nascent United Auto Workers union, GM executive Charles Stewart Mott and Frank Manley, a local educator, launched a community education program that remade all of the city's public schools into neighborhood civic centers. Under the leadership of the Mott Foundation, Charles Stewart's private charity, civic leaders forged a close partnership with the board of education that brought millions of dollars in corporate funding to Flint's public schools. Through that collaboration, local officials pioneered a system of community education and "neighborhood schools" that hundreds of cities across the country copied during the postwar period.

Community education boosters sought to build a vibrant public sphere that valued social intimacy, personal improvement, economic productivity, and community spirit in an increasingly privatized, diverse, and divided urban space. Mott and Manley thus envisioned community schools as "lighted community centers"—places where trained professionals could educate rural migrants, promote civic unity, combat communism, and, above all else, boost labor productivity. Open nearly around the clock, Flint's community education facilities served as both neighborhood schools and civic centers. By the thousands, Flint residents descended daily upon their local schools for organized recreation, adult education, and a variety of other activities and services. Located no more than one-half mile from every home in the city (at least in theory), Flint's community schoolhouses were the concrete and steel manifestations of a sweeping urban renewal strategy. However, unlike the so-called slum clearance projects that would later tear through hundreds of inner-city neighborhoods in the 1950s and 1960s, community education derived from an emphasis on people rather than physical spaces. Over the course of its nearly forty-year existence, the Mott program provided valuable educational and recreational programming, delivered vital health care services, and created a bustling public culture.

Nevertheless, the Mott initiative also institutionalized patterns of administrative racial segregation, educational disadvantage, and economic inequality that remained long after its demise. Believing that civic harmony and community solidarity could take root only in socially homogeneous environments, school board members and other education leaders implemented a variety of exclusionary policies that sorted and separated citizens by race and class. Like local realtors and federal mortgage underwriters, the designers of community education thoroughly embraced segregation. However, they did so not simply to protect real estate values and "split" the metropolis along racial lines. They also sought to strengthen community solidarity. School board members and representatives of the Mott Foundation wanted local residents to bond and connect with others in the community, and they believed that the publicly owned schoolhouses scattered throughout the city were the most sensible and cost-effective venues for such activities, but they also believed that people of different races were so incompatible that they could not forge meaningful ties across those racial divides. School and foundation officials thus looked to racial segregation as an indispensable component of community education.[1]

The Roots of Community Education

The Vehicle City's community schools program bore the unmistakable imprint of its primary benefactor, a man known locally as "Mr. Flint, First Citizen." Like most of the pioneers of the automobile industry, Charles Stewart Mott was a product of the Gilded Age, born in 1875 in Newark, New Jersey. His father, John Coon Mott, owned the Genesee Fruit Company, a successful cider and vinegar business, as well as a portion of the Weston-Mott Company, a manufacturer of wheels and axles. After graduating from the Stevens Institute of Technology in 1897 with a degree in mechanical engineering, the younger Mott joined the US Navy and served briefly in the Spanish-American War. Upon his return from battle, Mott accepted an executive position with Weston-Mott in Utica, New York. In 1903 he became Weston-Mott's president and general manager after purchasing the company from family members. Under Charles Stewart's leadership, Weston-Mott doubled the size of its factory and gained new business from automobile manufacturers across the country. News of the company's successes eventually reached Flint, where Billy Durant and others were organizing the Buick Motor Company. In the summer of 1905, Mott received a business proposition from Durant requesting that he move his company to Flint. Hoping to establish a wheel and axle plant adjacent to his proposed Buick facilities, Durant offered Mott an enticing deal that included a large parcel of land, local stock subscriptions in the amount of one hundred thousand dollars, and an exclusive sales contract. After accepting Durant's proposition, Mott relocated to Flint in 1906, just as the local automobile industry was beginning to boom.[2] In 1913 Durant purchased the Weston-Mott Company, forever tying C. S. Mott to General Motors and Flint. Mott accepted Durant's buyout offer in exchange for GM stock and a seat on its board of directors, which he occupied until his death in 1973. Although Mott loathed discussing his personal fortune, his holdings in excess of fifty million dollars made him GM's single largest stockholder and one of the richest men in the nation.[3]

Flush with a sense of obligation toward the city that had given him so much, Mott, a staunch Republican, became an active participant in Flint's public affairs. As a three-time mayor—elected in 1912, in 1913, and again in 1918—Mott worked to establish a business-friendly municipal government by modernizing the city's accounting procedures, cutting expenses, improving infrastructure, and implementing a new building code.[4] Executives from the Chevrolet Motor Car Company expressed their appreciation for such improvements by announcing in 1912 that they would base their manufacturing operations on Flint's west side. Not all of Mott's accomplishments were in the sphere of industry, however. At around the same time, he began donating money to construct a modern hospital and to support citywide recreation programs for children. Through his early experiences in politics and philanthropy, Mott came to an understanding of his civic duties that ultimately drove the community education program: "It seems to me that every person, always, is in a kind of informal partnership with his community. . . . The institutions of a community . . . are the means by which . . . individuals express their faith, their ideals, and their concern for fellow men."[5]

After losing a bid for governor in 1920, Mott returned to Applewood, his sprawling estate in Woodlawn Park, where he immersed himself in philanthropic endeavors. In 1926 he established the Charles Stewart Mott Foundation, a decision based in part on his desire to take advantage of federal tax incentives for charitable giving.[6] Initially the foundation funded a broad range of institutions and programs including the YMCA, the Flint Community Fund, and the American Red Cross. That all changed in 1935, however, when a fellow New York transplant named Frank Manley persuaded Mott to focus his philanthropic efforts on building a community education program in the Flint Public Schools.

Although Manley and Mott often took credit for inventing community education, the idea was clearly not their own.[7] Rather, it was the product of the combined efforts of John Dewey, Clarence Perry, and other reformers and scholars of the Progressive era who sought to place schools at the center of urban life. By the time that Manley arrived in Flint in 1927, Perry's influential writings advocating the "wider use" of schoolhouses and other education-centered urban development strategies had already been circulating widely in Michigan for over two decades.[8] In 1912, for instance, the Flint Socialist Party's formal electoral platform included a plank stating, "School buildings shall be open for the use of the public, when not in use for school purposes." For their part, local business leaders, many of whom wanted to expand adult

FIGURE 2.1. Community education leaders, 1960. This photograph captures a meeting between General Motors executive Charles Stewart Mott (center), local educator Frank Manley (right), and Harlan Hatcher (left), president of the University of Michigan. Between the 1930s and the 1970s, Mott and Manley helped to establish community education programming in Flint and hundreds of other cities. The two men also played an important role in maintaining administrative segregation in Flint's schools and neighborhoods. Courtesy of the Genesee Historical Collections Center, University of Michigan–Flint.

vocational programs at the public's expense, also saw the benefits of community education.[9]

At around the same time that Flint's socialists were endorsing community education, Manley's graduate mentor Wilbur P. Bowen, a physical education professor at Michigan State Normal College in nearby Ypsilanti, became a strong proponent of opening up schools for public use. Drawing extensively on the works of Perry and others, Bowen maintained that local policy makers should transform public schools into around-the-clock neighborhood civic centers. In their 1923 monograph *The Theory of Organized Play*, Bowen and a colleague named Elmer D. Mitchell argued that public school buildings ought to be the "dwelling place for a large community family."[10] As a graduate student of Bowen's during the 1920s, Manley must have been familiar with his mentor's work on community education as well as the broader progressive discourse on the wider use of school buildings. Regardless, there can be little doubt that when Manley decided to accept a teaching position with the Flint Public Schools in 1927, he was already a proponent of community education.

Shortly after Manley's arrival in Flint, the onset of the Great Depression triggered a massive crisis in the city's education system. As in other cities, the decline in tax revenues that occurred in Flint during the 1930s forced board of education officials to curtail both the length of the school day and the duration of the academic year. The result was a spike in juvenile delinquency and child accidents. Angered by the situation, Manley sought support for a citywide recreation program.[11] It was not until 1935, however, that his idea took hold. That year, Manley persuaded members of the Flint Board of Education and Charles Stewart Mott to cosponsor a pilot program in community education in five of the city's all-white elementary schools.[12] An immediate success, the program attracted well over five thousand participants per week in its inaugural year and helped to reduce juvenile delinquency in the city by approximately 70 percent.[13]

Initially Manley and Mott focused exclusively on the issues of juvenile delinquency and child safety. However, events in 1936 and 1937 compelled the two men to reassess the nature of the city's problems. On December 30, 1936, the Great Flint Sit-Down Strike erupted in GM's Fisher Body 1 plant and spread quickly throughout the city. The strike, which turned out to be one of the most consequential labor struggles in American history, pitted members of the newly formed UAW against GM in a battle for union recognition. After forty-four grueling days, the dispute ended on February 11, 1937, when GM agreed to recognize the union and engage in collective bargaining. Inspired by the victory, workers throughout the nation launched a wave of industrial organizing drives in the aftermath of the Flint sit-down. The Flint strike also

FIGURE 2.2. The Great Flint Sit-Down Strike, 1937. For nearly two months during the winter of 1936–37, thousands of union members, including these workers from the Fisher Body 1 plant, photographed on January 12, 1937, occupied GM factories in the Flint area in order to win collective bargaining rights. Their victory helped to solidify the power of the UAW while inspiring workers in other industries to pursue collective bargaining. The upsurge in union organizing that occurred during the 1930s also motivated Charles Stewart Mott and Frank Manley to launch their community education initiative. Courtesy of the Genesee Historical Collections Center, University of Michigan–Flint.

marked the birth of the UAW and its parent organization, the CIO, as a force in both industrial relations and American politics.[14]

However, the battle in Flint also served to galvanize Mott and other industrialists. Mott reacted harshly to the news of factory occupations in Flint, viewing the CIO and the UAW as protosocialist threats to private property and the rule of law.[15] More broadly, though, he and other American business leaders found the urban milieu of the 1930s deeply troubling.[16] When Mott first arrived in Michigan in 1906, Flint was a city known more for its horse-drawn carriages, tree-shaded boulevards, and small-town atmosphere than for automobiles, skyscrapers, and highways. To Mott and many of his peers, it seemed to be a small, homogeneous, and harmonious city—a socially intimate place where urban planning, transportation technology, architecture, civic pride, and racial segregation combined to create a palpable sense of community solidarity. Indisputably it was also a company town, or what one UAW activist referred to as a "giant stronghold of feudal capitalism."[17] By the late 1930s, however, Flint had become a fractured and diverse city of big cars, big factories, big unions, and big business. As such, it was virtually unrecognizable to Mott and other civic elites.[18] Like Manley, Mott looked to community schools as a means to undermine the UAW's newfound political power while simultaneously restoring a sense of personal responsibility and civic solidarity among city residents. "We must build back to community activities to get people to know their neighbors and bring about a wholesome, small-town atmosphere in a big city," Mott claimed.[19] To achieve those goals, Mott and Manley made the decision to expand their inchoate community education initiative and become more active in the governance of the public schools.

As part of that effort, the two men began promoting their supporters for school leadership positions and endorsing sympathetic candidates for the at-large, citywide races for the Flint Board of Education. Their goal, in Manley's words, was to get "the real leaders of our community on the board of education and in other influential positions."[20] That strategy first began to bear fruit in the mid-1940s, when a sympathetic majority on the school board granted official representation to the Mott Foundation by appointing Frank Manley to the position of assistant superintendent of community education. Once in a position of power, Manley and other representatives of the Mott Foundation obliterated all but the most trivial distinctions between private interests and public policies as they spread community education and a GM-approved vocationally centered curricula throughout the city.[21] As noted in one survey of Flint schools, "Frank J. Manley is, for all practical purposes, the superintendent of Flint's schools and through him the influence of the foundation is applied . . . to every sector of school programming."[22]

The Birth of Neighborhood Schools

By the early 1940s Mott's funding had helped to bring Manley's vision of a "school-centered" community close to fruition. Yet it was not until 1947 when the first bona fide community school opened in Flint. In a move that broke sharply with precedent, Manley chose predominantly black Fairview Elementary School for the inaugural demonstration program. The Mott program had operated exclusively in all-white schools through the first decade of its existence, serving at most a handful of African Americans who had slipped across Flint's color line. A devastating race riot that erupted in Detroit in 1943 forced Manley and Mott to consider a different approach, however. Hoping to prevent an outbreak of violence in Floral Park and St. John, members of the Mott Foundation designed the Fairview Project as an experiment in community improvement from within a segregated neighborhood school.[23] Like the St. John district that surrounded it, Fairview was predominantly black by the close of World War II. In 1947 the school had an enrollment of 393 students, 92 percent of whom were African American. Although the area surrounding the school was economically diverse and socially vibrant, it was also among the most polluted, overcrowded, and dilapidated sections of the city.[24] With the prospect of riots and other civil disturbances looming ominously, Mott and Manley hoped that the Fairview Project would "lessen frustration and aggression" among residents by improving the social conditions in one specific neighborhood. More generally, though, the foundation's strategy at Fairview was to combat inequality by raising the living standards and intellectual capacities of a few select African Americans. As Manley noted, "We should put the Negro on our middle class level of thinking."[25]

Prior to the launch of the Fairview program, the Mott Foundation conducted a survey of the neighborhood and used the findings to create in-service trainings for teachers. Additionally, foundation officials purchased new furniture, lights, artwork, and educational materials for the school. Mott also funded a complete overhaul of the Fairview curriculum. For their part, Manley and members of the board of education approved a new school orchestra, a series of enrichment activities for students, an updated physical education program, longer class periods, regular intelligence and achievement testing, and an improved pupil-teacher ratio. Mott also subsidized medical exams and a free breakfast program for poor students, along with after-school activities for the children of working parents. Making Fairview a true community school, however, entailed bringing adults into the building. With that in mind, school officials provided parents and other neighborhood residents with free meeting space and offered a variety of courses and recreational programs. Among the most important of these, according to foundation officials, was the adult homemaking class, which taught mothers to plan and cook meals, clean house, and make clothing. Members of the school board and foundation officials actively recruited poor mothers for these classes, believing that education and moral uplift were the keys to eradicating poverty.[26]

A year after unveiling the Fairview Project, the Mott Foundation put together a major publicity campaign to advertise its successes. Because the foundation released almost no data regarding the experiment, it is unclear whether the project significantly improved health outcomes or reduced unemployment and poverty in the St. John Street community. Still, in creating an attractive, well-utilized community center for citizens of all ages, the

Fairview Project was a rousing success. By the close of the 1940s, hundreds of Flint's leading white citizens had crossed the color line to see the Fairview community school firsthand. As Manley later remembered, "Almost every organized group in the community had at least one speaker who discussed in detail what was taking place at Fairview."[27] In support of this policy of neighborhood-based community schools, C. S. Harding Mott, son of Charles Stewart and vice president of the Mott Foundation, said, "We cannot solve world affairs satisfactorily without first tackling our community problems. We believe we have the best chance for doing this at the neighborhood school."[28] With the Fairview experiment, the segregated neighborhood school had emerged as ground zero for community education.

Mott and Manley looked to capitalize on the positive publicity generated by the Fairview Project by organizing a fundraising campaign to expand community education. Although Mott had ample funds to support a citywide program on his own, he and Manley were frugal by nature and hoped that taxpayers would eventually shoulder the burden for financing community schools. Gaining that support was no small challenge in a city such as Flint, however. Compared with other school systems of similar size, the Flint Public Schools typically ranked near the bottom in pupil expenditures, teacher pay, and construction outlays. A combination of factors including the city's rapid growth, its business-friendly system of local tax limitation, and the school board's conservative "pay as you go" financing guidelines were largely to blame. Due to perennial budget shortfalls, the Flint Board of Education constructed no new public schools between 1929 and 1950 and operated a full school year only once between 1932 and 1946.[29] Supporters of the Mott program looked to reverse that trend by waging an aggressive bond campaign in 1950. At a June 1 event sponsored by the board of education, Mott announced a one-million-dollar donation to fund the construction of a four-year college in Flint. However, he made his offer contingent on voters' approving the new education bond. Several days after Mott's announcement, voters overwhelmingly endorsed the seven-million-dollar bond issue—the first such victory in the city's history.[30] Flush with desperately needed funds, the school board moved quickly to erect new schools and implement community education on a citywide basis.

The June vote marked the beginning of what one *Flint Journal* writer dubbed a "golden era" of public education. Not coincidentally, this period of immense public investment in Flint's schools coincided with the pinnacle of administrative school segregation. By the end of the 1950s, board members had used their new revenues to build eight new elementary schools and launch scores of new community programs. However, the board's investment strategy privileged the all-white subdivisions of the urban periphery over the increasingly black neighborhoods of the inner city. In fact, only one of the eight new schools constructed after the war, Stewart Elementary, contained African American pupils, and it opened in 1955 with an 83 percent black enrollment. The remaining seven schools, all located in the booming white neighborhoods ringing the city's core, enrolled no black students whatsoever.

The most ballyhooed of the new facilities was Freeman Elementary, the city's first school designed specifically for community education. Located in the all-white Farnumwood neighborhood on Flint's far south side, Freeman was the embodiment of community education. The school sat on five acres of land and contained forty-three thousand square feet of floor space, significantly more than the older multistory schools near downtown. Complete with a wooded grove for outdoor activities, picnic tables, and a large parking lot, Freeman's spacious grounds signaled the board's embrace of the suburban ideals of open space, single-story architecture, and automobile-centered transportation. Inside the building were ten classrooms as well as a gymnasium, a library, an auditorium, an arts and crafts center, a community room, a health clinic, a teacher's rest area, and administrative offices.[31] Reflecting the board's desire to bring people of all ages into the school, the gymnasium and auditorium included extra space for dances, town hall meetings, and other community events, and the library was open to the public. According to Manley, the design of the building was a clear "physical expression of an educational philosophy."[32]

Organizationally complex, the Mott program consisted of eleven overlapping divisions, all of them operating within the framework of "community education." By far, the recreation and adult education divisions were the most popular. However, Mott's initiative also included a dental clinic, a health center, a children's camp, and a variety of other agencies and services. To promote his efforts, Mott established a graduate training institute and a workshop series for visitors. By 1970 the foundation's publicity campaign had helped to spread community education programs to over three hundred school districts nationwide.[33] Regardless of the division in question, the principal aim of the program was to "bring people into the school so that their interest would draw them into projects which would lead to a better community."[34] Invariably, that involved teaching workers how to become more efficient and educating the poor about personal responsibility, free markets, and self-sufficiency.[35] Courses such as "Economics of Our Community," designed by scholars from the University of Chicago, offered conservative counterpoints to both Soviet-style communism and Keynesian liberalism.[36] Through this and other courses, Mott and Manley sought to construct a

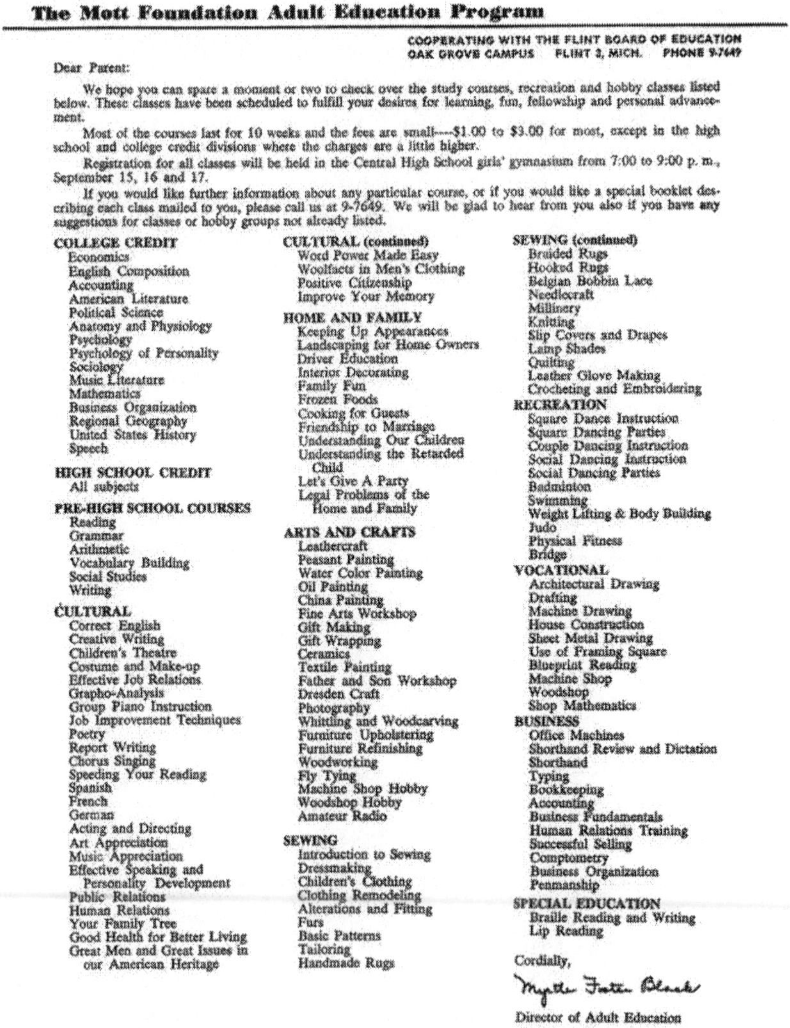

The Mott Foundation Adult Education Program

Dear Parent:

We hope you can spare a moment or two to check over the study courses, recreation and hobby classes listed below. These classes have been scheduled to fulfill your desires for learning, fun, fellowship and personal advancement.

Most of the courses last for 10 weeks and the fees are small----$1.00 to $3.00 for most, except in the high school and college credit divisions where the charges are a little higher.

Registration for all classes will be held in the Central High School girls' gymnasium from 7:00 to 9:00 p. m., September 15, 16 and 17.

If you would like further information about any particular course, or if you would like a special booklet describing each class mailed to you, please call us at 9-7649. We will be glad to hear from you also if you have any suggestions for classes or hobby groups not already listed.

COLLEGE CREDIT
Economics
English Composition
Accounting
American Literature
Political Science
Anatomy and Physiology
Psychology
Psychology of Personality
Sociology
Music Literature
Mathematics
Business Organization
Regional Geography
United States History
Speech

HIGH SCHOOL CREDIT
All subjects

PRE-HIGH SCHOOL COURSES
Reading
Grammar
Arithmetic
Vocabulary Building
Social Studies
Writing

CULTURAL
Correct English
Creative Writing
Children's Theatre
Costume and Make-up
Effective Job Relations
Grapho-Analysis
Group Piano Instruction
Job Improvement Techniques
Poetry
Report Writing
Chorus Singing
Speeding Your Reading
Spanish
French
German
Acting and Directing
Art Appreciation
Music Appreciation
Effective Speaking and
   Personality Development
Public Relations
Human Relations
Your Family Tree
Good Health for Better Living
Great Men and Great Issues in
   our American Heritage

CULTURAL (continued)
Word Power Made Easy
Woolfacts in Men's Clothing
Positive Citizenship
Improve Your Memory

HOME AND FAMILY
Keeping Up Appearances
Landscaping for Home Owners
Driver Education
Interior Decorating
Family Fun
Frozen Foods
Cooking for Guests
Friendship to Marriage
Understanding Our Children
Understanding the Retarded
   Child
Let's Give A Party
Legal Problems of the
   Home and Family

ARTS AND CRAFTS
Leathercraft
Peasant Painting
Water Color Painting
Oil Painting
China Painting
Fine Arts Workshop
Gift Making
Gift Wrapping
Ceramics
Textile Painting
Father and Son Workshop
Dresden Craft
Photography
Whittling and Woodcarving
Furniture Upholstering
Furniture Refinishing
Woodworking
Fly Tying
Machine Shop Hobby
Woodshop Hobby
Amateur Radio

SEWING
Introduction to Sewing
Dressmaking
Children's Clothing
Clothing Remodeling
Alterations and Fitting
Furs
Basic Patterns
Tailoring
Handmade Rugs

SEWING (continued)
Braided Rugs
Hooked Rugs
Belgian Bobbin Lace
Needlecraft
Millinery
Knitting
Slip Covers and Drapes
Lamp Shades
Quilting
Leather Glove Making
Crocheting and Embroidering

RECREATION
Square Dance Instruction
Square Dancing Parties
Couple Dancing Instruction
Social Dancing Instruction
Social Dancing Parties
Badminton
Swimming
Weight Lifting & Body Building
Judo
Physical Fitness
Bridge

VOCATIONAL
Architectural Drawing
Drafting
Machine Drawing
House Construction
Sheet Metal Drawing
Use of Framing Square
Blueprint Reading
Machine Shop
Woodshop
Shop Mathematics

BUSINESS
Office Machines
Shorthand Review and Dictation
Shorthand
Typing
Bookkeeping
Accounting
Business Fundamentals
Human Relations Training
Successful Selling
Comptometry
Business Organization
Penmanship

SPECIAL EDUCATION
Braille Reading and Writing
Lip Reading

Cordially,

Myrtle Foster Black

Director of Adult Education

FIGURE 2.3. Mott Foundation adult education leaflet, n.d., circa 1950. As part of their community education initiative, representatives of the Charles Stewart Mott Foundation and local school officials sponsored numerous recreation and enrichment programs for residents of Flint. The adult education courses featured in this leaflet were among the most popular of the Mott offerings. Courtesy of the Richard P. Scharchburg Archives, Kettering University, Flint, MI.

national model for urban renewal rooted in the principles of self-help and free enterprise. The foundation also supported the Personalized Curriculum Program (PCP) for potential school dropouts and the Better Tomorrow for the Urban Child (BTU) initiative for black children. The BTU program emphasized vocational instruction, casting aside what Manley called the "intellectualism" of traditional education in favor of practical training for minority children. The goal in all of this was to create innovative local programs to counter the liberal policies of the New Deal while simultaneously fostering intraracial community solidarity and racial segregation.[37]

Owing to the Mott Foundation's special relationship with the Flint Public Schools, community education had an impressive reach. Of the nearly two hundred thousand people who lived in postwar Flint, nearly half participated in community education programming. And participants ranged across the demographic spectrum. "At our house," one resident noted, "we just take it for granted that we are all going to take Mott Foundation classes." According to another enthusiastic participant, community schools "are like country clubs for the working class."[38] In effect, Mott and Manley used their extraordinary influence to revolutionize the role of the school in the process of community building. They did so by successfully transforming ordinary schoolhouses into the neighborhood civic centers that leading intellectuals such as Clarence Perry had first envisioned during the Progressive era. With the citywide implementation of community education, Flint's public schools became the focal points of civic life.[39]

The Magic Lines of Segregation

Without a doubt, the Mott program provided important community services and helped to strengthen numerous associational bonds in postwar Flint. At the same time, though, Mott's initiative was a very clear manifestation of segregationist community building that ultimately hardened social inequalities. From the start, in fact, both administrative racial segregation and racist resource allocation policies were crucial, if unspoken, components of community education. As critics correctly pointed out, segregated housing often led to Jim Crow schools. Yet pupil segregation was not simply an unintended outcome of the turn toward community education and neighborhood schools in a residentially divided city. Beyond the student segregation enforced by redlining and other housing policies, the deliberate actions of school officials helped to preserve and extend Jim Crow in the Vehicle City.

Unlike large sections of the South, however, where legal segregation reigned until well into the 1960s, many northern and western cities such as Flint, Boston, and Seattle exemplified an administratively driven mode of school segregation.[40] Despite the existence of an 1867 state law forbidding racial segregation in Michigan's public schools, Mott Foundation officials and members of the Flint Board of Education repeatedly manipulated student transfer policies, built new facilities in segregated neighborhoods, and gerrymandered attendance boundaries—all in an attempt to maintain the color line. The board's policies left schools as segregated, or even more so, than the highly segregated neighborhoods they served. Moreover, because the Mott program operated its community programming through neighborhood elementary schools, the gerrymandering of attendance zones affected both children and adults. As much as subdivision lines, then, school attendance zones created neighborhood boundaries and established the rules of membership in the city's rigidly segregated public sphere. The public policies that maintained segregation in Flint's schools and neighborhoods reflected a broad ideological consensus among whites that racial segregation was essential to supporting property values, strengthening community ties, and improving the city.

Prior to 1954, when the Flint Board of Education formally adopted boundaries for elementary and junior high schools, district officials designed attendance zones using an ad hoc calculus that revolved around new housing construction, building capacities, travel times for students, natural barriers, safety factors, and race.[41] During that period, the board handled student transfers on a case-by-case basis. "In the absence of an official policy," a 1949 school board report noted, "applications for transfers have been considered more or less in accord with unwritten precedents stemming from a combination of past practice, custom, and administrative dicta."[42] Since almost all of Flint's black citizens already lived within the areas served by the four elementary schools of the St. John and Floral Park neighborhoods, local policy makers only seldom employed racial gerrymandering prior to the 1950s. For white children who resided in these mixed districts, the board routinely granted transfers to all-white schools. Attendance and transfer policies exhibited no such fluidity when black families moved into white neighborhoods, however. A case in point occurred in February 1935, when the board of education quietly altered the unofficial boundaries between Dort and Parkland Elementary Schools, two adjoining districts serving the North End, after a handful of blacks from the Parkland area moved into a neighborhood near the all-white Dort School.[43] The decision to shift the attendance boundaries one block southward resulted in the immediate return of all of Dort's black pupils to Parkland. The 1935 boundary move marked the first documented case of deliberate administrative segregation in the Flint Public Schools. Scores more would soon follow.[44]

In the decades preceding the Supreme Court's 1954 *Brown v. Board of Education* decision, a powerful combination of housing and education policies kept Flint's students and their families rigidly segregated. Out of the twenty-seven public elementary schools open in the city in 1950, nineteen were all white, two were over 95 percent white, and two were over 95 percent black. Of the four remaining schools, only one, Dewey, could claim an integrated student body, and that status proved to be fleeting.[45] Despite the city's large black population increases in the 1950s, the racial geography of local schools changed very little during that decade. In fact, in a number of instances, segregation actually increased during the postwar era. By 1955 the Flint Board of Education was operating thirty-three public elementary schools with an enrollment of over twenty-one thousand pupils. Among these schools, seventeen were all white, five were at least 95 percent white, and four were at least 95 percent black. Out of the seven remaining schools, three tallied black enrollments of 50 percent or higher. At decade's end, none of the city's 6,490 black pupils studied in a building that represented, within ten percentage points, the racial demographics of the system as a whole. Perhaps more revealing, over half of the elementary students in the city attended completely segregated schools. Despite the tremendous national significance of the court's *Brown* ruling, its moral and legal demands did not substantially affect public policy in the Flint Public Schools.[46]

School construction, location, and boundary decisions played decisive roles in maintaining the educational and residential color lines in postwar Flint. This was especially true in the 1950s, when student enrollments spiked across the city. Like their counterparts in other urban centers, members of the Flint Board of Education subsidized segregation by constructing new schools within all-white FHA-backed neighborhoods.[47] Wherever possible, the board sought to follow the tenets of community education by locating new elementary schools at or near the center of newly built white subdivisions. The goal, Manley and other education officials claimed, was to ensure that all the city's pupils lived within one-half mile of a neighborhood school. The result, however, was new formations of administrative segregation. On the strength of a long string of successful millage campaigns, the school board continued its neighborhood school construction policies throughout the postwar era, erecting over a dozen additional elementary facilities, almost all of them within segregated residential enclaves.[48]

Still, school officials accepted no blame for any type of segregation. Instead, board members, Mott Foundation personnel, and school superintendent William Early consistently maintained that educational segregation in Flint was solely the result of extralegal forces in the private sector. If pressed, school representatives alleged that "de facto" segregation in the schools stemmed directly from segregated housing.[49] Clearly, Early and his colleagues spoke the truth in linking popular attitudes, housing, and pupil segregation. However, citizen racism and residential segregation could explain neither

TABLE 2.1. Number of black pupils and total student enrollment in Flint's public elementary schools for selected academic years, 1950–60

Elementary school	1950–51	1955–56	1959–60
Civic Park	0/899	0/879	1/831
Clark	474/645	559/582	611/614
Cody	0/555	0/724	0/742
Cook	0/514	0/591	0/557
Coolidge	0/517	0/815	0/525
Cummings			0/664
Dewey	65/463	355/710	474/712
Dort	0/615	251/739	626/949
Doyle	0/402	37/530	404/623
Durant-Tuuri-Mott	1/363	8/402	21/474
Fairview	385/393	524/535	531/531
Freeman		0/761	0/908
Garfield	0/699	0/776	0/679
Gundry		0/655	0/1,103
Hazelton	0/282	13/321	8/260
Homedale	0/759	0/809	11/830
Jefferson	181/440	660/776	954/980
Lewis	0/800	2/499	1/712
Lincoln	0/627	9/388	47/379
Longfellow	0/746	0/916	0/936
Martin	0/494	115/639	750/1,007
McKinley	0/460	0/734	0/852
Merrill		0/800	0/757
Oak	0/437	2/489	37/435
Parkland	394/395	585/591	615/618
Pierce		0/479	0/484
Pierson	0/883	0/916	0/909
Potter		0/864	0/1,386
Roosevelt	428/516	592/623	570/718
Selby			0/811
Stevenson	0/361	0/366	1/371
Stewart		440/530	621/819
Walker	14/475	190/542	207/412
Washington	0/933	0/699	0/572
Zimmerman	0/725	0/907	0/591
TOTAL	1,942/15,398	4,342/21,587	6,490/24,751

Source: "Racial Distribution by School, K–12—1950–1968," n.d., box 23, 78-8.3-31f, Frank J. Manley Papers, Richard P. Scharchburg Archives, Kettering University, Flint, MI.

the policies that maintained the color line on the city's rapidly shifting racial frontiers nor the school board's involvement in preserving segregation within the local housing market.

School officials used a variety of tools to maintain administrative segregation. Sometimes they gerrymandered attendance boundaries, other times they used temporary classrooms, and in some instances they quietly approved illegal racial transfers—all out of a desire to hold the color line and maintain popular support for Mott programming.[50] Board members also used their partnerships with residential developers and federal housing officials to ensure the racial exclusivity of new subdivisions and school facilities. On at least several occasions, the board formally collaborated with local builders and representatives of the Federal Housing Administration in the planning and construction of new schools adjacent to racially restricted neighborhoods.[51] In Flint these segregated school-neighborhood units were a matter of official government policy.

Gerrymandering was among the most common and effective of all the methods used to maintain administrative segregation in Flint and other urban school systems.[52] One of the clearest examples of this practice occurred during the early and mid-1950s, when the Flint Board of Education repeatedly redrew attendance zones to maintain pupil segregation in the racially contested neighborhoods between Clark and Pierce Elementary Schools. In 1952 the school board opened Pierce with an all-white enrollment in one of the city's most exclusive and segregated east side neighborhoods. As was the case with several other schools that opened in the 1950s, the board had an opportunity to integrate Pierce from the outset by doing nothing more than adhering to a strict geographic policy of neighborhood schools. If the board had honored its own guideline that children, wherever possible, should attend the schools closest to their homes, then Pierce would have drawn a significant number of African American pupils from the racially transitional Elm Park and Sugar Hill enclaves adjoining Floral Park on the southeast. These two predominantly working- and middle-class residential areas were located just south of the new Pierce building. Prior to Pierce's opening, students from Elm Park and Sugar Hill attended Clark Elementary, a decaying, overcrowded structure with a black enrollment of over 95 percent. Despite the fact that many of the Sugar Hill and Elm Park students lived closest to the new building, school board members excluded them from the Pierce district.[53] Although Pierce did not reach full capacity until the early 1960s, the board also refused all transfer requests from black pupils. Moreover, over the loud objections of numerous black parents, board members repeatedly redistricted Pierce during the 1950s and early 1960s in order to maintain the school's all-white enrollment and keep its Mott programs segregated.[54] In several cases, the board's districting decisions left neighbors from the same street attending different "neighborhood schools." "They drew boundaries around houses," Ruth Scott remembered, "down the middle of the street. . . . When blacks moved onto a street, they would change the boundaries."[55] As a direct consequence of the board's decisions, Pierce remained all white at the end of the 1950s, while Clark's black enrollment soared to 99.5 percent. The rapidly shifting, oddly shaped boundaries that separated Pierce and Clark and other school pairings clearly eviscerated the notion that a colorblind ideal of proximity drove the community education concept. In recognition of the intentional racial gerrymandering that affected Pierce, Clark, and other area facilities in the 1950s and 1960s, the federal government later went so far as to acknowledge "the substantial duality of the Flint School System."[56]

When the school board announced the opening of Stewart Elementary in 1955, it again found itself facing the Pierce-Clark segregation issue. Located one-half mile south of Pierce and three-quarters of a mile east of Clark, Stewart served the residential areas south of the Lapeer Road color line, including sections of Elm Park and Sugar Hill. Prior to Stewart's opening, the board had blocked black pupils from transferring to Pierce, choosing instead to house nearly three hundred African American students from the overcrowded Clark building in dilapidated temporary structures. Through the careful drawing of Stewart's boundaries, the board relieved such overcrowding at Clark while maintaining racial homogeneity at Pierce. In order to exclude all blacks living in the area from the Pierce district, the board once again violated its policy on proximity by extending the Stewart boundary over a mile and a half to the north and east, resulting in an 80 percent black enrollment at Stewart while Pierce remained under capacity and all white. In effect, the construction of Stewart allowed the board to ease overcrowding at Clark by sending white pupils to Pierce and black pupils to Stewart, regardless of their proximity to either.[57] For those white pupils who still found themselves enrolled at Stewart, school officials established separate classrooms for white and black children.[58] When confronted with all of this news, Edgar B. Holt of the Flint chapter of the National Association for the Advancement of Colored People (NAACP) resorted to using supernatural metaphors in an attempt to communicate the geographic absurdity of the city's school districts: "There were magic lines for racial discrimination. The Flint Board of Education used these magic-racial lines in establishing school boundaries as if they were sacred."[59] As Holt no doubt understood, Flint's neighborhood schools were not neutral entities that corresponded to colorblind demographic forces. Rather, they were malleable, socially constructed institutions that reversed even modest breaches of the residential color line. In truth, Flint's neighborhood schools were imagined communities rooted in elaborate geographic fictions.[60]

On February 9, 1954, just three months prior to the *Brown* verdict, the Flint Board of Education jettisoned its secretive methods of drawing pupil

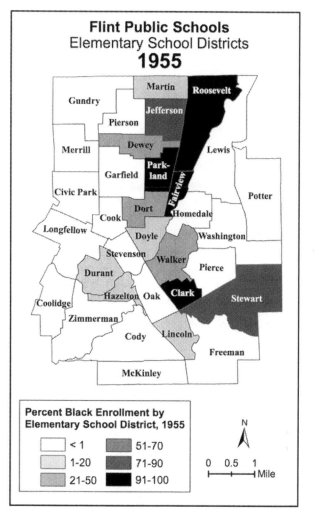

Flint Public Schools
Elementary School Districts
1955

Martin
Roosevelt
Gundry
Jefferson
Pierson
Dewey
Merrill
Lewis
Park-
Garfield   land
Fairview
Civic Park
Potter
Dort
Cook   Homedale
Longfellow
Doyle   Washington
Stevenson
Walker   Pierce
Durant
Clark
Hazelton Oak   Stewart
Coolidge
Zimmerman
Cody   Lincoln
Freeman
McKinley

Percent Black Enrollment by Elementary School District, 1955

< 1	51-70
1-20	71-90
21-50	91-100

N

0   0.5   1
Mile

MAP 2.1. Public elementary school attendance boundaries in the city of Flint and the distribution of black student enrollment, 1955. As an extension of a much broader philosophy of community building, members of the Charles Stewart Mott Foundation and the Flint Board of Education maintained rigid racial segregation in the city's public schools from at least the mid-1930s until the late 1970s. Of all the strategies used to impose administrative school segregation, the gerrymandering of student attendance boundaries, as shown above, was among the most common. By drawing oddly shaped attendance districts such as the 1955 zones for Clark, Dewey, Oak, Pierce, Stewart, and Walker Elementary Schools, board of education officials were able to maintain highly segregated facilities even as residential neighborhoods were becoming more integrated. *Sources*: Official Minutes of the Flint Board of Education, 1955, Flint-Genesee County Reference Collection, Flint Public Library, Flint, MI; Minnesota Population Center, *National Historical Geographic Information System: Version 2.0* (Minneapolis: University of Minnesota, 2011).

attendance zones by formally establishing new school boundaries throughout the city. Though it would take nearly twenty years for the federal government to acknowledge the illegality of the 1954 boundary reorganizations,
the board's plan clearly expanded administrative segregation, particularly in
elementary schools. This was especially obvious on the northeast side of the
city, where the 1954 plan split the integrated enrollment of Roosevelt School
between the official school site on Thetford Road and a group of temporary
structures located a mile and a half to the north. For the main Roosevelt
building, located in a majority-black neighborhood near the intersection of
Stewart Avenue and Dort Highway, the board drew boundaries that resulted
in a 95 percent black enrollment. However, to the temporary structures north
of Roosevelt the board assigned fifty white pupils and perhaps one black student. In effect, the 1954 plan split the Roosevelt enrollment along both racial and geographic lines, establishing separate facilities for white and black
students from the same school.[61] Angered by these and other segregationist
boundary reorganizations, UAW officials Robert Carter, Norman Bully, and
Earl Crompton drafted a joint letter in September 1955 accusing the school
board of violating the US Constitution. Their letter and many others like it
went unanswered, however.[62]

In other parts of the city, board members practiced administrative segregation by allowing racial transfers, tweaking school boundaries, tracking
students, and erecting temporary classrooms.[63] On the northwest side, school
board members often turned to temporary structures known as "primary
units" to preserve administrative segregation in the face of racial transition
and student enrollment increases. Unlike the temporary classrooms erected
during the postwar era by many other boards of education, which often abut-
ted permanent school buildings, Flint's primary units sat in residential neighborhoods across the city.[64] The units resembled small single-family ranch
homes, with each structure housing one classroom, and they were virtually
indistinguishable from surrounding residential landscapes. School board
members promoted the units as fiscally sound solutions to the problem of
overcrowding, arguing that they could be resold as single-family homes upon
the final resolution of enrollment crises. Because primary units allowed the
board to maintain racial separation without shifting attendance boundaries,
they were also useful tools for upholding the color line. Between 1950 and
1966, the board of education oversaw the construction of 116 such structures,
many of them in newly integrated sections of the northwest side. Although
board members could have resolved overcrowding and pupil segregation
there through limited boundary shifts between Martin, Jefferson, Pierson,
and Gundry Elementary Schools, instead they opened forty-one primary

units. By 1959 the board's decision had resulted in an especially glaring dis-juncture between Pierson and Jefferson. Although only a half mile separated the two facilities, Pierson contained an all-white enrollment of 909 pupils in a building intended to house 1,301 students. By comparison, Jefferson—which claimed at least seven primary units—had an enrollment of 980 pupils, 954 of whom were African American, in a building constructed to hold only 906 children. Primary units made it possible to maintain such strict forms of ad-ministrative segregation even in the midst of serious enrollment crises.[65]

Black population increases between the close of World War II and the mid-1950s placed extraordinary pressure on residential color lines in Chi-cago, Atlanta, Oakland, and other urban centers. Flint, as was often the case, proved to be no exception.[66] Prior to the war, Saginaw Street formed a nearly impermeable barrier between the solidly black and transitional neighbor-hoods of the North End and the all-white sections of Flint's west and north-west sides. However, as the dilapidated St. John Street neighborhood reached its capacity in the 1950s, black families began seeking out housing west of Saginaw Street. Occasionally, African American homebuyers—especially business owners, professionals, and others with ample funds—were able to pierce through the color line by purchasing homes outside of the ghetto in cash or through informal, often deeply exploitative private real estate agree-ments known as "land contracts."[67] These residential movements generated many new opportunities for educational integration, especially in the junior and senior high school districts that served larger areas of the city, but the board responded to such population shifts by continuing to gerrymander. In April 1954 board members unveiled an official boundary plan for junior high schools. The plan included an oddly shaped, zigzagging attendance bound-ary on the northwest side that pulled handfuls of white students into the Longfellow district while pushing black pupils into the neighboring Emer-son zone. The board's hard-to-follow boundary line resulted in a 100 percent white enrollment at Longfellow that held throughout the 1950s. At Emerson, by contrast, the boundary created a black student body that constituted 23.9 percent of the school's overall enrollment. As in other instances, the board allowed white students from Emerson to switch to Longfellow while deny-ing black students the same privilege. Spatially, the Longfellow-Emerson boundary represented another clear violation of the geographic principle of neighborhood schools, with both black and white students often traveling to the more distant of two facilities. According to a 1958 editorial that appeared in the *Bronze Reporter*, the city's leading black newspaper, "If children were simply sent to the schools nearest them, we would find much fewer all-Negro and all-White schools."[68]

As black families from Floral Park and the North End began to move into white neighborhoods in the 1950s, the board of education worked feverishly, if less successfully, to maintain administrative segregation. For its junior and senior high schools, which drew students from larger, more diverse sections of the city, the board could not impose strict segregation and thus managed boundaries and transfers to minimize interracial enrollments. At the high school level, for instance, the board's attendance boundaries served to concentrate black pupils at Northern and Northwestern High Schools while diluting African American enrollments at Central and, later, Southwestern High Schools. For younger children, however, the board continued to plan for rigid segregation wherever possible, even if that meant requiring elementary pupils to cross major traffic thoroughfares and attend schools outside of their residential neighborhoods.[69] The racial imbalances between schools were especially visible in the North End. There, board policies left the neighboring schools of Lewis and Fairview with stark demographic contrasts even though the two buildings were only one mile apart. Located west of the Flint River in the St. John district, Fairview had a 100 percent black enrollment during the 1959–60 academic year. That same year, Lewis School, located just east of the river in a working-class white neighborhood, contained only one black student out of 712 pupils. With a capacity enrollment of only 661 students, Lewis was severely overcrowded in spite of its proximity—and easy access, via vehicle and pedestrian bridges—to Fairview. However, the school board refused to allow transfers between the two buildings.[70]

The pattern of overcrowded white schools adjacent to integrated and all-black facilities repeated itself across the city as the color line shifted to the north and west. Yet the school board would not relieve pupil overcrowding by breaching the color line. Such refusals often resulted in severe student-teacher mismatches. At Fairview, for instance, the outward migration of white students reduced the student-teacher ratio to 27:1. The ratio at nearby Kearsley, an all-white school, was 36:1, however. When parents criticized education officials for their inability to solve school overcrowding in the white neighborhoods surrounding St. John, board members drew from a pre-*Brown* political discourse that framed segregationist policies as a form of compensatory education for underprivileged black students: "Occasionally," a board report asserted, "socio-economic factors make a lower pupil-teacher ratio desirable for a given school."[71]

By the dawn of the 1960s, Flint's public schools—like those in Chicago, New York City, Detroit, and other urban centers—were deeply segregated.[72] Of the thirty-five public elementary schools operating in the city in 1960, sixteen contained all-white student bodies, while four additional schools

FIGURE 2.4. Racial segregation at Fairview Elementary School, 1960. By 1960, when an unnamed photographer captured this image of students from Fairview, Flint's public schools were deeply segregated. In Flint and elsewhere, the educational color line derived from a combination of popular, legal, and administrative forces. Courtesy of the Alfred P. Sloan Museum, Flint, MI.

claimed fewer than ten black pupils each. Meanwhile, four of the city's elementary schools had black student majorities of greater than 90 percent, with one facility posting an all-black enrollment. With a total elementary student population of 24,751 children, 26 percent of whom were African American, the Flint Public Schools could not claim a single integrated primary school, defined as a building with pupil enrollments approximating, within ten percentage points, the racial demographics of the system as a whole.[73]

The numbers told only part of the story, however. Stark as they were, the glaring contrasts between facilities at times masked the segregation of pupils that occurred within school buildings. By 1960 black families had begun to move east across the Flint River. Like Lapeer Road on the south side, the Flint River was a powerful symbol of the city's rigid segregation. Homedale Elementary School, which served the working-class white neighborhoods east of the river, gained its first black pupils in the late 1950s. In 1959 eleven of its 830 students were African American. One of those children was a second-grader named Wesley King. For King and perhaps other black pupils at Homedale, integration proved to be a frightful experience. After receiving complaints from white parents who opposed integration, King's teacher decided to place the boy's desk inside a coat closet, physically removed from the other children. Once alerted by King's mother, Flint NAACP members responded with an unannounced visit to the school, where they confirmed the child's story. The sole black student in a class of thirty, King had experienced an especially humiliating form of popular segregation from within an ostensibly integrated classroom. There can be no way of knowing how many other children experienced similar forms of racism.[74]

The case of Wesley King may have been exceptional by virtue of its cruelty, but popular and administrative modes of segregation—or, in the vague language of the day, "homogeneity"—within seemingly integrated schools was common in Flint and other cities.[75] Beginning in 1958, the Flint Board of Education and the Mott Foundation officially adopted the "primary cycle," the Personalized Curriculum Program, and the "Strawman Procedure on Grouping," forms of pupil tracking that segregated students by ability levels and, often, race. Following the dictates of these three methods, Flint teachers ranked and separated children according to classroom performance, achievement tests, IQ scores, and other ostensibly colorblind measurements. Yet in a city where black children typically scored two grade levels below their white peers on standardized tests, tracking contributed greatly to pupil segregation.[76] Furthermore, by sorting students into the Talented Child and Better Tomorrow for the Urban Child programs, school officials were able to maintain pupil separation by race and student performance under the guise of a colorblind assessment of achievement levels. As NAACP activists pointed out on a number of occasions, black children simply could not gain access to the Talented Child program because the board offered it only in white schools. For majority-black schools, the board favored BTU, a strictly vocational initiative.[77] According to Woody Etherly Jr., who graduated from Northern High School in 1962 and later went on to become a member of the Flint City Council, black students received an inferior education in the public schools: "Most black folks they put on the track system. They'd train you to be able to go into General Motors but not to be able to go to college."[78]

In Flint school boundaries shaped segregated communities—both in space and in the minds of residents. By the close of the postwar era, the board's policies had remapped neighborhood boundaries throughout the city, undermining even the slightest shifts toward residential integration. Because schools served as the focal points for neighborhood and civic activities, the gerrymandering of student boundaries affected nearly everyone in Flint, regardless of age, gender, or race. The government policies that created segregated neighborhoods and gerrymandered schools nurtured an artificial sense of community solidarity among whites in the city. In a 1956 article on the Mott program, *Coronet* magazine cheered the city's esprit de corps and credited community schools with making Flint the "Happiest Town in Michigan."[79] In similar fashion, *Reader's Digest* announced that "Flint's Gone Crazy over Culture," while *Family Circle* magazine referred to Mott's "life-saving, blues-curing plan" that had people "dancing in the streets."[80] By the end of the 1950s, white support for community education had reached a pinnacle. However, as time would tell, the broad but shallow white consensus in favor of public schools stemmed as much from segregation as it did from the innovative programming of the Mott Foundation and the board of education.

During World War II and its aftermath, Mott Foundation leaders and GM executives often boasted that new homes, industrial jobs, and community schools made Flint a model city and an engine of American progress. For their part, trade union leaders and members of the city's liberal establishment also heralded the special opportunities available to local citizens, arguing that Flint was one of the most progressive and racially "open" cities in the United States. Although these claims were clearly exaggerations, they were not entirely fanciful. For many white autoworkers, the decades following World War II made up an unparalleled era of job security and consumer prosperity. Due to the philanthropy of the Mott Foundation, this period also witnessed the rise of an expansive public culture organized around community education. Yet discrimination and segregation were integral features of this so-called golden era of civic togetherness, and not just in schools and neighborhoods. Beyond educational and residential segregation, African Americans in Flint faced widespread employment discrimination. In response, black workers and their allies in the UAW and other organizations waged vigorous campaigns for workplace equity. Like the pitched battles that would eventually emerge over segregated schools and neighborhoods, struggles over employment were essential to the city's gathering black freedom movement.

Jim Crow, GM Crow

In or around 1964, an anonymous civil rights activist authored a satirical play about employment discrimination at General Motors. The short play features five characters: Mr. Smith, a GM personnel manager; Mr. Bryant and Miss Fuller, African American job applicants; Mr. Ed Thornton, a white job seeker; and Charlie, a black janitor. The action begins in an unidentified GM employment office, where Miss Fuller, an applicant for a secretarial position, greets Mr. Smith beneath a sign reading "General Motors—An Equal Opportunity Employer." Miss Fuller, a graduate of Bryant and Stratton College with two years of secretarial experience, presents her application to Mr. Smith. Upon reviewing her credentials, Mr. Smith concludes, "There's no question that you have the qualifications, Miss Fuller, but we had in mind a much older person, with more experience." Mr. Smith then abruptly bids farewell to Miss Fuller, though he promises to keep her application on file.[1]

The next applicant, Mr. Bryant, is a millwright with six years of experience who has seen an advertisement for a millwright's position. Although impressed by Mr. Bryant's experience, Mr. Smith does not offer him the job. "After summing it up," Mr. Smith announces, "you are exceptional. . . . Unfortunately the millwright jobs have all been filled. It just so happens that we have an opening in the foundry for an iron pourer. The job's a little hot and dirty but being a healthy Negro, you'll be able to stand the heat." Desperate for income, the chronically underemployed Mr. Bryant reluctantly accepts the foundry position.

Following Mr. Bryant's departure, Mr. Thornton, a white job applicant, enters Mr. Smith's office. After overhearing Mr. Bryant's interview, Mr. Thornton states, "I was thinking of applying for a millwright position, but I'll accept a position in the foundry."

Looking through Mr. Thornton's application, Mr. Smith responds, "Wait a minute, Ed. . . . I see you worked in a paper factory for six months—unemployed for over three years—then you worked for Minute Car Wash approximately one year. . . . What's wrong with the job you have now?"

"The job only pays a dollar an hour," Thornton replies. "A Negro can live on small wages even with a large family. I'm not married but it takes more for me to live like a white person's accustomed to."

Persuaded by Mr. Thornton's explanation, Mr. Smith decides to offer Ed the millwright position: "Ed, I can work it so that you can be placed in a millwright job. Let's forget what happened here today. This colored boy had a little experience but we like to hire white men for the skilled trades."

As Mr. Thornton departs the stage, Charlie, the black janitor, enters Mr. Smith's office carrying a broom. While Charlie sweeps the floor, Mr. Smith says, "I hear that a bunch of agitators from out of town are stirring up trouble for the company."

Charlie responds, "Are you referring to the demonstration at the Auto Show that the N–double A–CP is planning?"

"Charlie," Mr. Smith asks, "when are you people going to wise up and stop this nonsense? . . . You know, as well as I do, that there are Negroes in all kinds of jobs around here."

"Yes sir," Charlie replies, "but how many are there in the skilled trades and white-collar jobs?"

Reassuringly, Mr. Smith answers, "Things like this take time, Charlie."

"But it's been over a hundred years now, sir," Charlie exclaims.

"Charlie," Mr. Smith cries, "did it ever occur to you that all this agitation isn't making the U.S. look too good to the rest of the world? . . . This country has done a lot for you people. You ought to show a little more appreciation."

Stung by Mr. Smith's remarks, Charlie interjects, "Sir, no one loves America more than Negro people."

After debating the issue for several additional minutes, Mr. Smith ends the discussion by reminding Charlie to empty all the ashtrays. The scene concludes with Charlie quietly sweeping the floor of the office.

Though ostensibly fictional, the play offers an accurate depiction of the quotidian realities of racial and gender discrimination in postwar America. During the 1940s and 1950s, tens of thousands of African Americans moved to Flint and other urban centers in search of good jobs, better housing and schools, and first-class citizenship. The influx of black job seekers helped to drive Flint's African American population from 6,559 to 34,521, an increase of over 400 percent.[2] Yet when these migrants arrived in the Vehicle City, they quickly discovered that GM and other local employers were committed to maintaining the color line. Those same leaders were also quite serious about practicing gender discrimination, as the case of Miss Fuller suggests. From GM's founding in 1908 through the postwar era, the company's black male employees tended to work in either foundry or custodial positions. For their part, most black (and many white) women, if they were lucky enough to land jobs at GM at all, found themselves concentrated in low-paying secretarial, unskilled assembly, and janitorial jobs. Just as Charlie, the fictional custodian in the play, notes, GM officials generally refused to hire black workers for skilled trades and managerial positions, regardless of their qualifications. Because of GM's discriminatory practices, thousands of African Americans in the Vehicle City, much like Mr. Bryant and Miss Fuller, suffered from unemployment and underemployment even during periods of economic expansion.

In response to such conditions, black workers in the United States waged fierce campaigns for equal employment opportunities.[3] In Flint the struggle for job equity hinged on a two-pronged effort to create a municipal fair employment practices commission and desegregate local workplaces. Ultimately, however, supporters of fair employment legislation were unable to secure a majority on the GM-dominated Flint City Commission. Frustrated by the defeat of the proposed Flint Fair Employment Practices Commission (FFEPC), African American activists launched a direct action campaign against employment discrimination. Repeatedly, however, the quest for fair employment brought activists into conflict with union members, corporate directors, and elected officeholders who were, at best, only marginally committed to racial fairness and, at worst, staunch defenders of white privilege.

The Hardest, Dirtiest, Lowest-Paying Jobs

The conclusion of the Great Flint Sit-Down Strike of 1936–37 gave rise to an iconic moment in the Vehicle City's storied labor history. On February 11, 1937, after learning that GM officials had formally recognized the UAW and agreed to participate in collective bargaining, strikers from Plant 4 of Flint's Chevy in the Hole complex triumphantly exited the factory for a long-awaited victory celebration. Leading the procession was an American flag–wielding janitor named Roscoe Van Zandt, perhaps the only black participant in the strike. Captured by a nearby photographer, an image of Van Zandt and his white coworkers standing together on Chevrolet Avenue garnered newspaper headlines throughout the world. This was not an impromptu moment, however. Prior to the strike's conclusion, the workers from Plant 4 had specifically asked Van Zandt to carry the flag. Though only a symbolic gesture,

FIGURE 3.1. Roscoe Van Zandt and fellow UAW members celebrate the victorious Flint Sit-Down Strike, 1937. Following the February 1937 conclusion of the sit-down strike, UAW members from Plant 4 of the Chevy in the Hole complex selected a black janitor named Roscoe Van Zandt, pictured here carrying a large American flag, to lead the workers from the factory. Although the image of Van Zandt and his colleagues reflected the sincerest hopes of millions of trade unionists and antiracists throughout the world, such acts of interracial solidarity were rare in the Vehicle City. Courtesy of the Alfred P. Sloan Museum, Flint, MI.

the workers' decision demonstrated the very real ways in which industrial organizing undermined Jim Crow.[4]

In spite of its symbolic power and potential, however, the story of Van Zandt and his UAW comrades conceals much more than it reveals about the experiences of black workers in Flint and elsewhere. Prior to World War II, African Americans made up only a tiny fraction of the auto industry's labor force. In 1910 American automobile manufacturers employed fewer than six hundred black workers nationwide, less than 1 percent of the industry's total domestic workforce. Although by 1940 that number had grown to 23,015, black workers still accounted for just 4 percent of the national total that year. Among the "Big Three" automakers—Ford, General Motors, and Chrysler—Ford consistently outpaced its competitors in the hiring of black workers, employing four times as many African Americans as any other firm in the industry. Of Ford's ninety thousand workers in Michigan in 1940, 12 percent were African American. By comparison, black employees at GM's

Michigan and Indiana facilities—where the vast majority of the company's African American wage earners worked—constituted only 2.5 percent of the total laboring force.[5]

The disparities resulted from a variety of factors, one of the most significant being GM's decentralized management structure. At GM local leaders from the company's numerous divisions made most hiring decisions.[6] Consequently, black employment opportunities varied widely among cities and even individual facilities, depending upon factors such as past practices, the prerogatives of local managers, public opinion, union militancy, the existence of fair employment legislation, and the status of the labor market. Nevertheless, across GM's many divisional boundaries, personnel managers tended to hire black workers only for the least desirable and lowest-paying positions. According to Lloyd Bailer, who conducted extensive surveys of the auto industry during the 1930s and 1940s, "The vast majority of Negro automobile workers are employed in the foundry, paint, and maintenance departments or as unskilled labor." In fact, of GM's twenty-five hundred black employees in Michigan and Indiana prior to World War II, at least two thousand worked in foundries. Most of the remaining five hundred worked in onerous low-status jobs: paint chippers, car washers, sanders, machine cleaners, scrap handlers, and janitors.[7]

Like African Americans, women also suffered from employment discrimination and job segregation during the interwar decades. Prior to World War II, women accounted for less than 10 percent of the nation's automobile manufacturing workforce. The gender gap stemmed in large part from the fact that personnel managers at GM and other firms hired women only for a limited number of low-paying "pink-collar" jobs, primarily those involving small parts assembly, clerical work, upholstery sewing, and janitorial labor. Women working in such positions endured a wage scale that was deeply discriminatory. According to one study conducted in 1925, women workers in the automobile industry earned an average of forty-seven cents per hour, whereas men took home almost twice as much. In Flint and other automobile manufacturing centers, both Jim and Jane Crow work arrangements were harsh facts of life.[8]

During the 1920s and 1930s, GM supervisors maintained almost complete racial segregation in Flint's plants. At Fisher Body and AC Spark Plug, where the color line was starkest, the workforce was nearly all white with the exception of several black custodians. As late as 1941, AC employed approximately thirty-five hundred workers, yet only twenty-three were African American, and all of them janitors. During the same year, Flint's two Fisher Body plants employed only six black workers as either janitors or car washers. On the eve

FIGURE 3.2. Black foundry workers at Buick, n.d. Prior to World War II, rigid forms of employment discrimination severely limited the job prospects of African Americans in the Flint area. At GM black job seekers found it all but impossible to obtain employment outside of custodial and maintenance departments or the Buick foundry, where the two men shown here worked. Although World War II helped to open up many new opportunities for black workers, employment discrimination remained a persistent feature of the local labor market throughout the twentieth century. Courtesy of the Alfred P. Sloan Museum, Flint, MI.

of the war, approximately three hundred African Americans worked at Flint's Chevrolet division complexes, but all of them served as janitors, truck drivers, or machine cleaners. African Americans fared slightly better at Buick, but discrimination was still rampant at the massive North End complex. Outside of the custodial department, black job seekers at Buick could find employment only in Plant 70, the foundry, or in Plant 71, the cleaning room. In 1941 Buick employed approximately six hundred black workers within its foundry division.[9] "That [foundry work] was the black man's job," remembered Charles Skinner. "That's where you hid, that's where you were supposed to be, in the foundry."[10]

Foundry jobs were among the most unpleasant, dangerous, and difficult positions available in the industrial world. In Skinner's words, "Those were the hardest, dirtiest, lowest-paying jobs that they had."[11] Beyond the onerous nature of their jobs, black foundry workers suffered from rigid racial segregation. According to Harry Rolf, who held the decidedly (Frederick Winslow) Tayloresque position of "time study" official at Buick, GM flatly refused all transfer requests from black foundry workers until the 1940s: "No one was allowed to transfer out of the Foundry. If you were in the Foundry, that's where you stayed."[12] Reflecting back upon his experiences as a foundry supervisor at Buick, Melvin "Pudge" Van Slyke noted, "It was always hot, dirty work. That's why it was ninety percent colored. If a black person went to the employment office looking for a job, he would be sent to the Foundry."[13] Once in the foundry, black workers often encountered hostile managers who policed their behavior and movements. According to former worker Roger Townsend, some Buick foremen went so far as to confine black employees to their work stations and forbid them from exiting the foundry, even to use the bathroom: "Blacks were limited to the foundries part, they were not permitted to go beyond the imaginary line [that separated black and white spaces] even to use the bathroom. They had certain bathrooms that were off limits to them. Whites could naturally go there."[14]

On the surface, the starkly discriminatory arrangements that reigned at GM's local plants and in other area workplaces prior to World War II appeared to be straightforward instances of popular segregation in which personnel managers and other individuals acted alone to perpetuate the color (and gender) lines. In reality, though, job segregation and employment discrimination had their roots in both the private and public sectors. Clearly, racist managers operating in the private marketplace—especially at GM, with its decentralized corporate structure—played a major role in maintaining Jim Crow work arrangements. However, in the absence of municipal, state, or federal fair employment legislation, the state effectively bestowed the right to discriminate on all of the nation's employers—at least until the early 1940s, when federal officials made the first of several major policy shifts designed to combat employment discrimination.[15]

World War II and the Double V Campaign

The first major cracks in GM's color bar began to appear during the Second World War, when a variety of new opportunities opened up for black autoworkers. In Flint, Detroit, and other "arsenals of democracy," the outward migrations of soldiers and other service members combined with the unprecedented federal demand for matériel created massive labor shortages.[16] Just in Genesee County alone, GM added twenty-five thousand new jobs during the conflict.[17] Viewing the wartime buildup as a prime opportunity to secure new civil rights protections, black socialist A. Philip Randolph quickly sprang into action. In January 1941 Randolph launched a nationwide March on Washington Movement to demand a federal ban on employment discrimination. In June of that year, just a week before Randolph's planned march in the nation's capital, President Roosevelt, concerned that protests would undermine the government's military preparedness, issued Executive Order 8802, which prohibited GM and other government contractors from practicing racial discrimination in defense-related industries.[18] Although Roosevelt's enforcement agency, the Fair Employment Practices Commission (FEPC), had only limited authority to combat discrimination, government negotiators from the Office of Price Management (OPM) and other federal agencies were able to persuade automobile executives to relax the color line at most defense facilities. By September 1941, OPM agents had signed an important agreement with automobile manufacturers and the UAW guaranteeing seniority rights for all workers and mandating that existing employees, regardless of race, receive priority treatment for all war-related job openings. At Buick, where the conversion to defense work resulted in the temporary closing of the foundry, the OPM agreement allowed hundreds of African American workers to transfer to less arduous, higher-paying positions on the assembly line. Spurred into action by a nationwide campaign for a "Double V" victory over fascism abroad and Jim Crow at home, federal officials played an important role in weakening long-standing proscriptions against black employment at GM and other companies.[19]

Although wartime reforms left many forms of discrimination and popular segregation untouched, signs of progress were still evident in numerous American workplaces.[20] During the conflict, GM recruited thousands of southern blacks to work in Genesee County's auto factories. From 1941 to 1945, GM's African American workforce in the Flint area more than doubled. At the city's two Fisher Body plants, for instance, the number of black workers rose from six to ninety-four. Desperate to fill the area's labor shortages, personnel managers also began hiring large numbers of women during this period. For example, at Fisher Body, eighty of the ninety-four new black employees working at the close of the war were female "janitresses."[21]

The war also opened up new possibilities for black workers, especially men, seeking jobs on the assembly line. Such was the case for Layton Galloway, a black migrant who arrived in Flint in 1942 to work as a janitor at GM but went on to work on the assembly line. "I went out to Plant 3 and went to work the same day," Galloway remembered. "They needed workers. I started sweeping Plant 3. I did that for about four months, then went into production." In 1943 Annalea Bannister shattered a major barrier—and raised quite a few eyebrows—by becoming the first African American to hold an office job at the AC Spark Plug facility on Flint's east side. "There were people walking past my office in droves trying to see the black girl," Bannister later recalled.[22] In Flint and across the nation, the war generated numerous possibilities for black workers to breach the color line. As Anna Howard remembered, "It took WWII to put black men in the shop on production. Before, they were janitors and did menial work. My husband got a better job. He came out of that foundry and got to be a finisher. That's what he always wanted to be."[23]

With few exceptions, job opportunities for urban black workers continued to expand after World War II. During the late 1940s, residents of the St. John neighborhood coined the term "the rush-in" to describe the Second Great Migration of black job seekers to Flint and other big cities.[24] According to black migrant John Rhodes, "Flint was crowded. Plenty of jobs. You could get a job anytime."[25] Nevertheless, the postwar era was also a period of rising expectations within the black community. After defending liberty abroad, thousands of black veterans returned to Flint and other cities with a renewed commitment to challenging Jim Crow at home.[26] As Flint resident Dolores Ennis recalled, "Those young [black] men who served in the war, they were not going to come back and take less."[27] Still, African Americans in postwar Flint faced a corporation, a city government, and a white community that had always offered less to black people. By war's end, the stage was set for a fierce battle over employment equity.

Civil Rights at Work (and Play)

Flint's status as a UAW stronghold may have been an important pull factor for the tens of thousands of black migrants who participated in the postwar rush-in. Formally organized in 1935 during the lead-up to the Flint sit-down strike, the UAW gained a great deal of visibility during its first two decades as one of the nation's most militant and racially progressive unions. The reputation was mostly well deserved, particularly at the national level.[28] Notably, the UAW's inaugural contract, signed at the close of the sit-down in February 1937, contained clear antidiscrimination language that protected all GM workers, regardless of race, from any retribution that company officials might take against prounion employees. The contract also guaranteed seniority privileges for all workers. As future UAW president Walter Reuther—himself a staunch racial liberal—would often proclaim, these were the first instances in which automobile executives formally recognized the rights of black workers.[29]

During the war, UAW representatives officially endorsed fair employment legislation and advocated for the abandonment of discriminatory hiring practices. In 1946 delegates to the UAW convention followed Roosevelt's lead by voting to create a Fair Employment and Anti-Discrimination Department while ordering the establishment of Fair Practices Committees within each of the union's local branches. To head the national department, Reuther selected himself and William Oliver, who had gained acclaim in 1941 as the first black person to serve on the executive board of UAW Ford Local 400. Beyond enforcing the union's antidiscrimination policy, the leaders of the new departments implemented human relations programs within plant cities and advocated for the passage of equal employment and fair housing laws at the federal, state, and local levels. Delegates to the 1946 convention also mandated that the treasurers of all locals deduct a small percentage of each member's monthly dues to fund the new fair employment divisions. Furthermore, union officials succeeded in persuading many industrial employers to enact voluntary nondiscrimination clauses. Between 1946 and 1957, the percentage of the UAW's contracts containing nondiscrimination language skyrocketed from 20 to 80 percent. Moreover, on numerous occasions union leaders used their power to invoke sanctions against rank-and-file members who participated in so-called hate strikes, or labor stoppages launched by white workers in response to the hiring of blacks. These and other efforts in support of racial equality helped to make the UAW one of the most powerful institutional supporters of the black freedom struggle in the United States.[30]

Nevertheless, the UAW's racial progressivism at the national level went unmatched within many individual locals, plants, and departments.[31] This was definitely the case in the Flint region, where white unionists routinely ignored, reinforced, or even created new racial inequalities. Criticism of the UAW's stance on civil rights often revolved around the union's weak position on corporate hiring policies. Because the union played no formal role in the hiring process, local personnel managers practiced discrimination against blacks, Latinos, women, and other groups with virtual impunity. Such activities, many critics argued, stemmed from the UAW's unnecessarily narrow focus on the shop floor concerns of already employed workers, which left little room for the crucial battle against discriminatory hiring practices.[32]

Even with regard to existing employees, however, Flint's UAW locals amassed very spotty civil rights records. On numerous occasions throughout the postwar period, rank-and-file white unionists reacted violently to the hiring of black workers and the desegregation of workspaces. In the worst cases, whites organized hate strikes or physically assaulted black workers. More often, though, white workers subjected their black peers to a daily barrage of harassment by doing things such as breaking their tools, issuing verbal threats, or vandalizing their cars. Although Walter Reuther and other UAW leaders in Detroit took a firm position against such activities, black workers, both men and women, often complained that union stewards and other officers refused to take their grievances seriously.[33]

Complaints against the UAW also stemmed from acts of discrimination committed outside of the shop. One of the most noteworthy of such examples occurred in 1946, when four black unionists confronted rigid forms of popular segregation at the Flint CIO Building. Located near the heart of downtown at the corner of Saginaw Street and Second Avenue, the union-owned ten-story CIO Building housed assorted professional, commercial, and business establishments as well as several government agencies, UAW offices, and a handful of private clubs. Representatives from the board of directors of the CIO Building—all of them agents of the UAW—solicited donations from local unions while assessing a fraction of each member's monthly dues in order to purchase and operate the building. Even though such assessments were mandatory for all area workers, including women and African Americans, the CIO Building operated on a segregated basis. In fact, many of the business establishments located in the building flatly refused service to blacks and women of any race.[34]

The CIO Building also housed a segregated bowling alley where UAW members and union-affiliated leagues competed. Following the guidelines enacted by the American Bowling Congress (ABC), the sport's official governing body, which stipulated that league play was open only to "individuals of the white male sex," the managers of the alley denied women and blacks the opportunity to participate in local competitions. Moreover, in a decision that went beyond the exclusionary policies of the ABC, local managers also denied African Americans and women the chance to use the facility during nonleague "open bowling" times. On September 8, 1946, black UAW members L. T. Heller and Silas Horton tested the ban by turning out for open bowling, but they were denied access to the lanes. Two weeks later, Heller and Horton returned to the lanes with Silas Hogan and R. C. Moore, also black union members, and were turned away once again. In response, the four men filed a complaint with the UAW-CIO's International Executive Board of the Fair Practices Committee. Their petition was but one of a large number of postwar grievances filed by black UAW members across the country who could not participate in union-sponsored bowling events.[35]

Upon receiving the complaints from Flint and other cities, UAW board members publicly implored the ABC to change its policies. Walter Reuther and other top leaders from the group's headquarters in Detroit also launched a national public relations campaign in which they challenged ABC officials to "destroy undemocratic sportsmanship."[36] For their part, some left-wing union members in Flint organized pickets in front of the CIO Building and other facilities that housed Jim Crow bowling alleys. Victory did not come overnight, though. It was not until 1950, four years after Heller and his peers had filed their complaint, that the ABC ultimately rescinded its segregationist policies. Still, the campaign against Jim Crow bowling in Flint could not formally conclude until the CIO Building's board of directors and other local union officers abandoned their own discriminatory guidelines. Consequently, in their ruling in support of the four black bowlers in Flint, the UAW-CIO's Fair Practices Committee condemned both the ABC's policies and the popular segregation practiced by the Flint membership. In the summer of 1950, shortly after the ABC abandoned its discriminatory policies, the managers of the CIO Building relented and fully desegregated their bowling lanes.[37]

However, new fractures within the UAW would soon develop. To be sure, just as the battle over segregated bowling was ending, another conflict erupted that illustrated both the complex fissures within local political coalitions and the deep racial divisions that beset the national union movement. This time, though, the contestation was over the merits of municipal fair employment legislation. Predictably, the struggle for equal employment reform centered on a series of clashes between civil rights supporters and the leadership of General Motors. But those battles also brought black workers and liberal-minded UAW officials into direct conflict with rank-and-file white unionists and union-backed city commissioners. These splits first surfaced during the early 1950s, when civil rights activists and left-wing trade unionists fought for the passage of municipal fair employment legislation and the formation of a Flint Fair Employment Practices Commission modeled after President Roosevelt's wartime creation.

Despite the deep racial divide within Flint's labor movement, supporters of the FFEPC elected to press their case by forming an alliance with the local leadership of the UAW-CIO. Their goal was to gain a favorable majority on the Flint City Commission. The strategy backfired, though, when the UAW failed to maintain control of the commission. After preserving a slim majority throughout the war, the UAW-CIO lost its hold on the commission in 1946 to a bloc of GM-backed candidates, only to win it back again in 1948. Two years later, GM's brass united with Michael Gorman, the staunchly Republican editor of the *Flint Journal*, to support a probusiness slate of office seekers. In a bitter election that pitted the UAW and its supporters against the GM-backed Civic Research Council of Flint (CRCF) and the Manufacturers Association of Flint (MAF), labor once again lost control of the commission. Then in 1952 antiunion commissioners consolidated their hold on local politics with yet another victory at the polls. Although the UAW continued to

FIGURE 3.3. "Bowl with the Ball of Fair Play" leaflet, 1950. During the late 1940s and early 1950s, leaders from the UAW organized a nationwide movement to ban segregated bowling. Like other antiracist campaigns that the UAW sponsored, the attack on Jim Crow bowling brought racially progressive officials from the union's headquarters in Detroit into conflict with segregationist members in Flint and other cities. Courtesy of the Walter P. Reuther Library of Labor and Urban Affairs, Wayne State University, Detroit.

endorse candidates and maintain an impressive voter turnout operation during the 1950s, supporters of the FFEPC were never able to gain the majority
necessary to implement municipal fair employment legislation.[38]

The difficulty stemmed in large part from GM's near-obsessive efforts to
dominate local politics.[39] Because Charles Stewart Mott and other GM leaders had succeeded in winning a 1930 campaign mandating nonpartisan municipal elections, the UAW faced a perennial legal obstacle in securing adequate institutional support for labor-friendly candidates. The nonpartisan
election rules also made it difficult for UAW leaders to galvanize the city's
heavily working-class, Democratic-leaning electorate. Corporate directors
exacerbated the UAW's electoral problems by quietly funneling large amounts
of money and resources to the CRCF, the MAF, and other groups that supported probusiness commissioners and policies.[40] In the words of one especially frank MAF document that circulated internally during the 1950s, "The
future program of the Manufacturers Association of Flint is aimed toward
finding a good answer to the question: Who's going to be sitting in those
Commission chamber chairs? . . . We like to feel that we should and do have
a hand in installing capable and fair-minded people in positions of public
trust."[41] As a result of GM's efforts, the city commission and other local governing bodies routinely acted as arms of the corporation—even during the
UAW's supposed postwar heyday.[42]

The UAW's inability to maintain control of the city commission had grave
consequences for supporters of workplace equity. In October 1951 Commissioner Robert Carter, who also served as a regional director for the UAW-
CIO, introduced a fair employment bill for the commission's consideration.
Probusiness legislators promptly rejected the proposal, however, in a six-to-
three vote. Mayor Paul Lovegrove defended his negative vote by asserting
that the FFEPC would drain the city's coffers. He and other opponents of
the legislation also claimed that it was impossible to legislate against racial
prejudice. Angered by the defeat, Carter submitted a blistering editorial to
the *Searchlight*, the newspaper of UAW Local 659, in which he charged GM
with controlling the city commission. Still optimistic, though, Carter predicted that Flint voters "will not let these commissioners get away with this
type of reaction."[43]

Carter proved to be an extremely poor judge of the Vehicle City's electorate, however. Throughout the early 1950s, Flint's registered voters, over forty
thousand of whom were UAW members, consistently elected city commissioners who opposed fair employment legislation. In 1954 FFEPC supporters rejoiced when prounion forces regained a majority on the commission.
Anticipating an improved political climate, advocates for workplace equity

introduced a second nondiscrimination proposal. On March 14, 1955, the commission, by a six-to-two vote, endorsed the proposal on its first reading. In response, AC Spark Plug manager Joseph Anderson and G. Keyes Page of GM's Civic Research Council led a vigorous campaign to kill the proposed legislation that included the leadership of the Flint Chamber of Commerce and the MAF as well as the editors of both the *Flint Journal* and the *Flint News-Advertiser*. On April 7, 1955, in an unmistakable show of force, GM supervisors packed the city commission chambers to witness—and, presumably, influence—the final vote on the ordinance. This time, though, the FFEPC went down to defeat by virtue of a four-four deadlock.[44] When pressed to defend their votes, Mayor George Algoe and the other prolabor commissioner who had changed his position argued that education, not legislation, was the most appropriate means of combating segregation and other discriminatory practices. In addition, all of the anti-FFEPC commissioners maintained that color-conscious antidiscrimination legislation was an unjust remedy for racial prejudice.[45] As the FFEPC vote highlighted, GM was consistently able to exercise state power in Flint—even when its candidates failed at the ballot box.

In the absence of strong civil rights protections, black job seekers in the 1950s confronted near-daily reminders of employment discrimination.[46] At the Flint Trolley and Coach Company, which operated the city's public transportation system, there were no black drivers.[47] Likewise, out of the hundreds of long-haul truck drivers based in the Flint metropolitan region, none was black. The same pattern held true at the Flint headquarters of the state police department, where only white officers held commissions. As in virtually all other cities, jobs in Flint's building trades industry were generally off limits to black workers. Blacks could occasionally find employment in the city's numerous banks and financial services firms, but only in menial, entry-level positions. As a rule, in fact, blacks living in postwar Flint had almost no chance of obtaining skilled, technical, or supervisory positions, irrespective of industry.[48] Furthermore, there were no black high school teachers within the city's public school system. Although there were a small number of African American teachers working in Flint's public elementary schools in the 1950s, virtually all of them taught in buildings with large black student majorities. Even at city hall, blacks could obtain employment only as custodians. Employment segregation was so thoroughly implanted in the city's culture that the *Flint Journal* continued to publish race-based classified advertisements throughout the 1950s.[49]

Flint's downtown business district proved to be an especially hostile environment for black job applicants. In July 1958, members of the Flint NAACP

picketed in front of the Smith-Bridgman department store, the city's largest and most prestigious downtown retail establishment, for refusing to hire
black workers in positions other than maintenance. The campaign received a
great deal of publicity because the Mott Foundation owned the store. Foundation representative Frank Manley, one of the leaders of the community
education initiative, responded to the pickets by forming the Flint Citizens'
Study Group (FCSG), a secret organization that brought together leading
members of the city's black and white communities. The FCSG convened
once per month for quiet negotiations over workplace discrimination, especially in the downtown business district.[50] A product of the severe economic recession of 1957–58, the FCSG also sponsored job training programs
for unemployed black workers.[51] Although the leaders of the FCSG rejected
protests, they defined themselves as "an action group."[52]

Participants in the FCSG relied heavily upon referrals from the Urban
League of Flint. Like their counterparts in other cities, officers from the local Urban League maintained a long list of black job seekers, complete with
recommendation letters and résumés, which they routinely submitted to the
FCSG.[53] After evaluating candidates, members of the FCSG quietly negotiated with employers on behalf of "deserving" black applicants. At the FCSG's
June 17, 1959, meeting, for instance, Dr. Donald Swank, who administered
the group's "testing" program, provided his assessment of a candidate named
Ellen Marie Burton, noting, "Has good test results and recommendations,
is excellent typist, attractive appearance, wants clerical or selling position.
Recommendation: Dr. Swank will tell Miss Burton to see John Stout regarding possibly being employed at [J. C.] Penn[e]ys."[54] If applicants performed
well on Swank's tests and possessed stellar qualifications, white members of
the FCSG would negotiate with local employers on their behalf. In most instances, however, members of the FCSG refused to act out of fear that the
placement of an unacceptable applicant would undermine their cause.

As an "action group," the FCSG proved to be unsuccessful. To test the ef-
ficacy of the group's program, Urban League functionaries commissioned a
survey of downtown employers during the 1959 Christmas shopping season.
After investigating twenty-six downtown retailers, Richard A. English found
only eleven black sales clerks, all employed at just four establishments. Notably, English discovered that Smith-Bridgman, the target of the NAACP's 1958
protest, had still not hired any black clerks.[55] Much to Manley's embarrassment, the survey confirmed that the FCSG's policy of gradualism and private
negotiation generated few employment opportunities for black workers. Even
for members of the Urban League, who were genuinely committed to quiet,
negotiated racial change, the pace of progress in the Vehicle City seemed to

be depressingly slow. In her preface to English's study, Edwyna Jones, chair of the Flint Urban League's Vocational Services Committee, wrote pessimistically, "We must face the fact that Flint continues to progress feebly in the development of better race relations." "Certainly," she continued, "this study points out that that 'every man at his best job and for every job the best man' is at best a catchy phrase lost in a hodge podge of bias and bigotry."[56]

The Battle at General Motors

As with the campaign to desegregate downtown workplaces, local efforts to combat discrimination and popular segregation at GM faced enormous obstacles. In the spring of 1953, representatives of the UAW conducted a demographic survey of the automobile plants located in the union's Region 1C district, an area that included metropolitan Flint, Lansing, and several other mid-Michigan communities. The results of the study were a sobering reminder of the barriers to fair employment in the automobile industry. Within the region they uncovered four plants that had no black employees at all and two facilities in which managers refused to hire women. Nine of the factories they visited employed African Americans, but only for unskilled positions, and two of the plants provided separate Jim Crow accommodations for black and white workers. In the Flint area, personnel supervisors at three of GM's major facilities—AC Spark Plug, Fisher Body 2, and Buick—hired blacks only for unskilled positions.[57]

Several months after the UAW released the survey, Earl Crompton, director of the UAW Region 1C Fair Employment Practices Committee, provided additional evidence about discrimination in local plants. According to Crompton, blacks in Flint simply could not gain entry into GM's apprenticeship programs. At Chevrolet's Tool and Die Department, for example, Crompton discovered that managers categorically barred blacks from becoming apprentices.[58] In most cases, though, company leaders preferred subtler approaches to maintaining the color and gender lines. Often personnel managers employed racially and gender biased aptitude tests and "emotional" assessments to measure a candidate's worthiness for an apprenticeship. The widespread use of openly sexist and nepotistic recruitment policies further served to maintain racial and gender homogeneity among apprentices. As stated in GM's official apprenticeship handbook, in fact, "[p]reference is given to qualified sons of employes, and to other capable young men of the community whose high school records and recommendations are satisfactory." Under GM's highly restrictive apprenticeship policies, the high-paying world of the skilled trades was in reality a closed universe in which special, life-altering perquisites passed primarily between white fathers and their sons.[59]

The company's exclusionary practices created particular hardships for black women, who, since they could claim neither the privileges of whiteness nor the benefits of maleness, often occupied the lowest rungs of the employment ladder. As a matter of fact, members of the UAW's various fair employment committees nationwide received thousands of reports of discrimination against black females during the 1950s.[60] Complainants in a number of instances alleged that GM managers reserved the highest-paying pink-collar positions exclusively for white women. For example, in the summer of 1953, Mrs. Lola Coleman applied for a temporary position as a nurse at the AC Spark Plug complex without mentioning her race. Soon after receiving the application, Dave Nichols from the AC Employment Department phoned Coleman and told her that her qualifications were satisfactory and she should set up an interview at once. Upon discovering that Coleman was black, however, Nichols immediately backed off, telling her that the current nurse—a white woman—would return soon enough. Even within the pink-collar employment ghetto, black women faced major obstacles in their quest for greater equity.[61]

Racial and gender biases also shaped the processes through which workers at GM achieved seniority. On at least two occasions in the 1950s, women of color in the Flint area reported being terminated just prior to achieving seniority. In 1958, after years of failed attempts, Sam Duncan, president of UAW Local 598, persuaded managers at the Fisher Body 2 facility to hire a black woman. Supervisors at the plant kept this unidentified woman on the payroll for exactly eighty-nine days, just twenty-four hours short of the ninety days needed to achieve seniority, before laying her off. Two years later, Duncan succeeded again in convincing Fisher executives to hire a nonwhite female worker. Like the woman before her, however, the new employee, an unnamed Chicana, remained at her post for eighty-nine days before receiving a layoff notice. Unable to obtain the job security that seniority bestowed, these and other women of color were little more than temporary workers who met GM's short-term employment needs while managers searched for permanent white male replacements.[62]

In response to these and other acts of discrimination, Duncan and other racially progressive UAW members repeatedly pressed company officials to reverse their policies. However, GM executives flatly refused to meet with union officials to discuss discrimination. According to Nat Turner, recording secretary of UAW Local 599, the company would not even answer letters that addressed the issue of workplace equity.[63] Activist Edgar Holt, who would

FIGURE 3.4. Protest against employment discrimination at General Motors, n.d. As part of the local movement for equal employment opportunity, activists such as these NAACP members, who picketed in front of a Buick showroom in downtown Flint, consistently challenged the racist hiring practices that prevailed within General Motors and other area businesses. Courtesy of the Alfred P. Sloan Museum, Flint, MI.

later become president of the Flint NAACP, concurred, noting, "Most of the time we do not get common courtesy. We try without success to relay the corporation's [formal] policy of fairness and equal opportunity but they think this applies to Mars."[64]

Although GM's long record of exclusion and popular segregation at times seemed to stand out in relation to the practices of other businesses, in reality, workplace discrimination was an inescapable fact in postwar American society.[65] This was especially true within the automobile industry, one of the nation's leading economic sectors. According to federal census records, there were 54,590 professional and technical workers employed in the American motor vehicle industry at the close of the 1950s. Of those highly paid skilled workers, however, only 327, or 0.6 percent, were nonwhite. Furthermore, all of the nearly four thousand lawyers, public relations professionals, labor relations experts, and social scientists employed in the industry were white. The numbers were equally skewed at the level of management, where nonwhites accounted for just 0.4 percent of the workforce. Discrimination was pervasive even within lower-paying job categories such as clerical work, where 97 percent of the industry's employees were white. As evidenced by the census data, vehicle production, like most other vocations in postwar America, was largely a Jim Crow enterprise.[66]

Of course, corporate managers were primarily responsible for employment inequities and the maintenance of popular segregation in the workplace. But employment discrimination was also in many instances a function of public policy. Nowhere was this as clear as in the technical training program for high school seniors cosponsored by GM and the Flint Board of Education. The initiative began in 1948 in response to concerns from GM managers that graduates of the city's public schools were entering the workforce without adequate skills. Local GM chiefs and other employers attempted to remedy the problem by working directly with teachers, administrators, and board members to make Flint's public school curricula more technically and vocationally oriented. In addition, GM managers persuaded members of the board of education to approve a cooperative training program in which participating students earned course credits for spending part of the school day working as interns and apprentices in training at local auto plants. The goal of the initiative was to create a more able workforce while helping GM meet its growing need for skilled laborers. Only white male students could apply, however. Although the program was nominally colorblind, members of the Flint Board of Education allowed only students from the all-white Flint Technical High School to participate. Moreover, in deference to GM's male-only apprentice policy, school officials prohibited Tech's female students from applying for the program. By excluding blacks and women from apprentice training programs, elected officials from the Flint Board of Education and their corporate colleagues participated directly in the maintenance of employment discrimination and administratively driven workplace segregation.[67]

Beyond the struggles over exclusionary hiring policies, contestation over popular segregation and discrimination within workplaces was also quite common during the postwar era.[68] This was certainly the case at Buick in the 1950s, where racial resentments and Cold War paranoia combined to ignite near-daily conflicts. On August 4, 1950, a hate strike involving over fifty employees erupted in Department 12 of Factory 40 when a supervisor temporarily assigned a militant black worker named Berry Blassingame to an all-white division. In addition to objecting to Blassingame's race, the white workers claimed that he was a communist. When a white supervisor named Bruce Batchlor arrived on the scene to investigate, Blassingame, a member of the Michigan Civil Rights Congress and vice president of the Flint NAACP, vehemently denied the charge. Unconvinced, Batchlor immediately suspended Blassingame. Upon investigating the matter, Batchlor found no evidence that Blassingame was a communist. Nevertheless, a week later Batchlor fired Blassingame for violating shop rules that prohibited "threatening, intimidating, coercing, or interfering with employees" and "distracting the attention of others." Although Blassingame ultimately won reinstatement after filing a successful grievance, managers refused to offer any back pay for the time he missed from work. Furthermore, upon Blassingame's return to work, Batchlor ordered him back to his old unit, a move that resulted in the resegregation of Department 12.[69]

Several years after the Blassingame incident, a violent wave of anticommunist hysteria swept across the United States, targeting many of the nation's most outspoken proponents of racial equality.[70] Locally, the trouble began in the spring of 1954, when US Representative Kit Clardy, a virulent anticommunist and ally of the redbaiting US Senator Joseph McCarthy, hosted two separate House Committee on Un-American Activities (HUAC) hearings, one of them in Flint, regarding alleged communist activities in Michigan.[71] During the proceedings, witnesses publicly identified dozens of local UAW members as members of the Communist Party, which in turn triggered a rash of violence in plants and neighborhoods throughout the region. In the weeks following Clardy's show trials, gangs patrolled the streets looking for suspected communists, stoned the homes of persons named in the hearings, and physically ejected left-wing UAW members and racial progressives from local plants.[72] On May 15, just two days after the Flint hearing, thirty workers at Fisher Body 2 attacked a suspected communist named Charles Shinn. Before removing Shinn from the plant, the workers threw him to the ground and attempted to toss him over the railing of a stairwell.[73] Several weeks later, a gang of workers at the Chevrolet manufacturing plant attacked a reputed communist named Howard Foster, beating him with brass knuckles and blackjacks. Flint, as Foster claimed, had descended into a "reign of terror."[74]

The attacks on alleged communists continued throughout the region in the months following Clardy's hearings. Eager to eliminate suspected radicals from their employee rolls, GM managers summarily fired all but a few of the individuals named in the HUAC proceedings. In response, many of the workers filed grievances with the UAW. However, local union bosses flouted the UAW's formal condemnation of the firings by refusing to pursue grievances on behalf of suspected communists.[75] Instead, UAW officials in Flint, much like their peers in other American unions, responded to the Red Scare of the 1950s by systematically purging suspected communists from their ranks. Yet in the process of doing so, anticommunists helped to strip the UAW and other workers' organizations of their most racially progressive members, a fact that would have important implications for the future of Flint and the American labor movement.[76]

Notwithstanding the significance of Cold War–related conflicts, the campaigns for workplace equity that erupted in postwar Flint usually revolved around more localized plant- and department-specific concerns. Regardless of the corporate division or plant location, day-to-day shop floor struggles over race often centered on the complex and at times contentious relationships among white and black line workers, GM supervisors, and UAW shop stewards. Such was the case at the Chevy in the Hole complex, where supervisors developed a nasty reputation for harassing black workers. In March 1953 Raymond Walker, a progressive white steward, reported, "Management of Plant 2 is trying to run their plant like a concentration camp. . . . They are using every tactic known to man, including a member of supervision over each employee with stop watches, pencils, and paper breathing down the men's necks; making catty remarks, threats, intimidations, reprimands and penalties using the most vile Jim Crow and discriminatory tactics." Nearly every day, Walker found, supervisors from Plant 2 and other factories invoked Shop Rule 29, a vague prohibition against "restricting output," to harass black workers. The case of Paul Kenny is instructive. Just a few weeks after Walker complained about racial harassment, a white supervisor from Plant 2 suspended Kenny, a black honing worker, for restricting output. Kenny received a three-day suspension as punishment for the offense. The following week, managers penalized Kenny yet again for violating Rule 29, but this time they imposed a harsher one-week suspension. On both occasions, Walker determined that Kenny had produced a higher than average output for the day. After fielding similar complaints from other black workers, Walker further concluded that Kenny's case was part of a much broader pattern of racial harassment, noting, "[T]here are dozens more of such cases that I for one could report."[77]

The infernal conditions that often prevailed in the Buick foundry were another source of conflict between black workers and management. Upon entering the foundry, workers faced a barrage of unhealthy irritants including foul-smelling smoke and silica dust, ear-splitting noise, and intense heat. Once they arrived at their stations, many foundry workers spent their shifts handling and pouring large vats of molten metal into mold cavities, an extremely dangerous process. Because of the harsh conditions, foundry employees suffered more than twice the number of disabling work injuries as their peers in other manufacturing sectors. According to one study sponsored by the UAW's Foundry and Forge Department, an average worker who spent a thirty-year career in the foundry could expect to lose approximately fifteen years of life expectancy. One of the principal factors driving such poor health outcomes for this class of workers was the high incidence of respiratory diseases such as lung cancer, asthma, silicosis, and other conditions caused by the inhalation of foundry dust and smoke.[78] In response to the dangerous conditions, Albert Christner, chair of the UAW's Buick Shop Committee, filed a number of grievances asking for the installation of new fans, more custodial support to keep the foundry clean, and upgraded safety equipment to protect workers, but managers rarely took such complaints seriously. As Christner noted in an undated complaint from the 1950s, Buick managers refused to act even though "these problems have been brought to management's attention on several occasions."[79] For foundry workers and other black employees at GM, the grievance process often proved to be a dead end.

Unable to obtain justice through formal channels, frustrated black workers often found themselves in direct confrontations with other employees and supervisors. On January 3, 1956, black activist John Hightower, a member of the UAW's FEPC committee, came to blows with Terry Wilk, a white foreman at the Buick foundry. Prior to the January confrontation, Wilk had called Hightower a "nigger" on at least one occasion and had routinely subjected other workers to "vile language and harsh treatment." Hightower reacted by filing a grievance, but officials from Buick and the UAW ultimately determined that there was not enough evidence to punish Wilk. Remembering the initial incident, Hightower later commented, "I turned and went around the corner and cried and became emotional because I'd just gotten out of the war and I said, 'How could this person . . . be allowed to call me a nigger?'"[80] During the January 3 confrontation, Wilk went even further, calling Hightower a "dirty, black, stinking nigger." This time, however, Hightower took matters into his own hands by delivering a "smashing right cross to the jaw" of his supervisor. Years later, Hightower remembered the details of the confrontation, claiming, "I beat him severely. . . . I used to smoke a pear-witted pipe and I took my pipe out and I told him for the rest of your life here was a nigger you'd never forget." As punishment for the incident, Buick officials fired both men.[81]

The experiences of Berry Blassingame, Paul Kenny, John Hightower, and other black workers in Flint's automobile plants stood in stark contrast to GM's official proclamations on employment discrimination.[82] Throughout the postwar era, GM executives in Flint and Detroit, like the fictional character Mr. Smith, formally endorsed the principle of equal opportunity, but their actions suggested they were committed to upholding the color line. According to local NAACP leader Edgar Holt, "GM like Rip Van Winkle has slept too long. The world and events have passed them by. Of course their machines are modern and pretty but their concern for humanity stinks worse than the manure of a thousand dinosaurs." What GM directors needed, Holt asserted, was to "open their doors and let freedom winds blow through."[83]

In the two decades following the close of World War II, African Americans in Flint fought fierce battles against employment discrimination and poor working conditions. The struggle for employment equity in Genesee County, as in other communities, took place on multiple fronts including the shop floor, the city commission chambers, corporate boardrooms, union halls, and the streets. Yet for Edgar Holt, John Hightower, and other activists, those battles seemed to bring too few victories. Angered by the lack of progress, Holt and other organizers gradually jettisoned the Urban League's quiet negotiation campaign in favor of more confrontational tactics for addressing discrimination. As part of that shift, local activists in the late 1950s and early 1960s began organizing picket lines in front of area factories and coined a new slogan, "Jim Crow Must Go, GM Crow Must Go," that reflected the widespread frustrations of black workers.[84]

The new phrase also signaled a growing awareness among civil rights activists that the campaign at GM was closely linked to the broader struggles for social justice erupting across the region, nation, and world. In sprawling metropolitan areas such as Flint, those larger campaigns necessarily took on a spatial dimension, bringing the racial politics of suburban development to the forefront of fair employment debates. As part of the effort to secure access to better-paying jobs, black workers campaigned vigorously for better schools, new housing, and the right to live and work in segregated suburban areas. Proponents of civil rights were not the only community members involved in these struggles, however. By the close of the 1950s, the rapid but uneven growth of Flint's suburbs had produced political polarization on both sides of the color line. Over time, this fragmentation manifested itself in a new political consciousness among advocates of suburban capitalism who sought independence from both the central city and its increasingly black citizenry.

4

Suburban Renewal

Signs of the dramatic rise of American suburbia were visible all over metropolitan Flint during the decades surrounding World War II. In 1930 just over a quarter of Genesee County's 211,641 residents lived outside of Flint. By 1960, however, well over half of the county's 374,313 people lived beyond the city limits. Over the same period, the percentage of Americans living in suburbs jumped from just 13.8 percent to over 30 percent.[1] The ascendance of the suburbs during the twentieth century was without question one of the most consequential developments in American history. The nationwide migration of people, jobs, wealth, and power from cities and the countryside to suburbs radically reshaped local economies, fostered new types of inequalities, spawned a variety of important political and social movements, and generated novel forms of cultural production. Together, these changes in American society marked the transformation of the United States into a predominantly suburban nation.[2]

None of these shifts would have been possible, though, without policy interventions in the market economy. Over the course of the postwar era, the Federal Housing Administration exerted a powerful influence on the real estate, construction, finance, and banking industries. Through its mortgage insurance program, the FHA induced private banks to grant millions of long-term low-interest fully amortized loans for existing and especially new homes. Not all Americans benefited equally from FHA investments, however. With few exceptions, FHA underwriters maintained administrative forms of Jim Crow by refusing to guarantee mortgages in racially integrated areas and withholding insurance from most black buyers. Moreover, FHA guidelines on minimum lot and home sizes, construction standards, and "neighborhood appeal" privileged low-density residential suburbs over high-density mixed-use urban areas.[3] According to the FHA's explicitly antiurban and prosegregation standards, crowded and diverse city neighborhoods were intrinsically less desirable than all-white suburban subdivisions with large lots and new single-family homes.

Nevertheless, federal housing policies did not reward all suburbs with new investment. On the contrary, from the mid-1930s until well into the 1950s, the FHA systematically redlined suburban areas that possessed inadequate building and zoning regulations, utilities, services, and schools. Lenders and federal underwriters also withheld funds from communities that seemed to reject the secessionist politics of suburban capitalism. In Flint and elsewhere, the communities that suffered from suburban redlining were dispropor-tionately populated by poor and working-class residents, many of them self-builders with limited funds who wished to keep taxes—and, by extension, municipal services—at a minimum. Because the FHA required suburbs to meet numerous economic, governmental, and service standards in exchange for gaining federally backed mortgages, only the most developed sections of the metropolis—typically those with large middle-class majorities—were eligible for federal support. Yet at the close of World War II, only exclusive city neighborhoods such as Woodlawn Park and a small number of Flint's most urbanized middle-class suburbs satisfied the FHA's requirements. Con-sequently, local builders operating during the 1930s, 1940s, and early 1950s fo-cused their efforts on a handful of segregated, well-developed suburbs such as Davison and Mt. Morris and fully-serviced, all-white neighborhoods within the city such as Glendale Hills and Pierce Park (see table 1.1.). During the postwar era, suburban officials—especially those representing poorer, more undeveloped districts—raised taxes, developed a flurry of zoning and build-ing codes, constructed new roads and schools, and worked to provide water and sewer services in their attempts to improve living conditions and attract new investment. Ultimately the FHA rewarded those revitalization efforts by agreeing to insure mortgages in most suburban areas.

Over time, however, the FHA's embrace of suburban capitalism led to greater political fragmentation in the region. To secure the FHA's approval, suburban policy makers raised taxes, funded expensive infrastructure proj-ects, and added many new layers of government. Suburbanites also turned increasingly to municipal incorporation, which made it difficult for Flint to grow. Furthermore, as a byproduct of the modernization and incorporation processes, residents of the out-county began to develop new identities as suburbanites. Eventually these identities fueled the development of a fiercely independent suburban political consciousness that often manifested itself in hostility toward Flint.

Suburban Blight

During the 1940s and 1950s, tens of thousands of families living in Flint and the rural areas around it packed their belongings and headed off to the rigidly segregated suburbs of Genesee County. This mass migration caused the populations of the four townships bordering Flint to grow by over 50 percent.[4] As in other metropolitan regions, migrants set out for Flint's suburbs for a host of reasons. Many moved out of a desire to obtain more land, new homes and businesses, and economic security. Others relocated in order to gain more privacy or because they had grown tired of urban pollution, crime, and traffic. Still others made the move in order to fulfill cultural aspirations for a more pastoral existence or to be closer to work and the new retail establishments opening outside of the city. Although whites seldom spoke openly about issues of race during this era, the desire to relocate to racially segregated subdivisions also played a role in driving the exodus to the suburbs.[5]

Upon arriving in their new communities, many of the nation's suburban homesteaders quickly found that life on the urban fringe was not at all what they hoped it would be.[6] In 1946 researcher Walter Firey published an especially critical study of housing conditions and unplanned development in Genesee County that spoke to the many frustrations of new migrants. "All around Flint," he wrote, "there are hundreds of acres that stand idle. . . . Billboards notifying of 'Lots for Sale—$1 Down, $1 a Week,' street signs that have no streets, sidewalks that have no houses, great expanses of land grown up to weeds, squatters living in tents and trailers—these are the tangible symbols of what has resulted from unguided settlement in the fringe area." Like others who had surveyed the suburbs before him, Firey discovered deep poverty and substandard living arrangements in neighborhoods throughout the out-county. In one especially egregious instance, he encountered a family of eight persons living in a one-room "paper shack" with no water, electricity, or car. While Firey conceded that arrangements such as that represented "the extremity of blight," thousands of suburban families lived under similarly difficult conditions.[7] As late as 1940, over half of the homes in the suburban fringe had no running water, while nearly three-quarters lacked indoor toilets.[8] According to geographer Peirce Lewis, who surveyed the region in the 1950s, "the edge of the city is marked by what is euphemistically called 'substandard housing'—in fact rural slums."[9]

The harsh living conditions that predominated in many of Flint's outlying areas were a reflection of the economic diversity that prevailed in many American suburbs.[10] Prior to 1950, well over half of the Flint region's suburban dwellers were unskilled or semiskilled factory workers, and a disproportionately large share of unemployed persons and subsistence farmers lived outside of the city.[11] As a result, living conditions in many of Flint's suburbs tended to be austere. In the Dayton Heights and Thornton neighborhoods of Flint Township, for instance, two-room houses, basement dwellings, and even tents were common.[12] Nearby, in the Fentonlawn and Romayne Heights subdivisions, many residents kept livestock and engaged in subsistence gardening in order to make ends meet during economic downturns. "I've planted some trees and some berries and we've got a hundred chickens," claimed a suburban homeowner in Flint Township. "We're counting on this after I lose my job."[13] Beyond their qualms about housing standards and economic uncertainty, residents of these and other working-class suburbs often complained about poor utilities, unpaved roads, weak building and zoning codes, and the lack of public transportation.[14]

Misgivings about suburban living were even common among middle-class homeowners. Among residents of Burton Township's Witham Place subdivision, for example, many indicated that they felt isolated from friends and families living in Flint. Some complained that their homes were too distant from grocery stores, movie theaters, and other consumer and cultural attractions in the city. Still others worried that weak building and zoning codes would allow nearby homes and businesses to detract from their property values and quality of life. According to Firey, even carefully restricted communities such as Witham Place needed zoning protection in order to defend against the blighting influences of "tiny dwellings, service stations, and roadside stands." Without zoning, he argued, "property values depreciate all around, stable residents desert the neighborhood, and physical deterioration sets in." Even in completely segregated suburban areas that catered to middle-class professionals, blight was an ever-present concern.[15]

For urban transplants who had grown accustomed to big-city services and infrastructure, the paucity of amenities in the out-county was a source of great frustration. At midcentury most residential areas in the Genesee County suburbs lacked paved roads, modern schools, garbage collection services, public water and sewers, zoning and building codes, and plat regulations.[16] Additionally, police and fire services were almost nonexistent in Flint's postwar suburbs.[17] During a July 1948 meeting of the Genesee County Board of Supervisors, Argentine Township supervisor Charles Markley maintained that the postwar surge in suburban home building had created "a very serious problem . . . as it relates to drainage, sewers, water, fire control, schools, and other facilities."[18] The author of a 1957 study of suburban development concurred, noting, "The fringe is deficient in virtually all of the services required for reasonably efficient and sanitary urban living."[19] To many of Flint's unhappy suburbanites, it seemed as though their communities had grown too quickly and chaotically and that unchecked growth was undermining neighborhood tranquillity, lowering property values, and creating infrastructural crises.

Problems with waste disposal were especially frustrating to suburban residents. At the close of the 1950s, only two of the four townships surrounding the city offered rubbish collection services. In many areas, homeowners burned their garbage, buried it, or simply dumped it in open fields or wooded areas. Furthermore, sewers were extremely rare in the suburbs. Outside of Flint, Burton Township, and the northern suburb of Beecher, where sewer services were widely available, most homeowners relied on outhouses or private septic systems. As suburbs grew, septic system failures and effluent seepage became pressing public health problems.[20] In a 1947 study of local government in suburban Flint, Israel Harding Hughes claimed, "Septic tank effluent flows from the townships through Flint in nearly every direction." "Frequently throughout the year," admitted a local journalist, "Flint literally is surrounded by an open sewer."[21] Septic seepage occurred for a variety of reasons including insufficiently large drainage fields, nonabsorbent soils, runoff, and improper maintenance. The relatively impervious clay soils that predominated in many parts of Genesee County presented a special problem for septic system users. According to one report, "the prevalence of heavy clay soils results in surface drainage of much of the effluent and subsequent seepage into wells." Even when the systems worked correctly, though, unrestrained development brought wells and other water sources into dangerous proximity with raw sewage.[22] By 1954 the local sewage problem had become so severe that the Michigan Water Resources Commission ordered governments throughout the out-county to install modern sewer systems by the beginning of 1958 or face stiff fines.[23]

Concerns over the water supply were also widespread among suburbanites. At the end of the 1940s, only about half of Flint's one hundred thousand suburban residents possessed both running water and indoor toilets.[24] Among those who had access to running water, most relied on private backyard wells.[25] Throughout the out-county, even in areas served by government-owned wells, residents often complained about effluent contamination, poor water pressure, and smelly, briny-tasting water. Over time, water shortages also began to emerge. In 1954 a water shortfall in the northwestern suburb of Flushing triggered a civic emergency. The crisis grew out of the growth of the 1940s, when Flushing's population surged by 25 percent.[26] As new homeowners dug private wells and tapped into the village's antiquated community water system, the water table fell rapidly. On January 29, 1954, after months of warnings, Flushing's public water system shut down completely. Hoping to avert a public health crisis, members of the village council searched for new wells, began a rationing program, and implemented "Operation Tanker," a plan to purchase and transport water from Flint via large trucks.[27] On November 23, councilpersons moved to prevent future emergencies by passing a major new water infrastructure bill. The three-hundred-thousand-dollar project, which included a new community well, water mains, pumping equipment, a 750,000-gallon storage tower, and a well house, was the largest and costliest development in village history. In order to fund the new system, the village council issued revenue bonds and increased fees for water customers.

Although Flushing's problems were more severe than those of other areas of the county, many suburban governments across the region and nation confronted similar resource, utility, and service crises during the postwar era. Just a few years after the trouble in Flushing, researchers from the University of Michigan warned of "an acute water shortage" looming on the horizon.[28] Nationwide, emergencies such as the one that unfolded in Flushing enraged millions of homeowners and forced suburban policy makers to develop expensive new infrastructure projects.[29] Without fail, these projects drew financial and political support from the Federal Housing Administration and other government agencies. By linking mortgage insurance to the provision of services, federal officials played a major role in transforming the working-class "shack towns" of the early postwar period into modern middle-class suburbs.

Making Suburbs Modern

During the 1940s and 1950s, suburban policy makers worked closely with federal officials, lenders, and builders to boost property values in new subdivisions and resolve the mounting service and utility crises in Genesee County. Following the FHA's explicit recommendations, suburban legislators developed numerous master plans, zoning codes, and building regulations while working to improve government services.[30] Throughout the metropolitan region, the changes wrought by the FHA's suburban redlining policies were impressive. In south suburban Fenton, for instance, village leaders invested millions of dollars in new infrastructure improvements in order to obtain federal backing for mortgages. Just in 1947, members of the village council purchased a road grading machine, a two-ton truck, and a gravel loader. During the course of the year, workers used the new equipment to pave two miles of roads, lay over a mile of new sewers, and build six blocks of new sidewalks. They did so in large part because the FHA required all of its new developments to include sidewalks, paved roads, sewers, and other utilities.

If policy makers could not meet these conditions, then the FHA canceled its backing for new residential projects. This almost occurred in 1952, when FHA administrators threatened to redline the entire Fairfield subdivision in northwest Fenton until village officials had provided sewers, sidewalks, and water service. According to a local journalist, "the FHA demands that these services be available before construction starts."[31] In order to comply with the FHA's demands, authorities in Fenton and numerous other suburbs abandoned "pay as you go" financing and enacted new taxes to support development. However, the FHA rewarded such strides toward suburban capitalism by offering mortgage insurance to the Fairfield subdivision and other new residential developments.[32]

A similar situation occurred in Flushing in 1954, just after the resolution of the water crisis. That year, local builders set a new record by erecting fifty-nine new homes in and around the village.[33] Because the village council had acted quickly to resolve the emergency, new homeowners were able to enjoy virtually unlimited quantities of clean drinking water. Sewers, however, were another matter. Like most other suburbs in Genesee County, Flushing lacked a public sewer system. Instead, homeowners used outdoor privies or septic systems. Some even relied on private drainage lines that emptied untreated waste directly into the Flint River. In the fall of 1946, the Federal Works Administration approved a $130,000 grant to fund a sewer system for village residents. However, between 1946 and 1950 Flushing voters narrowly rejected two ballot proposals to raise the village's portion of the cost for the new sewers.[34] Despite the setbacks, members of the Flushing Village Council persisted in their efforts to gain the two-thirds majority of voters needed to approve the project. In March 1952 local voters narrowly endorsed a $165,000 bonding proposal to construct sewer lines and a waste treatment plant.[35] Village officials waited for two years to act, however, convinced that they lacked a strong public mandate for the project. In response, members of the Michigan Water Resources Commission moved to take punitive action, while FHA representatives threatened to withhold mortgage insurance throughout Flushing until the village council had built a modern sewer system. Within days of the FHA's announcement, the Flushing Village Council had agreed to construct a new system—including sewer mains, an interceptor, and a disposal plant—by no later than January 1, 1957. Satisfied by the council's pledge, the FHA rescinded its redlining threat in January 1955.[36]

Builders and investors responded to the rise of suburban capitalism in Flushing by launching a number of new residential developments in and around the village. In 1956, the year after the resolution of the sewer crisis, the village council issued construction permits for forty-one new homes, the second-highest total in Flushing history.[37] Throughout the summer of that year, full-page advertisements in the *Flushing Observer* announced that new homes were available for sale in the Labian Terrace subdivision. The ads reminded buyers that the subdivision had the FHA's full backing and that new homes in the neighborhood featured "all the conveniences of the Big City," including city water and sewers, school bus service, and paved streets. In a matter of years, federal housing and development policies had helped to transform the sleepy village of Flushing into a modern, independent suburb with a variety of urban amenities. The decision to incorporate Flushing as a home rule municipality would be just a few years away.[38]

Occasionally suburban officials balked at complying with the FHA's infrastructure, service, and economic development regulations. One such instance occurred in the far northern suburb of Clio. Like other suburban areas, Clio grew rapidly during the first half of the twentieth century. Between 1925 and 1951, enrollment in Clio's schools grew by 103 percent.[39] Members of the Clio Area Study Committee addressed the problem of school overcrowding by repeatedly recommending new taxes to fund two new elementary schools and major repairs to the district's high school. In 1956 committee members succeeded in getting a bond proposal on the ballot, but voters— some of them working-class homeowners who opposed tax increases on their self-built dwellings and others small farmers who wanted to keep the area rural—decisively rejected the initiative.[40] That same year, residents also voted down new taxes to build water and sewer lines. Construction and development within the city came to a virtual standstill in the wake of the votes. In January 1957, representatives from the Clio Chamber of Commerce organized a meeting with Oren Stone, director of Flint's FHA office, to discuss the situation. During the meeting, Stone claimed that Clio had fallen behind Fenton and other booming suburban areas because the city did not provide adequate schools, water, sewage disposal, and storm drains. At one point during the discussion, one of Stone's associates scolded Clio's residents for neglecting to modernize the city. If Clio's citizens did not want to pay taxes for improvements, the FHA representative offered, "[t]hey should sit down and get their horse and buggy back and keep on living in the past." Under the FHA's stringent guidelines, Stone averred, new development would simply not occur without urbanized infrastructure and other growth-inducing improvements.[41]

For those who sought evidence of what the combination of local initiative and federal investment could produce, they needed to look no further than the FHA-approved suburb of Davison, which many suburban capitalists held up as a model for development. Located eight miles east of Flint, this small, predominantly middle-class city grew faster than any other area of Genesee County during the two decades bracketing World War II. On April 19, 1946, the *Davison Index* newspaper ran a special column trumpeting the area's growth so far that year. The author listed the many construction projects that were under way in and around the city: a new roller rink and bowling alley, a school and convent, a business office operated by the Leffler Gravel Company, a Davison Dairy outlet, a new grocery store, Munsell Restaurant, and several new medical offices. To serve these new developments, the city launched its first major road-paving initiative.[42]

The author of the article also described the city's booming housing market. One of the year's most significant real estate events occurred when developers Robert Wisler and Harry Hill platted land and sought FHA approval for a new subdivision on the site of the Davison fairgrounds. The development was to feature new streets, sidewalks, sewers, and water service. Later in 1946, the author noted, O. L. Adams offered lots for sale in a newly platted subdivision on Davison's west side. All of the new projects, the *Index* article conceded, had created a massive workload for members of the Davison City Council. Just to accommodate the new zoning requests that it had received in 1946, the council increased the frequency of its meetings from once to twice per month.[43]

In October 1946 the FHA approved the subdivision plans submitted by Wisler and Hill. Later renamed Rosemore Park, the new subdivision consisted of 133 "carefully restricted" residential lots. The developers agreed that before selling the properties, they would install sewer hookups and water lines, and they made plans with Consumers Power Company to connect each of the new homes to the gas and electric grids. In order to insulate the subdivision from traffic, the FHA required Wisler and Hill to construct only one entrance into the neighborhood and build a sixteen-foot planting strip between several homes and a nearby street. Additionally, the FHA insisted that all new homes in Rosemore Park face toward the center of the subdivision. Upon purchasing their new lots, buyers could either choose from a list of government-approved home designs or submit their own architectural blueprints to the FHA, so long as the new abodes cost a minimum of six thousand dollars. Also, according to the FHA's guidelines, Rosemore Park was to be a whites-only subdivision.[44]

Neighborhoods such as Rosemore Park proliferated rapidly in postwar Davison. Just between 1946 and 1949, developers added the Adams, Daniel Allen, and George P. Hill subdivisions to the city's growing list of new, racially segregated residential communities. On July 8, 1949, a writer from the *Davison Index* boasted, "There has been no other town in Genesee County that

FIGURE 4.1. A sewer expansion project in the suburb of Davison, 1958. Between the 1930s and the 1950s, suburban officials in Genesee County, often with strong support from local taxpayers, spent a great deal of time, money, and energy modernizing their governments and building urbanized infrastructures, primarily to improve living conditions and attract new investment. As a result of efforts such as this one, the Federal Housing Administration largely abandoned its suburban redlining policies. Courtesy of the *Flint Journal.*

has gone so extensively into the building project. . . . Day after day, another basement is being dug and a house frame erected here in Davison." When asked to explain the area's growth, civic boosters pointed to its convenient location near Flint, rising property values, good schools, and the many new businesses serving the city. Municipal leaders also highlighted Davison's well-equipped fire department, modern utility services, newly paved roads, and growth-friendly government. Unlike their counterparts in Clio, Davison's leaders embraced the FHA's suburban capitalist policies, even when they required tax increases and additional layers of government. "Each new resident in Davison means increasing prosperity for the town," the *Index* explained.[45]

Taxpayers and government functionaries did their part to support Davison's turn toward suburban capitalism by passing new building and zoning regulations, recruiting new businesses, funding educational improvements, and maintaining popular and administrative modes of racial segregation. In 1948 the Davison City Council developed a master plan and passed a revised zoning code that expanded the number of restricted land use districts from three to eight. Supporters of the new ordinance argued that it would protect property values by strictly limiting the areas of the city in which builders could erect multiple-family housing and other unwanted developments.[46] In 1951 local business owners and professionals organized the Davison Area Chamber of Commerce, which helped the city increase its profile among investors.[47] Three years later, area voters overwhelmingly endorsed a plan to combine Davison's schools with twelve rural school districts surrounding the city. The consolidation dramatically increased local tax revenues, which policy makers then used to fund an expansion of Davison High School, a new elementary facility, and an athletic complex.[48]

The annexation of unincorporated land was another important part of the Davison City Council's suburban capitalist growth strategy. In September 1953, for example, developers announced plans to build the Manford Heights neighborhood just south of the city. Within a year, city council members had voted to annex the entire subdivision. As part of the agreement, the council agreed to provide homes in Manford Heights with water, sewers, and roads.[49] In response, the FHA agreed to underwrite mortgages for qualified buyers in the new community. Within weeks of receiving the FHA's endorsement, builders from Rasak Home Builders and other local firms began erecting 190 new houses in the subdivision. In order to protect property values in the new development, builders adhered to strict construction regulations and, in deference to FHA underwriters, selected respected real estate agent Frank Hachtel to handle all marketing and sales. Although by the early 1950s the FHA could no longer advocate racially restrictive housing covenants or other forms of legal segregation, Hachtel—like all other realtors serving the Davison community—accepted as gospel the notion that racial integration would detract from neighborhood stability and property values. Accordingly, he and his colleagues refused to show the new properties to African Americans.[50]

Local growth initiatives and federal housing and development policies combined to spur an unprecedented economic boom in and around Davison during the 1940s and 1950s. In a 1954 editorial that spoke to the rise of suburban capitalism, writers from the *Davison Index* pointed out that the area's growth did not begin until "leaders changed from a village to a city, established a modern school system, set up a zoning ordinance and building code, built a modern, large capacity sewage disposal plant to make Davison the most logical building site in the county, and created a Chamber of Commerce."[51] By embracing the FHA's numerous guidelines for development, Davison's leaders transformed their city into an attractive place for white homeowners, shopkeepers, and investors. Between 1940 and 1960, Davison's population more than doubled to 3,761, while the number of businesses in the city rose 93 percent.[52]

Davison's growth was part of a much larger transformation occurring co-evally in suburbs throughout the United States.[53] During the decade follow-ing World War II, as the nation hurtled toward a suburban plurality, all but one of the communities surrounding Flint grew in population. Except in a few rare instances, Flint's innermost suburbs—especially those that met the FHA's requirements—grew much faster than the outlying areas lacking ur-banized governments and services. In the 1940s, for example, the populations of Mt. Morris and Burton Townships grew by 109 and 66 percent, respec-tively. Similarly, the townships of Genesee and Flint each posted population gains in excess of 40 percent during the decade. Even in north suburban Clio and other areas of the out-county where voters clung to "horse and buggy" lifestyles, the postwar era brought impressive growth.[54] By 1950 Flint's share of the county's overall population had declined to 60 percent.[55] The growth of the out-county continued apace during the 1950s as suburban Genesee County added seventy thousand new residents. Although the central city also gained new residents during this period, suburban growth was six times that recorded in Flint. As new housing starts in the suburbs began to outpace those in the city, such imbalances only grew wider.[56]

Across the country, suburban officials struggled mightily to meet the growing service needs of new homeowners and shopkeepers. Most subur-banites wanted better schools and roads, more shopping centers, stricter building and zoning regulations, clean water, modern sewer systems, po-lice and fire protection, and other costly services. Elected officials were only rarely able to provide all of the services that residents expected, however. Further complicating matters, other suburban dwellers preferred smaller, less intrusive government and despised the development that was occurring around them. These antigrowth farmers and homeowners, many of whom were impoverished, fought to keep their taxes low and restrict new construc-tion. Rarely did either side get what they wanted, though. During the 1950s suburban frustration reached a boiling point as new residents and problems continued to pour into the cities, villages, and townships surrounding Flint. Ultimately the resolution of these conflicts over suburban capitalism and development brought newfound power and independence to local govern-ments in the out-county.

Just as they had done in the 1940s, suburban policy makers launched ma-jor improvement campaigns during the 1950s in order to satisfy grassroots demands for services and meet the stringent regulations of the FHA. The most pressing of these were water and sewer projects. By 1956 there were nine

FIGURE 4.2. A modern residential subdivision in suburban Fenton, 1960. With support from the Federal Housing Administration, local builders erected hundreds of new residential subdivisions such as this one during the postwar era. Over time, however, the FHA's endorsement of suburban capitalist policies led to greater racial segregation and increased political fragmentation in the Flint region. Courtesy of the *Flint Journal.*

public water systems operating in Genesee County, and nearly a third of Flint's suburbs possessed sanitary sewers. Although the new utilities helped to solve many public health problems, infrastructure projects also drained the coffers of suburban governments. Between 1940 and 1956, government outlays in the four townships surrounding Flint skyrocketed from a yearly average of $8,000 to more than $161,000, with most of the increases earmarked for sewer and water projects and other capital expenditures.[57] Other major expenses included the construction of new schools and roads, the expansion of fire and police protection services, and the opening of new public facilities such as libraries and parks. Provision of these public goods and services necessitated a major expansion of suburban government.

For the sake of protecting property values and controlling growth, suburban capitalists and their FHA underwriters also called on local governments to pass restrictive zoning and building codes. Between 1946 and 1955, legislators in twenty-four of Flint's suburbs adopted such ordinances. Suburban villages and cities such as Fenton, Flushing, Davison, Grand Blanc, and

Mt. Morris were the first to adopt the new codes. By the mid-1950s, how-ever, rural and semi-rural townships throughout the county had followed suit.[58] Under the new codes, suburban governments established minimum home and lot sizes, land use guidelines, and basic construction standards for residential buildings, all of which helped to harden racial and economic segregation. While the zoning laws in Flint and Flint Township established minimum lot sizes of five thousand square feet for most residential struc-tures, the codes in the remainder of the county were much more stringent. For instance, the 1947 zoning code for the city of Mt. Morris required lot sizes of at least 12,075 square feet for new single-family homes, while in many rural areas residential lots could be no smaller than twenty thousand square feet.[59] Because most of the new codes either prohibited or severely limited apart-ment complexes and other affordable housing units, they helped to generate new forms of administrative segregation.[60]

Even with such restrictive policies in place, however, planners in Gen-esee County and other suburban areas around the country found it virtually impossible to control postwar population growth.[61] This unchecked growth presented special problems for local school officials, who found themselves in a near-constant state of crisis. At the close of World War II, the twenty-six school districts serving the out-county were still largely rural institutions. As late as 1947, in fact, there were ten one-room schoolhouses and twenty-eight two-room schools in the four rapidly urbanizing townships surround-ing Flint. All of them, according to one study, were severely underfunded and largely unprepared for rising enrollments.[62] To accommodate the new growth, education authorities rushed to build new facilities, increased class sizes, and orchestrated a large number of district consolidations. Between 1920 and 1955, the number of school districts in Genesee County declined from 137 to 36. The consolidations allowed school leaders to draw from larger tax bases and deliver more cost-efficient educational services.[63] Nonethe-less, the postwar baby boom brought massive enrollment spikes that over-whelmed virtually every suburban school system in the nation, including those in Genesee County.[64] By the mid-1950s dozens of suburban school su-perintendents in the Flint area had declared that their districts were in a state of emergency.[65]

In the spring of 1958, the crisis hit Flushing High School, forcing Princi-pal Bertrand Long to suspend new admissions. That year, the school had a pupil-teacher ratio of 34:1. In the building's home economics room, thirty-two students crowded into a space designed to hold only twenty people. The study hall facilities, built to house a maximum of 140 pupils, often contained in excess of 200 students. "We cannot adequately house our students now," Superintendent M. D. Crouse warned. "We are definitely in trouble."[66]

While Flushing's schools were among the most overcrowded in the county, they were by no means unique. In the Mt. Morris and Beecher systems, for instance, the student-teacher ratios were 50:1 and 44:1, respectively. To resolve the crisis, voters in Flushing and dozens of other districts approved new taxes to hire teachers and construct new school facilities. In Flushing voters endorsed a $1.5 million bond proposal to build a new high school. However, like many other suburban schools, Flushing's new building was full within a few years of opening. Easily visible throughout the region and nation, images of schoolchildren attending study halls in gymnasiums, taking shop classes in converted coal sheds, and learning to read in tents defied easy stereotypes of suburban plenty.[67]

During the 1950s a team of scholars from the University of Michigan's Social Science Research Project (SSRP) documented citizen frustration over Flint's burgeoning suburban crisis with a series of reports on local public opinion. The reports suggested that suburbs were still far from idyllic places and that residents of outlying areas remained unhappy with many aspects of their lives. Not surprisingly, complaints over the quality of the water supply, sewers, and schools were foremost in the minds of suburban dwellers. However, there were numerous other sources of discontent as well. For example, only 25 percent of suburban residents polled were happy with the quality of police protection in their neighborhoods. Similarly, under half of the suburbanites queried were "very satisfied" with the garbage disposal services offered in their communities. Road conditions were also a source of frustration, with just 17 percent of household heads reporting that they were "highly satisfied" with their local thoroughfares. Likewise, a minuscule 12 percent of suburban respondents were happy with the street lighting in their subdivisions. Residents were also deeply disappointed with their libraries and parks. Furthermore, one of every two household heads in suburban Flint believed that their neighborhoods lacked adequate public transportation. Regardless of the issue, suburban residents were generally displeased with the governmental services and amenities available in their communities. Most of these unhappy homeowners were desperate to bring the conveniences of urban living to their new neighborhoods—even if that meant higher taxes.[68]

Suburban homeowners' willingness to pay higher taxes stemmed in part from inequities in the existing tax structure. As in most other metropolitan areas, the tax system in Genesee County favored homeowners residing in outlying areas over those living in the city. Specifically, local governments in the out-county received unfairly large sums of revenue through sales tax returns from the state of Michigan. Residents of Genesee County living outside of Flint paid approximately 42 percent of the local sales tax receipts, yet they received nearly 50 percent of the county's sales tax returns. Because of this imbalance, officeholders in the out-county were able to keep residential property taxes low, especially in the villages and townships that provided only basic services. According to one study conducted during the mid-1950s, only 18 percent of suburban homeowners in Genesee County paid more than one hundred dollars per year in local property taxes. Even among wealthy homeowners whose residences were worth more than fifteen thousand dollars, nearly 60 percent paid less than one hundred dollars in annual property taxes.[69]

In the late 1940s and early 1950s, however, suburban policy makers in Genesee County began adopting the Davison and Flushing models by raising taxes and fees in response to FHA strictures and the growing service needs of residents. Because most suburban homeowners had been spending only a small fraction of their income to support local government, opposition to new taxes was minimal.[70] In fact, nearly 70 percent of fringe residents actually supported tax increases to fund new schools, sewers, roads, and the like. "Contrary to popular belief," one researcher wrote, "the fringe residents are much more receptive to the payment of higher taxes than are central city residents."[71] The overwhelming support for higher taxes reflected the growing consensus among homeowners, investors, and other suburban capitalists that services and infrastructure in the out-county were simply insufficient to foster economic growth and a high standard of living.

Taxes in suburban Flint varied widely due to factors such as population size, business growth, and assessment rates. Even within local political jurisdictions, voters paid differently depending on the school district in which their property was located. Regardless of location, however, taxes rose dramatically in suburban Flint during the 1950s, due in large part to the cost of funding schools and utility projects. By the close of the 1950s, many suburban homeowners were shouldering higher tax burdens than city residents. In 1955, for instance, the owner of a twenty-thousand-dollar home in the city of Flint paid a total of $195 in local property taxes. That same year, owners of twenty-thousand-dollar homes in Burton Township's Atherton School District and the Dye School District of Flint Township paid just over $250 to fund local government. "When services are considered," journalist Homer Dowdy asserted, many suburban residents "find they're now paying as high taxes as their City friends." On a per capita basis, government in suburban Flint had become more costly than in the central city. Few suburban residents objected to that, however.[72]

The Birth of Suburban Identity

Before purchasing new homes in the suburbs, the most careful consumers conducted research on taxes, zoning restrictions, utility services, and other issues. Others visited local schools and inquired about property values. Most suburban migrants, however, knew only basic facts about their new communities prior to relocating. For the vast majority of home buyers, the search for real estate hinged primarily on the quality and affordability of homes, neighborhoods, and schools. Few buyers had either the time or the inclination to conduct extensive research on the governmental services available in the suburbs—or, for that matter, on the political structures of their new communities—before they moved. Even fewer understood with any precision what it meant to be a resident of a place such as Flushing, Mt. Morris, or Burton Township. Prior to the 1950s many residents of the out-county had yet to form explicit identities as suburbanites.

Things began to change in the 1950s, though, as homeowners—both on their own and collectively—worked to mitigate the unsatisfactory living conditions in the suburbs and overcome FHA redlining. Anger and disappointment over low-quality infrastructure, poor schools, and the lack of urban amenities brought people together to search for solutions, and those community interactions helped to foster a new sense of civic identity among suburban capitalist homeowners. Parents who were upset over pupil overcrowding joined their local PTA and worked to support new taxes to fund school construction. Homeowners formed neighborhood associations and block clubs in order to police their subdivisions, maintain racial segregation, and protect property values. Concerned residents united to demand new streetlights, sewer systems, and roads. As residents worked collectively in support of such efforts, they began to see themselves and their communities in a different light.[73]

In short, the urbanization of America's suburbs—whether it occurred in the late nineteenth and early twentieth centuries, as in Boston, or well after World War II, as was the case in much of the Flint area—spawned new forms of civic engagement and a new independent political consciousness among millions of suburban residents.[74] Locally, this nascent suburban capitalist identity manifested itself in a variety of complex and often contradictory ways. During the late 1940s and early 1950s, for instance, suburban retailers and boosters launched major campaigns to persuade residents to patronize local stores instead of shopping in Flint. In 1948 merchants in Fenton unveiled the slogan "Buy, Build, Believe in Fenton."[75] A year later, the *Fenton Independent* published a full-page advertisement informing readers, "You get more for the dollar you spend in Fenton." The advertisement offered seven reasons why Fenton residents should do business with local merchants. Chief among them was that shopping locally kept wealth inside the community. "Increased retail business," the ad maintained, "will either lower your contribution to any given Fenton project or make the project better financially." Like other suburban boosters, the architects of this campaign believed that local consumption strengthened civic bonds and made the community more coherent. "Beyond the money advantage," the advertisement concluded, "you have the satisfaction of being on the team that is Fenton."[76]

Like the grassroots agitation for new sewers and school taxes, suburban shopping campaigns were important components of community building. They were also indispensable to the broader effort to establish suburban capitalism. Therefore they were exceedingly common during the postwar era. Through participation in such efforts, suburban residents gained a heightened sense of civic pride. At the same time, local mobilizations reminded suburbanites that their civic and financial interests were no longer synonymous with those of Vehicle City residents. In fact, community solidarity in the cities, villages, and townships of suburban Genesee County derived as much from the feeling that suburbanites shared a distinctive identity vis-à-vis Flint residents as it did from any positive sense of civic pride. Though suburban dwellers often fought among themselves about how best, or whether at all, to modernize their new governments, residents of the out-county could agree on at least one thing: by choice, Flint was no longer their home.

Still, many proponents of suburban capitalism envied Flint's community schools and the high-quality services available to Vehicle City residents. By advocating new improvements, suburban renewers hoped to replicate the services available in the city. Most of those same individuals were vehemently opposed to joining the city, though. During the 1950s academic researchers conducted several local studies that highlighted the strength of the growing suburban independence movement. According to one 1955 survey, a scant 9.1 percent of suburbanites endorsed annexation, while just 28 percent of suburban residents expressed support for cooperative regional government. In 1957 University of Michigan scholars Basil Zimmer and Amos Hawley reported the startling fact that nearly two-thirds of suburbanites in Genesee County would never, under any circumstance, vote for annexation. Whenever they were asked, suburban capitalists made it clear that they wished to obtain big-city services and a measure of urbanization without actually becoming part of Flint. With opposition to annexation so deep and widespread, remedies for metropolitan problems would have to come through other, more creative means. The quest for such solutions led directly to the controversial New Flint proposal of 1958.[77]

The Metropolitan Moment

Between 1940 and 1960, executives at GM implemented a new investment strategy that hinged on the suburbanization of automobile production. In pursuit of that goal, corporate leaders supervised the construction of eight new plants outside Flint while shifting millions of dollars in capital and tens of thousands of jobs from the city to the suburbs. Not at all unique, the capital transfers that occurred in Genesee County were part of a nationwide migration of businesses and jobs from cities to suburbs that unfolded during and after World War II. By opening new facilities in outlying areas, corporate chiefs hoped to reduce production costs, tap new markets for their products, and take advantage of government subsidies designed to promote economic growth outside of central cities. Working in tandem, progrowth public policies and the profit-driven strategies of GM and other companies helped to restructure the geography of American capitalism.[1]

Although many urban dwellers resented GM's capital shifts, the decision to build new plants in suburban areas was not motivated by a desire to abandon Flint. Instead, GM's suburban investment strategy was part of a broader agenda of metropolitan capitalism within which the city of Flint was essential. At its core, metropolitan capitalism was a vision and plan for economic growth that revolved around the suburbanization of manufacturing and the creation of regional governments. Unlike their secessionist-minded counterparts at Ford and other companies, who looked to escape large urban centers by establishing new enterprises in politically independent suburbs, metropolitan capitalists wanted to do business inside enlarged cities.[2] Consequently, when proponents of metropolitan capitalism from GM and other American companies opened new facilities in the suburbs, they did so with the hope that the city would eventually gain authority over them, either through annexation or by other means.[3] This commitment to urban production grew out of corporate decision makers' belief that larger regional governments were more efficient economically than fragmented jurisdictional structures. But business leaders' support for metropolitan capitalism was also inseparable from their confidence that they could control municipal politics in order to gain competitive advantages over their rivals. In most cases Flint's political leaders were more than willing to cater to the company's wishes. Believing that Flint's future prosperity depended upon the construction of new factories, even if they initially had addresses outside the city, municipal policy makers subsidized GM's suburban strategy by providing water and sewer lines to outlying plants, offering discounted utilities, and building new roads to accommodate industrial traffic.[4] Like GM's directors, Flint's elected leaders viewed growth through a regional lens and believed that the city and its suburbs were part of an organic metropolitan whole.

Corporate officials pursued their metropolitan capitalist expansion strategy by supporting annexation bids and other efforts to enlarge Flint's boundaries. Moreover, in the late 1950s GM helped to launch New Flint, a plan to create a single Sunbelt-style "super government" for the entire metropolitan region.[5] Although GM managers actively supported the plan, suburban capitalists—who wished to shield their resources and investments from urban officials—united to defeat the proposed megacity. During the decade following New Flint's demise, the urban-suburban rift grew as municipal officials, with the support of GM, began "strip-annexing" factories and other businesses near the city limits. Suburban legislators responded by rushing to incorporate their communities, taking advantage of the state's lenient municipal formation statutes. The incorporation of Flint's suburbs left the Vehicle City extremely isolated from its neighbors. At the same time, the passing of Flint's metropolitan moment forced GM's leaders to reassess their commitment to the city. Ultimately the postwar battles between metropolitan capitalists and advocates of suburban capitalism produced a fractured metropolis with a rigidly segregated, rapidly deindustrializing, and increasingly black and poor city at its core.

Suburban Strategies

During the 1940s and 1950s, GM executives built an impressive network of modern single-story manufacturing, assembly, and warehouse facilities surrounding Flint. The new plants, eight of them in all, formed an arc around the city stretching from the southeastern community of Grand Blanc, through Swartz Creek and Flint Township on the southwest and west sides, to the northern metropolitan district of Beecher and the eastern suburb of Burton Township. In contrast to the densely arranged, inefficient multistory structures that lined the city's industrial corridors along the Flint River and rail lines, the new suburban factories signaled GM's preference for larger single-story plants and freeway-oriented production. Embracing an aesthetic ideal that celebrated the harmony between manufacturing and suburban home ownership, GM consciously modeled its suburban plants after surrounding residential landscapes.[6] According to a *Detroit Free Press* writer, "The new plants being built today all over the country resemble modern country club structures, with low, sleek lines graced by landscaping."[7] Although corporate architects and city planners hoped that the compatibility of suburban factories and residential subdivisions would translate into new forms of civic comity, their utopian dreams quickly gave way to a more contentious reality as local residents competed over the spoils of industrial decentralization.[8]

The decision to build new factories outside the city marked a significant spatial shift in the local economy. Prior to the Second World War, all of GM's major industrial facilities in Genesee County were located inside the city. As late as 1940, in fact, under 3 percent of GM's area employees worked in the suburbs.[9] Beginning in the 1940s, however, GM's directors replaced their strategy of industrial urbanism with a new philosophy of metropolitan capitalism. Evidence of this shift first became visible in 1942, when GM began producing Sherman tanks for the US Army in a federal defense facility in the south suburb of Grand Blanc. After the war, GM purchased the tank factory and reopened it as a metal-stamping plant employing three thousand workers.[10]

Corporate officials continued to invest heavily in suburban Flint during the postwar era.[11] In 1953 leaders from GM's Ternstedt division took possession of a federal aircraft engine facility in north suburban Beecher and converted it to a massive two-million-square-foot auto parts factory that employed nearly seven thousand workers.[12] Three years later, AC Spark Plug announced a major addition to its parts plant on Dort Highway. With the new addition, GM's AC facilities on the east side crossed the Flint border into suburban Burton Township.[13] Then in 1957 Chevrolet relocated its service and parts distribution warehouse from the Chevy in the Hole complex on Flint's west side to the southwestern suburb of Otterburn, later renamed Swartz Creek.[14]

Although signs of the emerging metropolitan capitalist order were easy to spot across the region, they were most noticeable in the southwest suburb of Flint Township, where GM constructed an ultramodern manufacturing and assembly complex near the intersection of Van Slyke and Bristol Roads.

MAP 5.1. The major General Motors facilities in metropolitan Flint and their opening dates, 1900–1960. During the early decades of the twentieth century, most of the General Motors factories in the Flint area were located inside the city limits. Beginning in the 1940s, however, company officials pursued a metropolitan capitalist growth strategy that revolved around the suburbanization of automobile manufacturing. These investment shifts were not examples of corporate abandonment, however. Rather, the decision to decentralize production was part of a much larger effort to expand Flint's boundaries and create a regional government. *Sources: Flint Journal*; Michigan Center for Geographic Information, *Michigan Geographic Framework: Genesee County* (Lansing: Michigan Center for Geographic Information, 2009).

This large compound of factories first took shape in 1947, when GM's Chevrolet division opened a body manufacturing facility and an assembly plant. Equipped with employee locker rooms, cafeterias, and other modern conveniences, the two plants were the showpieces of GM's postwar investment campaign and a source of great civic pride.[15] Six years later, in 1953, Chevrolet continued its decentralization campaign by opening a V-8 engine plant at the Van Slyke site. Shortly after that, GM rounded out its Flint Township compound by completing a fourth plant—this one a stamping and manufacturing facility.[16]

The opening of the new suburban plants set off a hiring frenzy that drove GM's countywide employment to nearly eighty thousand, an all-time high. But GM's investments in Flint Township and other suburbs came at the expense of major cutbacks and plant closures inside the city. Within the Chevrolet division, for instance, the opening of the Van Slyke facility in 1947 triggered the termination of body production and assembly work at Chevy in the Hole, one of GM's oldest industrial sites. Similarly, the opening of the highly

FIGURE 5.1. Aerial view of GM factories in Flint Township, 1954. In pursuit of their metropolitan capitalist growth strategy, GM officials built numerous suburban facilities such as this one following World War II. The opening of new suburban plants often triggered intense jurisdictional conflicts between suburbanites and city dwellers. Courtesy of the *Flint Journal.*

automated engine and stamping plants at the Van Slyke Road compound in the early 1950s rendered more labor-intensive factories at Chevy in the Hole obsolete. In a 1954 speech to the Flint Industrial Executives Club, Chevrolet general manager T. H. Keating reflected triumphantly on the new plants, claiming, "Flint is the fastest-growing center in the automobile industry and shows the way when big things are doing in America."[17] Missing from Keating's statement, though, was an acknowledgment of the fact that virtually all of GM's major postwar construction initiatives, both in Genesee County and in other US manufacturing hubs, were occurring in suburban areas. Yet this was significant, of course, because municipal officials in Flint and many other GM plant cities could not tax the new suburban plants.[18]

Still, escaping Flint was not one of GM's primary objectives. Rather, a variety of more complex political, economic, and social forces drove GM's metropolitan capitalist strategy. Chief among them were urban land shortages. In delineating the criteria that company leaders used to plan postwar plant construction, Frederick G. Tykle, the director of GM's Argonaut Real Estate division, explained the urban land limitations that led to industrial sprawl: "First of all, when we plan a new plant we look for a site of 250 or 300 acres. That automatically puts us more or less out in the country. There are no such sites left in cities."[19] On that point even UAW president Walter Reuther agreed, once noting, "One major problem that has defied solution at the local level involves provision of sufficient space for the location of modern industry."[20] Beyond such land shortages, a mix of corporate production interests and public policies drove the nationwide shift toward capital decentralization.[21] Cheap land in outlying areas, the growth of consumer markets outside cities, lower suburban tax rates, federal subsidies for "industrial dispersal," trade union demands for large employee parking lots, readily available city services in the suburbs, and, crucially, the construction of interstate highways all shaped the nation's sprawling patterns of industrial development.[22]

Express highways were the linchpin of GM's suburban strategy. In order to link its decentralized network of plants—and, of course, to boost automobile sales throughout the United States—GM aggressively lobbied federal, state, and local officials for additional highway construction. As part of that effort, GM officials actively campaigned for the passage of the 1956 Federal-Aid Highway Act, also known as the National Interstate and Defense Highways Act.[23] With its guarantee of large federal subsidies to fund a national system of interstate highways, the landmark legislation received a hearty endorsement from the Manufacturers Association of Flint, the Flint Industrial Executives Club, and other GM-backed lobbying groups. Locally, company leaders hoped to use the freeway system to increase sales and connect the county's

FIGURE 5.2. Homes bordering a GM factory in Flint, n.d. By the close of World War II, most of GM's industrial facilities in Flint were surrounded on all sides by residential and commercial developments. The lack of industrial space in cities such as Flint played an important role in driving GM's shift toward suburban production. Courtesy of the *Flint Journal*.

plants both to one another and to big cities such as Detroit and Chicago. Of special importance to plant managers was an urban freeway route that would link the Buick assembly facility on Flint's North End to the Fisher Body supply plant on the city's southern fringe.[24] Without such a route, executives maintained, the numerous trucks moving between the Fisher 1 factory on South Saginaw Street and the North End Buick plants would cause major road deterioration and unrelenting traffic snarls.[25] Representatives of the Michigan State Highway Department (MSHD) and the Michigan Good Roads Federation concurred, arguing, "Highway transportation is a vital element in the automobile production process. The highway itself is as much a part of the assembly line as the cranes that lift motor blocks onto chassis."[26]

Federal defense and taxation policies also played a role in driving the suburbanization of industry. During the late 1940s, as concerns mounted about the possibility of an atomic attack against the United States, federal lawmakers began advocating industrial decentralization. In 1947 President Harry Truman signed the National Security Resources Act, a law designed to foster "the strategic relocation of industries, services, government and economic activities."[27] By moving industrial facilities away from cities, proponents of the law hoped to create new spatial buffers between military targets and heavily populated urban neighborhoods.[28] In order to speed the dispersal process, federal officials offered a variety of tax and loan incentives to GM and other defense contractors who could demonstrate that their facilities were distant enough from central cities to render them safe in the event of a nuclear attack. For smaller urban areas such as Flint, federal guidelines stipulated that fifteen miles from the city center constituted appropriate dispersion. Yet Department of Defense officials generally granted manufacturers a great deal of discretion over site selection, which allowed firms such as GM to construct most of their suburban plants well within the fifteen-mile perimeter.[29]

In 1950 members of the US Congress created an additional incentive for industrial suburbanization by passing an accelerated tax amortization law that allowed defense contractors to deduct up to 100 percent of their capital expenditures on newly constructed suburban facilities. Within months, federal administrators had granted nearly five billion dollars in tax breaks to GM and other defense firms operating new suburban plants.[30] Three years after approving the new tax plan, Congress dealt yet another blow to cities such as Flint by approving the so-called Maybank Amendment to the 1953 Defense Appropriations Bill. The amendment, named for Burnet R. Maybank, a Democratic senator from South Carolina, helped to reshape the spatial organization of the American defense industry by requiring the government to purchase all of its matériel at the lowest possible price. Due to the new rule, employers in Flint and other high-wage, heavily unionized cities found it virtually impossible to land new defense contracts. As a result, much of that work shifted toward lower-wage nonunion facilities in suburbs and rural areas, particularly in the South and West.[31]

Federal incentives for industrial dispersal dovetailed seamlessly with GM's metropolitan capitalist strategy. In 1945 GM unveiled a plan to construct a nationwide network of branch assembly plants.[32] Several years later, in a 1948 report to stockholders, spokespersons from the automaker elaborated on their desire to move from centralized economies of scale to decentralized economies of place, noting, "[W]herever possible, operations are decentralized with manufacturing plants in areas convenient to sources of material supply and manpower, and assembly plants in the vicinity of the principal markets for General Motors products."[33] Executives were especially interested in eliminating the high cost of shipping assembled vehicles to distant dealerships. The economic calculus behind GM's policy was simple: while a standard rail car could hold only four fully assembled automobiles, that same container could transport twelve partially built vehicles.[34] For GM chiefs, then, it was much more profitable to construct new assembly plants throughout the nation—and, increasingly, the world—than to concentrate facilities in Flint and Michigan.[35] By 1955 the company brass had planned for new assembly plants in California, Georgia, Texas, Delaware, and Massachusetts. The dispersal of assembly work helped to reduce Michigan's share of the nation's automobile industry employment from 56 to 40 percent.[36] At the same time, the decentralization process drove suppliers to close Michigan factories in favor of new plants closer to branch facilities. In a 1959 article, economic analyst Neil Hurley interpreted such developments as a sign that Michigan was "losing its historic position of dominance" within the automobile industry.[37]

Many UAW members went even further, arguing that GM's policy of decentralization was part of a broader "southern strategy"—a pernicious plan to increase profits by shifting automobile production away from the high-wage prounion cities of the North to the South and West, where wages were lower and trade unions weakest.[38] At first blush, the charges seemed to make sense. Of the five new branch assembly plants that GM constructed between 1945 and 1955, all but one were located in the South or West.[39] Nevertheless, the southern strategy critique obscured an even greater truth about GM's policies—namely, that virtually all of the corporation's new facilities were either in predominantly white rural areas or within racially segregated suburbs. In fact, the shifting of resources from cities to suburbs and rural places was even more pronounced than the so-called Rust Belt to Sunbelt migration.[40]

Regardless of the industry or the area of the country in question, municipal public policies were indispensable to the suburbanization of capital. In
order to operate their facilities, business managers from GM and other firms
required sewers, large quantities of water, and other services that were often
unavailable in suburbs and rural areas. Representatives from GM thus aggressively lobbied Flint's city commissioners to extend water and sewer lines
to their new suburban plants.[41] By the close of the 1950s, their efforts had
resulted in new water and sewer hookups for at least seven of GM's suburban
plants.[42] The commission's decisions provoked intense opposition among
left-wing UAW members and other corporate critics, however. Even prior to
building its suburban factories, GM was responsible for over half of the city's
annual water consumption. According to Flint's public works and utilities director Ted Moss, the city did not have enough water to sustain GM's new suburban operations.[43] Others denounced the city's policies for helping to speed
the flight of industry from Flint. As city commissioner Robert Egan, a UAW
leader, lamented in 1957, the water and sewer extensions only helped to "send
the plants away from Flint."[44] To Robert Clark, director of Flint's CIO Political Action Committee, the city's probusiness policies seemed unfair because
they allowed GM "to enjoy all of the major services rendered by the city,
including fire and police protection; water and sewage disposal—everything
except the doubtful privilege of paying city taxes."[45]

The city's stratified rate structure for water customers amounted to an
additional subsidy for decentralization. Unlike Seattle, Chicago, and other
urban centers where customers purchased water at uniform rates, the city of
Flint maintained a complicated three-tier plan with separate costs assigned
to residential, commercial, and industrial users. Residential consumers purchased water at a rate of thirty-two cents per one hundred cubic feet for up to
10,500 cubic feet. Industrial users, by contrast, paid at the heavily discounted
rate of twenty cents per hundred cubic feet for all water in excess of 105,000
cubic feet. The policy, which rewarded the largest consumers of water with
significantly lower rates, amounted to a large subsidy for local manufacturers
as well as a major drain on natural resources. In the eyes of Temple Dorr of
the Taxpayer's Protective League, the city's water and sewer policies were little
more than corporate boondoggles. "We, of our organization, believe that we
should love our neighbor," Dorr conceded. "But I don't think we should love
our neighbor to the extent that if his house needs a coat of paint that we buy
the paint and put it on, too. And that's just what we're doing on this water
and sewer problem."[46]

When attacked for encouraging a corporate exodus, supporters of the
city's policies highlighted GM's commitment to the metropolis, arguing that

subsidies for suburban growth benefited the entire region. "We have to live here," Commissioner Craig asserted, "and without GM Flint wouldn't exist." "I don't believe in taxing or throwing them out of business," he added.[47] Craig and other leaders celebrated all of GM's new plants—even those located in Grand Blanc and Flint Township—because they subscribed to a brand of metropolitan capitalism that rejected distinctions between the city and its suburbs. Proponents of this nationally popular view believed that growth anywhere in a metropolitan region was a boon to everyone in that region.[48]

Not everyone accepted this logic, however. In fact, GM's commitment to metropolitan capitalism in the Flint area drew hostility from ordinary city dwellers and suburban capitalists alike. Many critics inside the city viewed the opening of the new suburban plants as the initial phase of Flint's urban-industrial crisis. Others asserted that the "new" positions created in the suburbs were merely replacements for existing jobs in the city.[49] Often, in fact, GM's suburban expansions coincided with sharp declines in employment, primarily due to automation.[50] At Buick, for instance, the number of hourly rated employees dropped from 27,500 to just 10,000 between 1955 and 1961.[51] To thousands of workers at Chevy in the Hole, Buick, and other GM facilities in Flint, the company's decentralization efforts looked a lot like a zero-sum game in which "new" highly automated jobs outside the city came at the expense of "obsolete" urban jobs.[52]

As GM's critics pointed out, without annexation or some form of metropolitan government, the lines between city and suburb mattered greatly. In the words of one researcher, "Industrial development is increasing disproportionately in the areas beyond the city limits, which means that taxable wealth is being added more rapidly in the fringe than in the city."[53] Even more than tax imbalances, though, the suburbanization of the economy helped to transform the spatial arrangement of power in metropolitan Flint. In addition to Flint Township, the county's primary beneficiary of industrial growth, the urbanized suburbs of Swartz Creek, Beecher, Fenton, and Grand Blanc—along with the townships of Mundy, Grand Blanc, and Burton—gained substantially from the decentralization of capital. After a flurry of postwar relocations, for instance, Burton Township could claim a portion of GM's AC Spark Plug plant on Dort Highway, Genesee Cement Products, Atlas Concrete Pipe, Moore's Iron Works, Dort Manufacturing, Mead Containers, and the General Foundry Corporation. In south suburban Fenton, where local chamber of commerce officials embarrassed their counterparts in Flint by creating the county's first industrial development committee, tool and die manufacturing and aircraft engine finishing operations helped to drive an economic renaissance.[54] Just between 1940 and 1955, the number of industrial plants in the townships surrounding the city increased from 7 to 119. Fueling this suburban manufacturing boom were nearly a dozen new highway-oriented industrial parks that opened outside the city after World War II.[55] From coast to coast, capital migrations such as those that occurred in metropolitan Flint brought newfound wealth and power to previously impoverished suburbs.[56] A prime example was Flint Township, where the opening of GM's seventy-eight-million-dollar Van Slyke Road complex brought a financial windfall to the Carman school district. By 1960 GM's Chevrolet division was paying 80 percent of Carman's school taxes, and the district could boast of a per pupil property tax valuation that doubled the city of Flint's.[57]

Evidence of capital decentralization was also visible in Flint's increasingly quiet central business district. Prior to World War II, downtown Flint was the retailing and services hub of mid-Michigan. Flint's downtown shopping district was home to major retailers such as Sears, J. C. Penney, Smith-Bridgman, Woolworth, and Kresge along with scores of locally owned restaurants, clothiers, and furniture shops. During the 1940s and 1950s, however, downtown merchants and professionals shifted their focus to the suburbs for new development. As that shift unfolded, the number of business establishments on Saginaw Street began a long, steep decline. From 1948 to 1954, the number of retail stores operating in the central business district decreased by thirty, a drop of 6 percent.[58] Replacing the old stores of Saginaw Street were increasing numbers of For Sale signs, boarded-up storefronts, and low-end discount shops. To some, the transformations occurring throughout the metropolis signaled the birth of a new suburban capitalist order — one in which big cities and their downtowns were rapidly declining in significance.[59]

In reality, however, the situation was much more complex than that. Despite the marked shift in capital investment patterns that occurred in Flint and across the United States after World War II, GM's leaders remained committed to the Vehicle City. Furthermore, although GM's suburban factories clearly drew capital and taxes away from Flint and other urban areas, the opening of new plants just across the city limits was not the equivalent of building them overseas or far away in the states of the South and West. For most workers living in Flint—where automobile ownership was nearly universal—the suburbanization of industry presented few geographic obstacles to employment. Therefore, when GM moved plants to suburban areas, most local employees could and did commute to work. Even among African Americans, who were well represented among the city's automobile owners, the distance between home and work proved to be far less significant than racist employment practices in shaping the suburban opportunity structure.[60] Because most of the area's workers were highly mobile, so-called spatial mismatches between the city and suburbs did not constitute major barriers to employment.[61]

And from the perspectives of local GM supervisors and Flint's political leaders, the implications of suburban development were even less consequential. To the creators of GM's metropolitan capitalist strategy and others who loathed the secessionist sentiments at the heart of suburban capitalist ideology, Flint and its surrounding communities were part of a regional whole.[62] That is why GM executives hoped that the city would one day take possession of all of the company's suburban factories. But this was not simply an empty wish. To actualize their vision of metropolitan capitalism, GM officials threw their support behind the New Flint plan of 1957–58, a proposal to create a unified regional government for the Vehicle City and its suburbs.

Corporate support for regional government was by no means unique to Flint. While the Flint model of corporate-backed regionalism seems not to have prevailed in Detroit, St. Louis, or other major industrial centers where business leaders found it difficult to maintain political power, local executives in Flint were not alone in their embrace of metropolitan capitalism.[63] On the contrary, in Jacksonville, Florida; Grand Rapids, Michigan; Charlotte, North Carolina; Indianapolis, Indiana; Phoenix, Arizona; and various other cities, particularly those in which either one firm or a small number of business interests thoroughly dominated local politics, members of the corporate establishment also endorsed metropolitan capitalism. In fact, throughout the postwar era, business leaders from companies and institutions as varied as Nations Bank (now Bank of America), the United States Chamber of Commerce, and even Walmart routinely looked to metropolitan capitalism as a means of increasing profits and fostering economic growth. Support for regionalism among such executives had little to do with civic pride and loyalty, though. Rather, business leaders embraced metropolitan capitalism out of a desire to reduce production costs and stimulate economic expansion through the creation of more efficient political economies of scale. The belief that such efficiencies could be achieved while operating inside of cities such as Flint and Charlotte flowed primarily from corporate leaders' confidence that they could effectively dominate urban politics. In Flint that optimism about the city and the surrounding metropolitan region remained high throughout the 1950s and early 1960s, long after the forces of capital flight and suburban secession had begun to take their toll on other cities. But such faith did not triumph everywhere—a fact that helps to explain the divergence between Flint and other cities.[64]

Executives from GM had many reasons to feel confident in their ability to get the New Flint plan implemented. Prior to 1957–58 company directors had

often played a decisive role in shaping the local policy process. Their involvement in Flint's municipal affairs reflected the growing influence of corporate
leaders in modern American society.[65] Although local executives occasionally failed to achieve their policy wishes, in most instances the Vehicle City's
municipal government served as an arm of General Motors. In fact, the connections between the city government and the company were often so close
as to obliterate any meaningful distinctions between public policy and private corporate interests. With New Flint, however, company leaders sought
to exercise power over less familiar governments and constituencies outside
the city's borders, a task that proved to be more formidable than controlling
Flint's municipal government.

The Battle over New Flint

The New Flint plan grew in part out of the economic fluctuations of the 1950s.
During that otherwise prosperous decade Americans suffered through two
painful economic recessions that brought the implications of GM's suburban
strategy into sharp relief. The severest of the downturns, occurring between
1957 and 1958, brought manufacturing in Flint, Detroit, and other industrial
centers to a virtual standstill. At Buick, for example, the recession drove annual vehicle production down from a postwar high of 800,000 in 1955 to just
232,000 in 1958.[66] General Motors laid off nearly thirty thousand workers in
Genesee County during the slowdown, causing unemployment to increase to
14.1 percent.[67] Among African Americans, many of whom had only recently
obtained positions at GM, joblessness rates approached 30 percent during
this period, largely because the company followed the UAW's seniority guidelines and laid off workers with the least experience first.[68] The situation was
so severe that in 1957 the federal government formally classified Flint and a
number of other industrial cities as economically depressed areas.[69]

Desperate to shift attention away from Flint's flagging economy, GM president Harlow Curtice adopted an outwardly carefree demeanor during the
meltdown. As part of that effort, he hosted a major citywide gala and minstrel
show in the fall of 1957. During his remarks to the crowd following the show,
Curtice went to great lengths to minimize the significance of the downturn, at
one point even adopting the persona of one of the "darky" minstrel perform-
ers to make light of "the hell of a mess we all is in."[70]

Behind the scenes, though, Curtice and his colleagues were hard at work
planning a major new policy solution for the region's deep economic crisis.
Unveiled in 1957, New Flint was a proposal to consolidate the entire region
under one government. Curtice and his associates hoped that the creation of

a much larger city would resuscitate the region's ailing economy in the short term. Yet New Flint was also part of a broader long-range plan to overcome the growing gulfs between the Vehicle City and its suburbs. At its core, New Flint represented a progrowth metropolitan capitalist approach to solving both urban underemployment and suburban overdevelopment. As such, it drew support from across the political and geographic spectrum. From the outset, in fact, New Flint was the product of a complicated partnership among academic researchers, corporate leaders, city commissioners, trade unionists, and concerned citizens.[71] The plan first took shape in the mid-1940s, when scholars affiliated with the University of Michigan's Social Science Research Project began producing dozens of reports on metropolitan development in Genesee County. Under the leadership of professors Basil Zimmer and Amos Hawley, researchers from the SSRP assailed the economic and political problems associated with sprawl and metropolitan fragmentation.[72] According to Zimmer and Hawley, Flint's "fringe problem" derived from an imbalance between the limited resources of out-county governments and the increasing desire for urban services among suburbanites. Residents and business owners wanted better schools and roads, clean water, and sewers, yet their arcane village and township governments did not possess the authority to fund new developments.[73] Fortuitously, however, the problems of the city and the urban fringe were complementary. Although the city possessed the services required to sustain new subdivisions, shopping centers, and industrial plants, it lacked the necessary land for expansion, while undeveloped land was still plentiful in the suburbs. "The city of Flint has the service and the Flint area has the space to form a perfect union for the benefit of all," proponents of New Flint claimed.[74] Zimmer and his colleagues sought to unite the city and suburbs over a shared set of problems by advocating a single government for the entire urbanized area.[75]

Like other regionalists, New Flint boosters viewed metropolitan fragmentation as one of the nation's most pressing problems.[76] By the late 1950s there were 174 metropolitan centers in the United States, but within those densely populated regions there were an astonishing 15,658 local governments. Although the entire greater Flint area covered only 150 square miles of land during the 1950s, far less than the area of metropolitan Chicago, Cleveland, and other large population centers, it contained at least forty-five separate units of government, each with its own governing authority and administrative structure.[77] Advocates of the New Flint plan hoped to combat such chaos by combining twenty-six political jurisdictions within the urbanized area of Genesee County into a single city with a unified school district and a regional planning agency. The proposed new municipality was over 162 square miles

in total area—significantly larger than Detroit, Philadelphia, and Atlanta—
and included the cities of Flint, Mt. Morris, and Grand Blanc as well as the
townships of Grand Blanc, Genesee, and Burton, along with portions of
Flint, Mt. Morris, Mundy, Vienna, and Thetford Townships.[78] By forming
this large new city, metropolitan capitalists hoped to stimulate economic
growth, modernize the county's infrastructure, deliver greater governmental
efficiency, and ensure parity for all taxpayers and schoolchildren. "With its
26 separate governments," a New Flint publication noted, "the present Flint
Community is like an automobile consisting of a Chevrolet engine—Cadillac
body—Ford transmission—Buick chassis—and Chrysler wheels. . . . But, it
can virtually be Aladdin and his magic lamp bringing progress—security—
opportunities for each and every one of us!"[79] The major challenge, of course,
lay in persuading all of the region's various constituencies that the seemingly
disparate fragments of the metropolis were in fact compatible.

To sell their plan, New Flint boosters appealed to city dwellers and suburbanites alike. Well aware of the suburban opposition to annexation, supporters of New Flint carefully framed their proposal as a cooperative endeavor.[80]
For residents of industrialized suburbs such as Burton Township, New Flint
advocates promised to deliver a better business climate and first-rate municipal services. In residential suburbs, where homeowners alone bore the costs
of government, proponents of New Flint took a different approach, offering a
unified school district with equalized expenditures and taxes. At a New Flint
rally in Mt. Morris, Zimmer spoke directly to the concerns of homeowners in
Flint's numerous bedroom suburbs, asserting, "Plants now located in townships belong to all the people in the community, not just to those who happen to live on one side of the road and therefore in one school district and not
another." Inside the city, the New Flint campaign centered primarily on the
issues of jobs, economic growth, and urban revitalization. "In the new city
of Flint there will be ample room and facilities to entice new industries and
the expansion of present industries for the benefit of all," supporters of the
plan claimed. "Industrial development has reached its peak in the Flint area
unless a more favorable atmosphere can be created." Other backers of metropolitan government adopted more of a moral approach, arguing that Flint,
a city whose products helped to drive suburbanization, had a special duty to
address the issue of sprawl.[81] Irrespective of the audience, supporters of New
Flint believed that they had something positive to offer just about everyone.

With its emphases on governmental efficiency and economic growth,
New Flint attracted strong support from GM directors and other advocates
of metropolitan capitalism. The leadership of the New Flint Planning Committee thus included Harding Mott, vice president of the Mott Foundation;

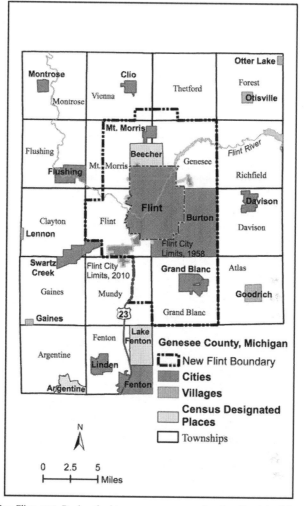

MAP 5.2. New Flint, 1958. During the late 1950s, supporters of regionalism joined forces with GM executives and other members of the Vehicle City's progrowth coalition in support of New Flint, a plan to create a single metropolitan government for the urbanized areas of Genesee County. Although the proposal drew widespread support from Flint's political and economic leaders, suburban capitalists and other opponents of the plan ultimately blocked the creation of the proposed megacity. *Sources*: Homer Dowdy, "Reasoning behind 'New Flint' Plan Given," *Flint Journal*, October 27, 1957; Michigan Center for Geographic Information, *Michigan Geographic Framework: Genesee County* (Lansing: Michigan Center for Geographic Information, 2009).

AC Spark Plug executive Joseph Anderson; and William "Oz" Kelly of the Manufacturers Association of Flint. According to Anderson, a tireless supporter of the proposal, "We need one government unit that should include all of Genesee County. . . . Just think of all the money that would save."[82] Predictably, members of the Flint City Commission went along with GM and offered their unanimous endorsement of the measure.[83] But so too did Robert Carter, regional director of the UAW, and Norm Bully, president of the Flint CIO, both of whom viewed New Flint as a potential source of jobs.[84] Noticeably absent from the New Flint coalition were African Americans, however. This was due in large part to the exclusionary practices of the Flint Area Study Group and the New Flint Planning Committee. Because the leaders of both organizations excluded African Americans from membership and largely avoided discussions of race, many black voters, even those who saw the promise of regionalism, had mixed feelings about New Flint. Another factor driving black skepticism toward the measure was the fear that the creation of a larger metropolitan city would increase the size of the white electorate, thereby diluting the already limited political power of African Americans.[85]

The New Flint plan elicited no such ambivalence in the suburbs of Genesee County, however. There, as in other parts of the country, proponents of metropolitan government faced stiff opposition from secession-minded suburban capitalists.[86] In industrial areas such as Flint Township, anti-Flint sentiment had been a part of the political landscape since at least the early 1950s, leading to several municipal incorporation drives.[87] Soon after the New Flint proposal's release in the fall of 1957, Flint Township supervisor John R. Dickenson denounced the plan, claiming, "The biggest majority out here are against [New Flint]. They moved out to get away from the city as it was." Echoing his colleague, Earl Swift, a county supervisor representing the booming south suburb of Grand Blanc, expressed his firm opposition to regional government. "I think we can run our own business without Flint coming in. The only reason Flint wants Grand Blanc is because of the [GM] tank plant. . . . This deal [New Flint] is all to their advantage, not to ours."[88]

Resistance to New Flint intensified in the winter and spring of 1958, when activists began circulating petitions to have the proposal placed on the fall ballot. In response to the canvassing, legislators and school district officials in Grand Blanc, Burton and Flint Townships, Beecher, and other areas with GM plants also spoke out against the initiative.[89] Zimmer and other New Flint supporters countered by organizing informational meetings throughout the county, but suburbanites turned out en masse to voice their opposition. On April 10, at a New Flint meeting hosted by Sheldon LaTourette in Mt. Morris Township, an unidentified woman excoriated the plan, stating, "Flint doesn't need us and we don't need Flint." Minutes later, a skeptic from Beecher, home of GM's Ternstedt plant, informed LaTourette, "There is not one thing in New Flint that can help us in Beecher."[90] At meeting after meeting, suburban residents from across the county showed up in droves to condemn the proposed new city.

By the summer of 1958, opponents of metropolitan government had formed the New Flint Resistance Committee (NFRC). Led by Louis Traycik, a resident of Grand Blanc, Carman School District superintendent Frank Hartman of Flint Township, and Joseph Parisi of the Michigan Township Association, members of the NFRC charged that New Flint would "subjugate our suburban people" through unmanageable government bureaucracies, substandard schools, and burdensome new taxes. "Super cities," NFRC petitions claimed, "mean super tax dollars."[91] Because many suburbanites had already voted to fund expensive FHA-mandated infrastructure projects, skepticism toward the prospect of additional taxes was especially acute. "We'll eventually have sewers and things anyway," Flint Township supervisor Dickenson maintained.[92] Other members of the NFRC argued that a metropolitan government would forever undermine their new civic identities as suburbanites. After attending a March meeting held in Grand Blanc, the site of GM's metal-stamping plant, *Flint Journal* reporter Lou Giampetroni described how the development of self-conscious suburban identities was undermining support for New Flint: "At the meeting there was not as much an indication of being 'against New Flint' as there was 'for Grand Blanc.' Those at the meeting indicated that they wanted to keep their identity as a community."[93] Whether they were prosuburban or anti-Flint, thousands of white homeowners united around the NFRC to block the proposed regional government.

New Flint generated an especially vigorous backlash in the southwestern suburb of Swartz Creek, where voters essentially seceded from Flint by incorporating as a city. Prior to 1958 the area that would become Swartz Creek, known as Otterburn, was an unincorporated, predominantly residential section consisting of parts of Gaines, Clayton, and Flint Townships. In 1957 GM provided a major boost to the local economy by opening a new Chevrolet parts warehouse and service center in the area. Fearful over the possibility of losing the facility, concerned homeowners and other advocates of suburban capitalism waged simultaneous campaigns for incorporation and against New Flint. New Flint boosters responded by pointing out that the boundaries of their megacity included neither the proposed city of Swartz Creek nor GM's warehouse and service facility. Unconvinced by such reassurances, pro-incorporation activists pressed on with their campaign. "If this city incorporation fails to pass," one resident warned, "there is a possibility the plant will be annexed to Flint. If this happens our backing per school child will decrease 50%. We need this plant! We must vote yes!" Ultimately fears over losing the Chevrolet facility and the desire for incorporation drove hundreds of residents to the polls. On August 5, 1958, these citizens declared their independence from both Flints, new and old, by creating the home-rule city of Swartz Creek.[94]

The Swartz Creek incorporation election stunned GM's local plant managers and other supporters of regional government. Nevertheless, because the newly formed city was not included in the New Flint boundaries, the campaign for metropolitan government continued. Despite mounting opposition in the suburbs, New Flint's well-funded supporters had little difficulty collecting the required signatures to have the initiative placed on the November ballot. In August advocates of New Flint submitted their petitions for certification to the Genesee County Board of Supervisors. The petitioners sought a single election on the issue to be determined by registered voters from the city and the suburban areas included in New Flint's boundaries. Led by Supervisor Thomas E. Bell of Mt. Morris Township, suburban opponents of the proposed new city formed a clear majority on the county board. The strongest opposition came from representatives of the townships, who feared losing electoral support, political power, and perhaps even their jobs if New Flint became a reality. Although the board's legislative affairs committee had determined that the petitions were valid and legal, Bell nonetheless urged all suburban supervisors to reject the request for an election. On August 12 suburban supervisors did just that by denying the validity of the New Flint petitions. Acting against the advice of their own legislative affairs committee and legal counsel, county supervisors, by a vote of twenty-five to thirteen, declined to place the New Flint proposal on the ballot.[95]

Following the board's vote, members of the New Flint Planning Committee filed an emergency appeal with the Michigan Supreme Court.[96] Members of the NFRC responded by invoking their constitutional rights as minorities. "The American concept of justice in government grants no given majority the power of indiscriminate transgression upon the rights of its ranking minorities," a NFRC statement maintained.[97] After agreeing to hear the appeal, the justices hosted a hearing on September 9. Attorney John G. David, speaking on behalf of the county board, argued that consolidation, rather than incorporation, was the only legal means of forming New Flint. Furthermore, David reminded the justices that New Flint violated a state law prohibiting the incorporation of already incorporated areas. Since New Flint would include already incorporated cities such as Grand Blanc and Mt. Morris, David urged the court to rule against the initiative.

On October 8 the justices granted David's wish by delivering a mortal blow to the New Flint plan. In a unanimous decision authored by Justice Harry F. Kelly, the court accepted David's arguments on consolidation and refused to order the November election.[98] Although Kelly's legal opinion kept alive the possibility of a new consolidation effort, the ruling marked the unofficial end of the movement for metropolitan government in Genesee

County. For regardless of the openings left by the court's decision or the future machinations of New Flint's backers, proponents of suburban capitalism had clearly articulated their opposition to metropolitan government. New Flint was dead, and to many observers Old Flint seemed to be dying.

Cities with Walls

The New Flint defeat left the Vehicle City in a difficult economic position. Absent a metropolitan government, the capital migrations of the postwar era had effectively diverted tens of thousands of jobs and millions of dollars in tax revenues from the city to the suburbs. Making matters worse, overdevelopment in the city had left the Buick, Chevrolet, Fisher, and AC Spark Plug plants with almost no room for expansion. To continue to prosper, Flint, a city of nearly two hundred thousand people, needed more space. Yet the city's boundaries had failed to keep pace with its population growth. In fact, between 1920 and 1960 the city's total geographic area remained fixed at just under thirty square miles. By contrast, booming cities in the South and West such as Houston and Los Angeles, whose leaders could more easily annex suburban land, contained hundreds of square miles of territory in 1960. In Michigan and many other northern and midwestern states, however, cities such as Flint suffered under a legal and political system that made it virtually impossible for them to expand.[99]

Municipal leaders continued to search for ways to grow in the wake of the New Flint debacle, but Michigan's antiurban annexation policies and liberal municipal incorporation statutes—similar to those in many other states— allowed most suburbanites to maintain their independence from the city.[100] Annexations in Michigan typically began with a lengthy petitioning process. State law required proponents of annexation to obtain signatures of support from at least 1 percent of the qualified property owners in the combined areas affected by the proposed boundary changes. Upon collecting the necessary signatures, city officials could forward them to the county board of supervisors, which held the power to order elections. In order to win an annexation election, a majority of voters in both the city and the unincorporated area in question, counted separately, had to endorse the plan. This could be an extremely difficult proposition. By requiring separate majorities in both city and suburb, state law provided suburban residents with veto power over most annexation proposals. In September 1961 residents of Flint Township exercised this authority by voting three to one against a proposed annexation to the city of Flint.[101] Although GM executives had offered their "wholehearted endorsement" of the plan—and though the total number of votes in favor of annexation far exceeded those against—the legal requirement of a separate majority in Flint Township proved to be decisive.[102] Urban growth would have to come through other means.

Enlightened by the failure of New Flint and the 1961 annexation defeat, Vehicle City officials next opted to circumvent suburban voters by "strip-annexing" factories, shopping malls, and other uninhabited or sparsely populated lands contiguous to the city. In such cases, state law required approval from only a simple majority of voters in the combined area, counted together, to proceed with annexation. This loophole provided legal space for city officials to annex several major industrial and commercial facilities that ringed Flint's border. On November 14, 1961, City Manager Thomas Kay filed four separate petitions to annex 5,422 acres of predominantly industrial and commercial property in parts of Flint, Mt. Morris, Genesee, and Burton Townships. The petitions included the four GM plants in Flint Township, the Ternstedt facility in Beecher, Bishop Airport, the South Flint Plaza, the Northwest Shopping Center, and a large parcel of vacant land north of the city slated for commercial and industrial development.[103] Like the New Flint proposal, Kay's plan received strong support from the Mott Foundation, the UAW, the chamber of commerce, GM, and other members of the progrowth coalition.[104] Speaking on behalf of GM, AC Spark Plug executive Joseph Anderson endorsed the drive. In doing so, however, he issued what proved to be a prescient warning about the consequences of suburban secession. "If a community holds a plant in a small area," he pointed out, "industrial development will be cut off. . . . GM can't afford to go into an area that's going to do that sort of thing."[105] Yet again, GM officials worked to solidify their metropolitan capitalist growth strategy by supporting Flint's expansion.

Civic leaders looked to gain support for the annexation bid by appealing once again to a shared sense of purpose among city and suburban residents. City Manager Kay even invoked images of ancient cities surrounded by walls to make his case, claiming, "History books have stories and pictures of cities with walls and, in American history, there were forts. The walls and forts fell as we became more civilized."[106] By employing the metaphor of the walled city, Kay hoped to highlight the divisive nature of suburban capitalism and resuscitate the metropolitan sensibility that had driven GM's local growth strategy.[107] Yet the response to his petition revealed that many suburbanites actually embraced the barriers that separated Flint from its neighbors. During the fall of 1961, residents formed antiannexation committees in each of the four townships surrounding the city. Predictably, the opposition was strongest in Flint Township, which stood to lose the four GM plants along with slightly less than two-thirds of its overall tax base of $112 million. Hoping to rally parents from the Carman School District, Flint Township supervisor Raymond Flavin charged that annexation would undermine the educational opportunities of suburban children by depleting the area's tax base. "Flint Township is being crucified," he lamented. "The people of my township have worked hard to give their children an education. Now you are trying to take the opportunity for education away from these poor children."[108] Like Flavin, many suburban populists believed that strip annexation made a mockery of American values by pitting greedy and powerful corporations and urban machines against innocent suburban families. Others, though, viewed annexation as a threat to white racial privilege. Feeding this perception was a widespread concern that annexation would somehow facilitate the integration of suburban schools and neighborhoods. In March 1961 Lawrence Rice, chair of the Carman School District Citizens Advisory Committee, appealed to such fears by predicting that annexation would bring an influx of "juke joints" to Flint Township.[109] Although racism often simmered quietly beneath the surface of annexation debates, it was a central feature of the suburban independence movement.

On February 27, 1962, Flint voters overwhelmed their suburban opponents at the polls. Needing to gain only a majority of votes cast in the city and suburbs, counted together, the proannexation forces easily won the votes for the four GM plants in Flint Township, Bishop Airport, the Northwest Shopping Center, and the South Flint Plaza. In all, the election yielded the city 1,370 acres of land and seventy million dollars worth of properties. The victory was a pyrrhic one, however. In Beecher and Mt. Morris Township, voters rejected the city's attempt to annex GM's Ternstedt facility and a parcel of developable land. Moreover, in the aftermath of the election, suburban school officials waged a successful counterattack against the city. Shortly after the February vote, members of the Genesee County Board of Education determined that all the school taxes paid by the newly annexed facilities would continue to fund the Carman and Westwood Heights School Districts and not the Flint Public Schools. Lawyers representing the Flint Board of Education fought the decision in court, but to no avail.[110] Beyond that major loss, the annexation battle delivered a terrible blow to the city's public image. Just a few weeks prior to the election, Spencer Carpenter, chair of the Flint Township Anti-Annexation Committee, voiced the frustrations of thousands of suburban residents when he announced that he planned to sever all ties with the city of Flint. "If this is the way the city of Flint handles its matters," Carpenter threatened, "I don't want to ever be a part of Flint."[111] Suburban bitterness over the 1962 annexation fight would linger for decades.

As part of the growing revolt against metropolitanism, suburban capitalists launched a number of municipal incorporation drives in the early 1960s. In April 1963 voters in Fenton and Flushing resoundingly approved incorporation proposals. Several years later, homeowners in Mt. Morris Township, Beecher, and Flint Township followed suit by waging their own incorporation campaigns. Propelling these and other suburban independence struggles were Michigan's extremely lenient incorporation statutes, which allowed suburbs to become home rule cities with a simple majority vote as long as they met a few basic density requirements. Once incorporated, suburban cities could levy higher taxes, build new schools, and provide better municipal services. As proponents of incorporation were quick to point out, the provision of such services resulted in improved living standards as well as new opportunities to receive FHA mortgage insurance. Such thinking was quite evident during the successful Flushing campaign, when full-page newspaper advertisements asked voters to choose incorporation to "end septic tanks . . . plug up the wells—get pure city water—qualify for low-cost FHA loans on mortgages."[112] Equally important, incorporation meant that suburbs could block all annexation attempts from adjoining municipalities. By the dawn of the 1960s, the era of suburban secession was already well under way.[113]

During the first half of the twentieth century, a combination of corporate expansion strategies, federal development initiatives, and municipal growth policies transformed the Vehicle City from an obscure mid-Michigan hamlet into one of the world's leading industrial metropolises. Intense racial segregation and jurisdictional division were by-products of that growth, however. Despite numerous efforts to remedy the area's growing fragmentation, Flint's metropolitan moment ultimately gave way to a rising tide of suburban independence. Although the significance of this development went largely unnoticed in the 1950s, the decline of metropolitan capitalism would have a profound impact on the fate of the Vehicle City. In the decade following the 1962 annexation fight, the gulf between Flint and its neighbors continued to widen as municipal leaders searched in vain for new opportunities to expand the city's shrinking economy. Making matters worse, as GM's political power and the prospect of regional government waned, company leaders began to reassess their commitment to producing vehicles in Flint. Meanwhile, regional battles over fair housing, school desegregation, and other civil rights issues helped to reinforce the growing urban-suburban rift. Indeed, as black activists mounted a direct assault on the Flint area's deeply entrenched system of Jim Crow, the political walls surrounding the city seemed to grow even stronger.

PART TWO

Fractured Metropolis

6

"Our City Believes in Lily-White Neighborhoods"

The spatial reorganization of metropolitan Flint engendered political up-heavals both in the all-white suburbs and among African Americans living inside the city. During the 1950s, suburban capitalists worked to consolidate their independence from the city by blocking the New Flint plan, resisting annexation, and organizing numerous municipal incorporation campaigns. Simultaneously, though, African Americans in Flint were coordinating their own uprising against the prevailing spatial order by challenging the region's rigid residential color lines. The decades-long battle for open occupancy grew out of the severe housing shortages, racial segregation, and neighbor-hood deterioration that plagued black Flint at midcentury. In the summer of 1967, as the city's leaders prepared to clear "blighted" housing in the black neighborhoods of Floral Park and St. John, riots exploded in the Vehicle City and many other urban centers. In response to the civil disorders and numer-ous protests against housing discrimination, city commissioners reluctantly passed a municipal open occupancy law. Several months later, open hous-ing supporters defended the new law in a hotly contested referendum. The 1968 election victory made Flint voters the first in the nation to endorse open housing at the polls. In the wake of the vote, civic boosters cheered the city's civil rights milestone, arguing that the new law made Flint one of the most progressive communities in the United States. In truth, however, the 1968 election did little to undermine either popular or administrative modes of segregation. By decade's end, the battles over urban renewal and open hous-ing had further fractured the metropolis.

Black Migrants and the Postwar Quest for Housing

The massive labor shortages that drove Flint's growth during World War II lingered into the postwar era. As GIs returned from battle and the nation's manufacturing firms rushed to meet rising demands for new cars and other consumer products, millions of high-paying jobs opened up in America's largest urban centers. Hoping to obtain economic security and first-class citizenship, large numbers of African Americans, Latinos, and others departed the rural South and Southwest during and after the war, many of them bound for northern and western cities such as Detroit, Chicago, Los Angeles, and Flint. In all, five million black southern migrants relocated to the North and West between 1941 and 1970. Joining these migrants of color were millions of white southerners in search of jobs and economic freedom. Collectively,

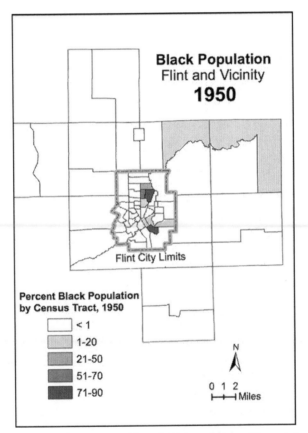

MAP 6.1. Black population by census tract, Flint and vicinity, 1950. *Source:* Minnesota Population Center, *National Historical Geographic Information System: Version 2.0* (Minneapolis: University of Minnesota, 2011).

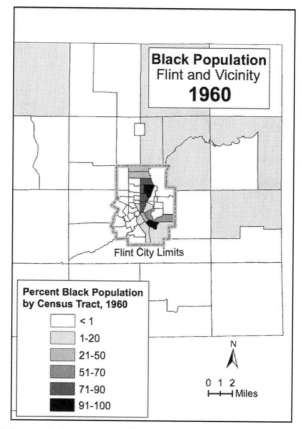

MAP 6.2. Black population by census tract, Flint and vicinity, 1960. *Source*: Minnesota Population Center, *National Historical Geographic Information System: Version 2.0* (Minneapolis: University of Minnesota, 2011).

these workers and their families radically transformed the nation's large metropolitan centers.[1]

From 1940 to 1960, Flint's African American population grew from just over six thousand to nearly thirty-five thousand.[2] Yet over the same period the boundaries of the city's black neighborhoods remained essentially fixed. With a residential segregation index that reached 94.4 in 1960, Flint ranked as one of the most racially divided cities in the country—more segregated, in fact, than Atlanta, New York, Chicago, or Los Angeles.[3] Beyond facing entrenched patterns of popular and administrative segregation, Flint's black population suffered from severe housing shortages. Although builders and developers used a combination of private funds and federal subsidies to construct thousands of new residences in the Flint area during the housing booms of the 1940s and 1950s, only a handful of those new homes were available to black buyers.[4] Consequently, all but a few of the city's black newcomers sought to acquire rental units, primarily in the overcrowded and rapidly deteriorating Floral Park and St. John neighborhoods.

The severe housing shortages that bedeviled African Americans at mid-century stemmed in part from the city's delayed efforts to construct public housing. Unlike New York City, Chicago, and other municipalities that adopted public housing in the 1930s, Flint did not inaugurate such a program until 1964, when citizen relocations due to urban renewal forced the city to provide low-income units.[5] On two occasions—once in 1939 and again in 1949—civil rights activists and UAW leaders persuaded members of the Flint City Commission to create a public housing commission.[6] Yet in each instance, builders and real estate developers launched polarizing referendum campaigns to block the city's efforts. Both times local voters, including large numbers of UAW members, voted to abolish the city's public housing program.[7] In the absence of a public housing system, residential shortages and neighborhood deterioration continued to plague the city and its people throughout the postwar period.[8] Nowhere was this truer than in the segregated, overcrowded, and heavily dilapidated black neighborhoods of Floral Park and St. John—the birthplaces of Flint's fair housing movement.

Floral Park and St. John

Prior to World War II, Floral Park had been a prized destination for black migrants to the Vehicle City. Beyond its proximity to downtown, the area was renowned and beloved for its diverse mix of businesses and residents, its relatively clean air, and its numerous recreational amenities.[9] Still, by the late 1940s Floral Park was in the midst of a dramatic transformation. The influx of black migrants during and after the war triggered several violent conflicts and a massive white exodus.[10] Over the course of the 1950s, waves of panic selling claimed all but a few of Floral Park's white residents. Meanwhile, members of the black middle class also began leaving the neighborhood as new housing options opened in Elm Park, Sugar Hill, Evergreen Valley, and other areas south and east of Floral Park.[11] By the end of the 1950s, Floral Park had fallen victim to the twin forces of popular and administrative segregation, and its once stable housing stock was beginning to show signs of significant deterioration.[12]

In 1960 Floral Park's black majority of 96 percent registered as the highest in the city. Though its housing stock included many homes in good repair, 28 percent of the area's residences were either dilapidated or deteriorating, a figure that nearly doubled citywide averages. At $8,900 per home, the median housing value in Floral Park was significantly lower than the $11,500 recorded for the city as a whole. In addition, the average family income in Floral Park, which totaled $5,167 per year in 1960, was nearly 20 percent lower than in the rest of the city. Along with poor housing, overcrowding, and poverty, many residents suffered from diabetes, hypertension, and other chronic ailments. With nearly a third of its residents under the age of nine, and a large number of households headed by single mothers, Floral Park's problems hit women and children with a special intensity.[13] Remarkably, though, the conditions there were not the worst in the city. In fact, living standards in Floral Park were generally far superior to those in St. John, the city's most isolated and impoverished black neighborhood.

By the 1950s St. John had become, ironically, the most deteriorated and most coveted section of the city. Home to Buick's world headquarters, St. John also contained the city's largest concentration of poverty and its second-highest proportion of African Americans. As in Floral Park, a combination of federal redlining policies, class tensions, and grassroots racism helped to transform the neighborhood during the 1940s and 1950s. In 1940 St. John had had a white majority of 60 percent, composed primarily of Catholic immigrants from southern and eastern Europe. Over the next decade, however, the number of black households in the area nearly tripled, while the neighborhood's white population declined by approximately 75 percent. At mid-century St. John contained 4,593 black residents and a white minority of 872. A decade later, when the city launched its freeway and urban renewal programs, African Americans made up over 95 percent of St. John's residents.[14]

The demographic transitions that swept across St. John and the broader North End after World War II paralleled those that occurred in Floral Park and many other urban enclaves in the United States.[15] As black job seekers from the rural South arrived in the area, shopkeepers, homeowners, and professionals left in droves. Their departures helped to create a decidedly poorer and more segregated North End. To Michael Evanoff, a white resident who moved away from St. John after World War II, it seemed as though "the community was virtually taken over by black people."[16] Although whites in St. John and other areas opposed integration with near unanimity, many middle-class black property owners also looked askance at their new neighbors. According to Ruth Owens Buckner, an African American resident of the south side who believed that blacks in St. John were a major cause of the city's problems, "Lots of those who came [to St. John] didn't have good morals so that carried on through the years."[17] In the eyes of Buckner and other middle-class observers, the Second Great Migration was responsible for an unwelcome invasion of unrefined black job seekers who threatened property

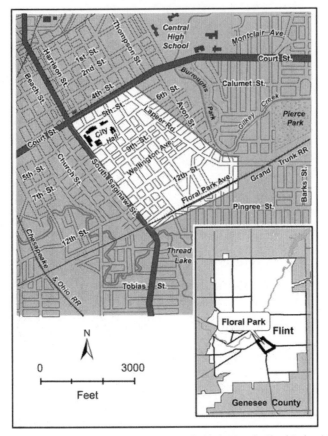

MAP 6.3. Floral Park, 1966. This map of Flint's near south side features the Floral Park neighborhood prior to the implementation of the city's urban renewal program. Although residents of Floral Park suffered from high rates of residential segregation, poverty, and housing dilapidation, the neighborhood's proximity to downtown and the Thread Lake recreation area made it a popular enclave among many African Americans. *Sources:* American Automobile Association of Michigan; Michigan Center for Geographic Information, *Michigan Geographic Framework: Genesee County* (Lansing: Michigan Center for Geographic Information, 2009).

values and public safety. Many middle-class blacks responded to such perceived threats by joining the postwar exodus from St. John.

The outward migrations of property owners and professionals created both hardships and opportunities for those remaining in St. John. Just as in Floral Park, redlining, disinvestment, and white panic selling took a severe economic toll on the area. Yet St. John was still the undisputed capital of black Flint. With the Columbia Theater, the St. John Community Center, and a host of black-owned businesses, St. John was the focal point of the city's segregated public sphere. "It was like being in a city of its own," re-

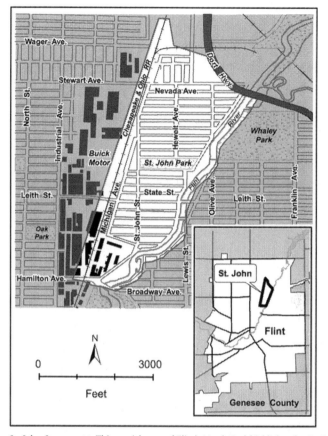

MAP 6.4. St. John Street, 1966. This partial map of Flint's North End highlights the St. John Street neighborhood on the eve of urban renewal. During the 1960s, African Americans composed approximately 95 percent of St. John's population. In the neighborhoods south of Whaley Park (and east of the Flint River), whites represented more than 99 percent of the population until well into the 1970s. *Sources:* American Automobile Association of Michigan; Michigan Center for Geographic Information, *Michigan Geographic Framework: Genesee County* (Lansing: Michigan Center for Geographic Information, 2009).

membered former resident James Blakely.[18] Moreover, like Floral Park and many other urban black neighborhoods in the United States, St. John housed a diverse mix of factory workers, domestics, small business owners, and professionals. Although its median housing values were among the lowest in the city, St. John was still predominantly a neighborhood of homeowners, many of whom found stable jobs as custodians or foundry workers at the nearby Buick plants.[19]

Nevertheless, St. John consistently ranked as the most impoverished, dangerous, overcrowded, and polluted section of the city. St. John had the dreadful distinction of containing Flint's largest rat population, and its hous-

ing stock—consisting primarily of aging wood-framed structures—was severely deteriorated. According to the authors of a 1960 housing survey, nearly
90 percent of the homes in the St. John district had "major deficiencies."[20]
At $3,900 per year, the median family income in St. John was less than half
of that for the county as a whole.[21] Although many black residents sought to
distance themselves from the area, a harsh mix of housing discrimination,
racial violence, and poverty severely circumscribed relocation options. Noted
one resident in 1963, "I loathe where I live, but I'm trapped."[22]

Residents of St. John Street had many reasons to feel trapped. Because of
its proximity to the Buick factories, the area received very little sunlight on
most days. When sunlight did arrive over the towering plants nearby, it had
to pierce through dense clouds of soot and ash that emanated from Buick's
smoke-belching foundry. According to Ailene Butler, owner of the Butler Fu-
neral Home in the North End and a future city commissioner, "When we get
a good sun-shiny day down there, we're lucky."[23] Unlike municipal leaders in
Pittsburgh, St. Louis, and other cities who enforced strict smoke abatement
programs during the postwar era, members of the Flint City Commission's
Air and Water Pollution Committee generally abstained from regulating industrial pollutants until state and federal officials forced them to do so in
the 1970s.[24] Instead, city commissioners and corporate officials presided over
nearly constant showers of smoke and ash that irritated lungs and depressed
real estate values in the North End. Throughout the northeast side, industrial
pollutants peeled paint from homes and cars, covered windows and wind-
shields, and caused innumerable respiratory ailments and cancers among
residents.[25] According to the authors of a neighborhood survey conducted in
the early 1960s, St. John was "an unfit environment for human habitation."[26]
On good days, when the factories hummed and the prevailing winds sent
Buick's smoke elsewhere, St. John residents could look beyond the pollution
and deterioration and take pride in their homes, schools, and churches. On
the worst of days, though, when clouds of toxicity hung like a pall over the
neighborhood, St. John was simply insufferable.

The Birth of the Fair Housing Movement

As living conditions deteriorated in Floral Park, St. John, and other inner-
city neighborhoods, members of the black community launched a grassroots
campaign against substandard housing and segregation. The activism in Flint
was part of a much larger nationwide movement against popular, legal, and
administrative segregation that erupted during the postwar era and culminated at the federal level with the passage of the Civil Rights Act of 1964,

FIGURE 6.1. Industrial pollution in the St. John Street neighborhood, 1971. This photograph features smokestacks from the Buick foundry along with a small section of the St. John neighborhood. Heavy pollution in the North End played a major role in driving demands for urban renewal and open housing reform. Courtesy of the *Flint Journal*.

the Voting Rights Act of 1965, and the Fair Housing Act of 1968.[27] In 1952 George Friley, an African American businessperson from St. John, addressed members of the Exchange Club at the Durant Hotel, where he noted that in spite of sharp black population increases, "[n]o provision has been made for additional Negro housing." "Negroes," Friley charged, "can walk into any bank in town and if they've got the down payment can receive funding for any automobile they want to buy. But let them try to borrow money to buy a house. In most cases they can't."[28] A year later, Arthur Edmunds of the Urban League delivered a formal request to the Flint City Commission to investigate Negro housing conditions.[29] "Today the supply of rental housing is so short," wrote Edmunds and fellow Urban League member Frank Corbett, "that it is virtually impossible to rent an unfurnished three room apartment, even in slum areas, for less than $18.00 per week." Well aware of the barriers to acquiring federally insured housing, the two men prevailed upon members of the private housing industry to tap the underappreciated African American housing market.[30]

Representatives of the real estate industry reacted harshly to such claims, however. During a 1955 citywide conference on housing sponsored by the Urban League, builder Eino Rajala attributed Flint's patterns of racial segregation to individual choices while flatly denying that there was a black housing shortage: "There is only one reason why every Negro in the city of Flint does not have a house, the one reason being they do not want to buy it."[31] Urban League director Rosa L. Kimp responded to Rajala's remarks with an emotional speech on the structural barriers to open housing. "We are living for the most part in compressed segregated neighborhoods whose boundaries are well defined," she charged. "Again I say without fear of contradiction that this segregated living pattern is forced!" Kimp then went on to indict the entire housing industry—including federal officials, lenders, builders, brokers, and landlords—for the "shameful" state of Negro housing in Flint.[32]

Housing inspections conducted in the mid-1950s revealed the extent of the city's residential crisis. In the spring of 1954, just prior to the Urban League's housing conference, members of the Flint City Commission ordered City Fire Marshal Elwood Rutherford to begin code enforcement inspections throughout the city. Significantly, Rutherford's inspections were the first carried out in Flint since the Great Depression.[33] As they toured the city, Rutherford and his fellow inspectors recorded housing and fire code violations in dozens of neighborhoods. The conditions in Floral Park, St. John, and the broader North End were especially grim, however. In December 1954 Rutherford encountered eleven persons, seven of them children, living in a North End basement on East Philadelphia Avenue that was accessible only by a trap door. That same week, inspectors visited a five-room bungalow on Elm Street that housed twenty-four people, including four separate families and one single adult male.[34] In October 1955, fire inspectors discovered nineteen people living in a five-room basement apartment at 1214 Hickory Street.[35] Several months later, fire marshals visited a North End home at 814 Lomita Avenue, where they found twenty-two persons living in a five-room house.[36] Infuriated by such discoveries, local UAW official Robert Carter denounced the Flint City Commission, noting, "They [city commissioners] have not done one single thing to clean up or eliminate slums in this area."[37]

As Rutherford's inspections confirmed, faulty wiring, leaky roofs, overcrowding, and other hazards were exceedingly common in Flint's densely populated black neighborhoods. Such conditions were not unique to the Vehicle City, however. According to federal census surveys, of the housing units in the United States occupied by African Americans in 1960, 23.8 percent were "deteriorating," 10.4 percent were "dilapidated," and 5.9 percent lacked basic plumbing facilities such as running water and flushable toilets. For whites, by contrast, the corresponding figures were 7.9 percent, 1.7 percent, and 2.1 per-

FIGURE 6.2. A dilapidated North End home, 1962. Municipal officials took this photograph of the home at 1111 Easy Street, located in the St. John neighborhood, as part of a citywide survey of housing conducted in 1962. Public policies played an important role in creating substandard living conditions such as these. Courtesy of the Alfred P. Sloan Museum, Flint, MI.

cent, respectively.[38] Flint's housing problems, while severe in their own right, were but a microcosm of the nation's plight.

Urban Renewal, City Services, and Civil Rights

By the late 1950s widespread anger over substandard housing, popular and administrative segregation, and the poor public health outcomes that accompanied those conditions had erupted in scores of black neighborhoods across the country. As part of the nationwide campaign for fair and affordable housing, thousands of African Americans living in St. John, Floral Park, and many other black urban enclaves called for the demolition and redevelopment of their neighborhoods.[39] By clear majorities, in fact, residents of these two segregated and impoverished communities supported neighborhood clearance. Although many community members felt a special attachment to their homes, churches, and businesses and were deeply suspicious of city leaders, for most residents the conditions in Floral Park and St. John were simply too awful to bear.[40] The issue was clear to St. John resident Lillian Huddleston: "We feel as if we were sewed up in a pocket, and we need Urban Renewal."[41]

Joining Huddleston and other black citizens in support of redevelopment were the Flint chapters of the Urban League and the NAACP as well as dozens of trade unions, African American block clubs, and civic groups. In 1955 the Urban League's Arthur Edmunds formally requested that municipal leaders "embark immediately on city-wide programs of slum clearance and urban redevelopment."[42] The following year, Urban League and NAACP members joined St. John and Floral Park residents in delivering three different petitions for slum clearance to the Flint City Commission.[43] For their part, liberal trade unionists and white integrationists also endorsed renewal. In 1957 an editorialist for the *Flint Weekly Review*, the city's leading left-labor newspaper, argued, "The real social need in Flint is to raze the ghettoes. . . . Until we do, none of us dare speak of the brotherhood of men."[44]

Beyond calling for neighborhood clearance, St. John and Floral Park residents also demanded more affordable housing, cleaner air, and improved city services.[45] To demonstrate their commitment to both quality services and redevelopment, residents of the two neighborhoods organized several large protests in 1958 and 1959. At around the same time, black renters and homeowners began attending numerous city commission meetings. Though black citizens brought a wide range of issues to the attention of city commissioners, complaints often centered on racial inequities in the city's street cleaning, public safety, garbage collection, pollution abatement, and street lighting programs. At a commission meeting held on March 23, 1959, for instance, Charles Murphy reported that the city had abandoned its street cleaning program in the northernmost section of Floral Park. According to Murphy, two years of municipal neglect had left Fifth Street and other roads littered with piles of rubbish.[46] Failure to deliver city services, he and other protesters asserted, contributed greatly to housing deterioration, blight, and disinvestment.

The city's failure to combat vice in the North End provoked additional protests. On scores of occasions during the 1950s, African American demonstrators attended city commission meetings to demand better police protection, especially from white johns and drug addicts. Because members of the vice squad focused their energies on arresting street-level drug dealers and prostitutes rather than their customers, white men in search of illicit substances and sex trolled the North End's red-light district with near impunity, often propositioning unsuspecting women who walked the streets. On August 8, 1955, St. John resident Roger Lewis informed commissioners that johns had become so aggressive that "women can't stand by their own fences or in their own yards without someone making vile remarks to them."[47] Without white johns, addicts, and suburban thrill seekers, North End activists maintained, the city's growing vice problem would virtually disappear.

A number of police brutality cases that surfaced in Flint and other cities during the late 1950s and early 1960s further fueled black resentments.[48] In September of 1958, tensions in St. John rose when two unidentified police officers assaulted and arrested an unarmed black girl returning home from choir practice. Operating under the incorrect assumption that she was a criminal, the officers "manhandled and injured" the young woman in front of a large group of onlookers. Outraged by the assault, protesters packed city commission meetings and picketed the Flint Municipal Center for four consecutive weeks.[49]

Shortly after the demonstrations died down, North End activists won a major victory when they elected black autoworker Floyd McCree, a member of the UAW's Local 599 FEPC Committee, to represent the Third Ward on the Flint City Commission. An employee of the Buick foundry and a well-known activist, McCree was the first African American to sit on the Flint City Commission and a harsh critic of police brutality, discrimination, and municipal neglect. Together, the protests against police brutality and McCree's election signaled that residents of St. John and Floral Park were fully committed to obtaining quality municipal services for all of the city's neighborhoods.[50]

Nonetheless, without fear of contradiction, many of these same activists also demanded the clearance of their neighborhoods. Although GM, the Mott Foundation, the Flint Chamber of Commerce, and other powerful organizations also supported such calls for redevelopment, Flint's economic and political establishment did not impose urban renewal upon unwilling African American communities. On the contrary, demands for urban clearance were an integral and organic part of the broader black freedom struggle in Flint. Fearing that redevelopment might exacerbate the problems of poor housing, popular and administrative segregation, and municipal neglect—and well aware of their small numbers and lack of political power—black activists joined the urban renewal coalition in order to shape its priorities. To thousands of African Americans, those priorities included safe and affordable housing, an end to all varieties of discrimination, improved public services, and clean air. Among members of the city's progrowth leadership, however, urban renewal was, above all else, an opportunity to foster economic development. As the city moved to implement its vast program of demolition and clearance, conflicts over these competing visions of redevelopment came to dominate Flint's fair housing movement.[51]

Rising Tide

Planning for Flint's urban renewal program began in earnest in the late 1950s, when city commissioners hired the firm of Ladislas Segoe and Associates to create a master plan. Released in 1960, the Segoe plan recommended complete demolition and redevelopment of all of St. John and a portion of Floral Park. Segoe and his colleagues proposed replacing "low-value housing" in the inner city with an expanded Buick facility, an industrial park in St. John, a freeway exit ramp in Floral Park, and a section of Interstate 475 to serve the central business district, the educational and cultural facilities east of downtown, and nearly all of the region's auto plants.[52] Soon after the release of the Segoe plan, the Michigan State Highway Department published *Freeways for Flint*, a report that contained the state's routing proposals for Interstate 475, M-78 (later renamed Interstate 69), and their connecting interchange. According to the MSHD document, I-475, a north-south freeway slicing through St. John, and M-78, an east-west route connecting Flint with Chicago and Canada, were to intersect via a large interchange in Floral Park. With the interchange and I-475, state and city officials hoped to link the assault on blight with increased automobile production and the resurrection of Flint's ailing downtown commercial core.[53]

Though the master plan and *Freeways for Flint* contained slightly different routing recommendations for the proposed highways and the I-475/M-78 interchange, both endorsed a massive reordering of the city's racial and economic geographies.[54] Flint contained a white majority of 83 percent at the time of the Segoe plan's release. Yet among the approximately three thousand families slated for relocation due to freeway construction, 60 percent were African American. Of the 1,193 St. John families that would have to move due to renewal projects, almost all were black. In addition, those slated for relocation were disproportionately poor and elderly, with approximately 85 percent of affected families earning less than six thousand dollars per year, one-third on public aid, and one-fourth senior citizens.[55] In St. John, which the Segoe plan designated as the city's top renewal priority, relocation presented extremely difficult challenges. "The trick [with relocation]," noted Olive Beasley of the Michigan Civil Rights Commission (MCRC), "is to prevent a new ghetto emerging."[56]

Although clear majorities of St. John and Floral Park residents supported clearance and relocation, the threat of losing precious dwelling units in Flint's already overcrowded ghettos triggered an intensified campaign for open housing.[57] Shortly after the master plan's release, as the city commission began weighing its options for North End and Floral Park redevelopment, members of the Flint NAACP and Urban League began negotiating with representatives from the *Flint Journal* over the explicit racial designations used in the newspaper's real estate advertisements. In the spring of 1963, *Journal* negotiators responded to a threatened boycott of the newspaper by agreeing to prohibit all such designations from classified advertisements.[58] Within months of the victory, Flint NAACP president Richard Traylor had drafted a municipal open occupancy ordinance banning racial discrimination in most real estate transactions. However, on October 3 activists suffered their first of many setbacks when Michigan attorney general Frank Kelley ruled that municipalities could not legally enact open occupancy ordinances under the state's constitution.[59] While fair housing advocates throughout the state worked to draft legal challenges to Kelley's opinion, supporters of open housing in the Flint area shifted their attention to combating popular and administrative segregation in the local real estate market. By the summer of 1964, the Evergreen Valley neighborhood in South Flint had emerged as a major flashpoint in that struggle.

Evergreen Valley was one of the many new subdivisions built in South Flint during the 1950s. Located two miles east of Floral Park near the city's southeastern border with Burton Township, the neighborhood contained 525 new single-family homes, most of them valued at between fifteen thousand and fifty thousand dollars. Like most of the new residential subdivisions erected in postwar America, Evergreen Valley began as an all-white community. Nevertheless, its mature trees, moderately priced homes, and location near downtown were appealing to many middle-class black families seeking to leave Floral Park and St. John. In July 1963 the neighborhood quietly received its first black residents, the result of a sale brokered by African American realtor Connie Childress. Within months, however, local realtors were waging an aggressive "blockbusting" campaign.[60] As part of that process, whereby brokers encouraged racial integration in order to profit from white "panic selling," realtors serving Evergreen Valley marketed homes only to black buyers. During the peak summer selling seasons of 1964 and 1965, white homeowners throughout the subdivision reported receiving anonymous phone calls urging them to sell, while others found leaflets on their porches describing the dangers of integration. On several occasions, realtors even showed up in person to warn white homeowners that they would soon be living in a predominantly black neighborhood.[61] Within two years of Evergreen Valley's initial desegregation, the area's black population had increased to 10 percent. As was often the case in newly integrated urban areas, the mere presence of black people in Evergreen Valley led to a rash of white departures. For most whites, even limited integration was a clear signal that the entire neighborhood was in decline and not worth defending.[62]

Whether in Evergreen Valley or elsewhere, realtors often played a decisive role in fueling the process of blockbusting. In most instances the decision to "break" a neighborhood stemmed from realtors' assessments of a community's "integrity."[63] If white residents expressed firm opposition to integration, then realtors generally prevented black home seekers from viewing properties for sale or rent.[64] However, if residents seemed more ambivalent, or if black renters or buyers were able to desegregate a neighborhood peacefully, then realtors often abandoned their traditional practices of exclusion.[65] Invariably, though, the loosening of racial restrictions led to panic peddling, mortgage disinvestment, and other acts of popular and administrative segregation. And because realtors targeted only a few of the city's neighborhoods for such treatment, the racial transitions that ensued were often quite rapid. Evergreen Valley, which registered a black majority within a decade of the onset of blockbusting, was a case in point.[66]

The struggles over integration in Evergreen Valley brought civil rights activists into direct conflict with realtors from the segregated Flint Board of Real Estate. Furious over blockbusting and the slow pace at which realtors were opening white neighborhoods to black buyers, Robert W. Rawls, chair of the Flint NAACP's housing committee, launched a boycott against discriminatory white real estate brokers in the fall of 1965. "How does one know which white real estate broker discriminates?" NAACP leaflets asked. "The answer is: They all do."[67] Members of Housing Opportunities Made Equal (HOME), a direct action group devoted to open occupancy, supported the boycott by picketing in front of the Flint Board of Real Estate offices and launching a test campaign to gauge the racial attitudes of white realtors.[68] As expected, the testing revealed that area realtors were staunchly committed to preserving the color line. With very little prompting, numerous realtors acknowledged that their firms had strict policies against breaking white neighborhoods. Others maintained that selling homes without regard to race was a clear violation of both the NAREB code of ethics and the policies of the Michigan Corporation and Securities Commission, the state agency responsible for issuing real estate licenses and regulating the conduct of brokers. After investigating the matter, HOME activists determined that on at least two occasions the state securities commission had officially censured Michigan realtors for either showing or selling properties to black buyers in all-white neighborhoods.[69] Throughout the county, open housing advocates found that realtors and government officials had constructed a seemingly impermeable web of popular and administrative barriers to integration.[70]

Although most of the 1960s battles over housing segregation occurred within the city limits, the suburban color line also came under attack during the course of the decade. In November 1965 Bill Hawkins, a member of HOME's board of directors, announced that his group was going to launch a new campaign of racial "scatterization" in Genesee County. In the first phase of the plan, HOME activists assisted thirty middle-class black families who wished to move from the city to the suburbs.[71] Modest though it was, Hawkins's plan generated strong opposition from real estate brokers and homeowners. Several months after Hawkins's announcement, realtors from the Hachtel-Pollock firm turned away a black woman who sought to buy a home in Davison Township, telling her that they were unable to show or sell homes in the area to Negroes. Members of HOME responded by picketing in front of the Belle Meade subdivision in Davison Township, where Hachtel-Pollock had a large number of listings.[72]

The spring demonstrations took township officials by surprise and created quite a stir among local homeowners and brokers. Within weeks, however, the Davison Township Board had come up with a plan to block future civil rights mobilizations. On May 9, board members adopted a parade ordinance requiring all persons who wanted to march, picket, demonstrate, or gather in public to obtain a permit from the township. The ordinance also stipulated, "No person shall intentionally lie down, stand, sit, or loiter in any public street, highway or sidewalk so as to interfere with the regular flow of pedestrian or vehicular traffic." Violators of the parade ordinance faced punishments of up to ninety days in jail and fines of up to one hundred dollars.[73] Though the board's actions infuriated HOME members, the demonstrations in Davison Township ended rather abruptly soon after the enactment of the new ordinance. Rather than face fines and jail time, HOME members sought out new opportunities in more hospitable areas.

Those would be extremely difficult to find, though. In addition to racist realtors and lenders, ordinary suburban whites were deeply committed to defending the residential color line. Although most suburbanites were peaceful, some of them resorted to violence to resist integration. In 1964 an African American teacher named Jerry M. Beaty and two black friends moved into a home in the all-white Candy Lane Estates subdivision in Mt. Morris Township. Neighborhood residents attacked Beaty and his housemates shortly after the move, however. According to a *Flint Journal* report, "An explosive situation developed as irate whites gathered around the house. Windows were broken and imprecations shouted. Once, an effigy was left hanging from the porch." After four difficult months in their new home, Beaty and his friends moved away.[74] Later that year, racial violence broke out again in Mt. Morris Township when Mr. and Mrs. Willie Mosley moved into a home in the Julianna subdivision. Hoping to run the Mosleys out of the neighborhood, white vandals broke the windows of their house and threatened the couple.[75]

In September 1966 violence against African Americans erupted in the Ottawa Hills subdivision of south suburban Grand Blanc. The trouble began when the Matthews family acquired a newly built house in the area through a "third-party" purchase—a somewhat uncommon arrangement in which sympathetic white purchasers bought and then immediately resold homes in segregated white neighborhoods to black buyers. An angry group of whites responded by picketing the property and harassing the homeowners. Upon learning of the duplicitous purchase, the developer who had built the house enlisted the support of black realtor Connie Childress, who offered to buy back the home and resell it to whites. When that plan failed, the builder told Mrs. Matthews that "she would be sorry she had moved" to Ottawa Hills. Far from isolated instances, these and other acts of intimidation and violence were fairly common occurrences in the suburbs of Genesee County and the nation as a whole.[76]

In response to such blatant acts of racism, the leaders of HOME and other local civil rights organizations publicly demanded that the government take action against residential segregation. After receiving signed statements from several victims of violence and discrimination, HOME president George F. Plum forwarded their complaints to Governor George Romney, Attorney General Frank Kelley, the Michigan Civil Rights Commission's Burton I. Gordin, and dozens of local journalists. In a June 7, 1966, letter to Gordin, Plum chided civil rights officers for procrastinating in response to discrimination complaints from the Flint area.[77] Hoping to deflect criticism of his agency, Gordin responded by agreeing to organize hearings on housing discrimination in Genesee County. Held from November 29 through December 1, 1966, the Flint hearings provided activists with a long-awaited opportunity to shine a spotlight on the public policies, administrative practices, and private racial animosities that had collectively thwarted equal housing opportunities in the Vehicle City.

Naming Names

The Michigan Civil Rights Commission's 1966 hearings were the product of many years' worth of protests, public relations efforts, and electoral campaigns waged by fair housing activists in metropolitan Flint. Although these organizers had already generated momentum in the months leading up to the hearings, an unexpected move by the Flint City Commission in the fall of 1966 helped to draw even more attention to the area's civil rights insurgency. Just a few weeks prior to the hearings, civil rights reformers won a surprising victory when city commission members selected popular Third Ward commissioner Floyd McCree as mayor. McCree thus became Flint's first black mayor and, as civic boosters proudly advertised, one of the first

FIGURE 6.3. Floyd McCree speaking to voters, 1954. McCree, an employee of the Buick foundry, won his first election to the Flint City Commission in 1958. After he served several successful terms in office, commission members appointed him to the position of mayor in 1966. One of the first black mayors of a major American city, McCree was a staunch advocate of open housing. Courtesy of the Genesee Historical Collections Center, University of Michigan–Flint.

African Americans selected to lead a major American city. Like most of his North End constituents, McCree was a strong advocate of both urban renewal and open housing. However, under Flint's commission-manager form of government, the mayoralty was essentially a symbolic position, which left McCree with no more power than that of other commissioners. Nevertheless, his appointment marked an important victory for fair housing activists.[78]

McCree and other supporters of fair housing used the hearings to explain how local citizens, members of the private housing industry, and public officials had worked to maintain both popular and administratively driven patterns of residential Jim Crow in Flint. During his appearance in front of the commissioners, the Flint Urban League's executive director John W. Mack testified that Flint, with its entire African American population confined to just twelve adjacent census tracts along Saginaw Street and Lapeer Road, was the most segregated nonsouthern city in the United States.[79] Witness George Plum, the president of HOME, attempted to shed light on the roots of these divisions by submitting seven sworn affidavits charging members of the Flint

Board of Real Estate, including GM's Argonaut Real Estate divison, with re-fusing to sell homes on the open market.[80] Deponents also charged that the real estate board practiced popular segregation by systematically excluding African Americans from membership, denying nonwhite realtors access to its multiple listing exchange, and levying substantial fines against member realtors who broke all-white neighborhoods.[81]

Builders also received a great deal of criticism during the hearings. In re-sponse to a question from one of the commissioners, Plum recounted an at-tempt that he and his wife had made to conduct a third-party purchase for a black couple who sought a home in a newly developed all-white subdivision. While negotiating over the purchase, the builder informed Plum that in order to prevent the house from being resold to Negroes, he would have to sign a purchase agreement giving the builder an option on the property should it be resold prior to occupancy. Other witnesses claimed that builders, like real-tors, steered African American clients toward all-black or integrated areas, "while the white market was diverted to all-white suburban subdivisions." Even the lone developer who testified at the hearings confirmed that, as a rule, "[b]uilders refuse to build or sell lots to Negroes in new subdivisions."[82]

Although many witnesses placed a great deal of blame on realtors and builders for maintaining segregated housing, the government did not escape criticism. During his conversation with the commissioners, Mayor McCree acknowledged that the city was woefully unprepared to handle the relocation of black citizens scheduled to move due to urban renewal. The city planned to uproot nearly three thousand predominantly poor Flint families, over half of whom were African American, to make way for Interstate 475, the Floral Park interchange, and the St. John industrial park. Yet according to McCree, municipal officials had planned for only two hundred units of low-income replacement housing, all of which would be located in segregated black neighborhoods. The Urban League's John Mack echoed McCree's charges, claiming that the city's newly opened Central Relocation Office (CRO) was already steering African Americans to all-black or racially transitional neigh-borhoods.[83] The FHA's Flint Service Office also faced harsh criticism dur-ing the proceedings. In addition to complaining about redlining, witnesses charged that the FHA subcontracted only with the Davis and Piper real estate firms, both of which refused to desegregate white neighborhoods, to market federally owned foreclosure properties in the county. The testimony regard-ing FHA properties highlighted a clear nexus between government adminis-trators from Washington and the local housing officials who helped to keep the city segregated.[84]

Residents of St. John Street and the broader North End provided some

of the most powerful testimony collected during the hearings. Beyond describing the harsh effects of redlining, residents pointed out that unregulated
industrial pollution and municipal neglect had made the North End virtually uninhabitable.[85] In her testimony, Ailene Butler, a throat cancer survivor,
spoke about her husband's death from the same disease, explaining, "There
is a heavy smog caused by the Buick factory which has been in existence for
about eighteen years. . . . The houses in this district are eaten up by a very
heavy deposit, something like rust. If you do not have a garage in this area
and buy a car, the insurance is higher because of these deposits. . . . You can
imagine what we go through down there breathing when this exists on just
material things."[86] Desperate to leave St. John, Butler and other witnesses all
but begged the commissioners for fair housing reform and urban renewal.

Much of the testimony from St. John residents revolved around the shortage of rental units. According to Cora Hollman and three other black women
who testified at the hearings, it was nearly impossible to find rental housing
in the north side "that is not overcrowded, unsafe, or unsanitary." Hollman
and her fellow witnesses, all of them public aid recipients, also charged that
landlords in St. John routinely refused to comply with code enforcement or-
ders and threatened eviction in response to even basic repair requests. In one
case, commissioners learned about a slumlord who had pledged to replace
his white tenants with black occupants in retaliation for their complaints
about substandard conditions. If the tenants did not cease their complaints,
the landlord vowed, he would evict them and "put niggers in the house," because he "could get more money per week out of the house if he had niggers
in it anyway, and the niggers wouldn't complain because they have no place
to stay." The city's bifurcated real estate market did more than simply trap
African Americans in the ghetto. It also undermined their ability to negotiate
with landlords.[87]

Flint fire marshal Elwood Rutherford, the coordinator of the city's code
enforcement program, confirmed many of the charges that witnesses made
regarding inner-city housing. "Complaints show by inspection that blight is
reaching into nearly every section of the city," Rutherford stated, "apparently
caused by a complacent public and lack of code enforcement." Rutherford
also informed the commission that there were "outhouses attached to shacks,
only minutes from City Hall," and scores of families citywide who lived in
"unremodeled attics and basements." Just between March and December of
1965, inspectors visited 1,806 dwellings, where they found nearly three thousand code violations, including 441 malfunctioning electrical systems, 147 inadequate plumbing facilities, and 107 illegal heating units.[88]

As part of the proceedings, commissioners looked carefully at residential segregation outside the city. Specifically, the commissioners wanted to know why only two suburban census tracts in the entire county contained black populations in excess of 2 percent. In response, witness Harold Draper, a representative of the Genesee County Board of Supervisors, readily acknowledged, "Equal housing opportunities do not today exist in Genesee County." However, Draper and other suburban witnesses claimed no responsibility for the color line. Articulating a classic defense of popular and administrative forms of housing segregation that echoed throughout suburban America during this period, Draper argued, "The slow movement of our non-white population to the county is a matter of economics. The vast majority of our [N]egro population cannot afford the cost of purchasing new homes in our county developments." Because "de facto segregation" was essentially a private economic matter, Draper maintained, "[t]he Board of Supervisors do[es] not have any legal power to act as to legislation."[89]

Seeking balance, representatives of the civil rights commission invited a host of GM executives, local builders and lenders, members of the Flint Board of Real Estate, and various civic leaders to testify on behalf of the local power structure. Most of the city's leadership abstained from the proceedings, however.[90] By boycotting the proceedings, Flint's corporate and civic elites may have sought to deny legitimacy to the open housing movement. In reality, though, their actions had the opposite effect. Following the hearings, MCRC members confirmed witnesses' charges of racial discrimination and substandard housing. Additionally, the commissioners issued a harsh rebuke to those who had declined to testify, noting, "The failure of financial, commercial, industrial and other private corporations and organizations to be represented at the hearing tends to show that that segment of the community is, at best, disinterested in the resolution of the problems of housing discrimination and segregation in the Flint area." The editorial staff of the *Flint Journal* concurred, suggesting that the city's social conscience had "dozed" during the hearings.[91] More than just a cathartic exercise for frustrated black citizens, the 1966 proceedings proved to be a powerful springboard for the fair housing movement.

The Battle for Open Housing

Open housing activists, many of them rejuvenated by the MCRC's findings, wasted little time in returning to the Flint City Commission. On February 13, 1967, a large group descended upon city hall to urge municipal lawmakers to reconsider the NAACP's initial open occupancy ordinance. After reminding the commissioners—at least five of whom were salaried employees of General Motors—that Flint remained profoundly segregated, HOME member Leroy Davis implored them to endorse equal housing opportunities. Speaking in opposition to the proposal, Harry M. Rapaport, executive vice president of the Flint Board of Real Estate, claimed that a fair housing ordinance would "create hostility because people don't like compulsion." City attorney Charles A. Forrest also criticized the bill, citing Michigan attorney general Frank Kelley's 1963 opinion that municipal open occupancy laws were unconstitutional. But Olive Beasley, Flint's MCRC representative, disagreed, informing commissioners that Kelley had since reversed his earlier decision. Seeking clarification on the matter, city commissioners voted to refer the bill to their legislative committee until they had received a definitive legal opinion on the NAACP's proposed ordinance.[92] On March 2, 1967, Kelley attempted to resolve the conflict by releasing an open letter endorsing the legality of municipal open housing laws.[93] Forrest refused to acknowledge Kelley's new directive, however, maintaining, "The attorney general can speak only though official legal opinions."[94]

By the spring of 1967, the issue of fair housing had become the most closely followed and volatile political issue in the region. Across the country, in fact, talk of open housing dominated newspaper headlines, legislative agendas, public meetings, and dinner table conversations.[95] Nevertheless, Flint's city commissioners continued to postpone a formal vote on the NAACP's proposal. Their delay served only to inspire fair housing activists, however. During an April speech to local union members, Herbert Hill, the NAACP's national labor secretary, scolded Flint residents for not forcing a vote on open occupancy. "Your town is a Jim Crow town," he claimed. "When are you going to bust out of the ghetto?"[96] On July 17 Marise Hadden warned commissioners of the consequences of inaction: "Unless elected officials take a firm and positive stand [on open occupancy], dissident elements come to believe that their only rewarding means of protest is open violence and rebellion. Under these circumstances, it is guaranteed that they are going to rebel against life in the ghetto."[97] Within a week, Hadden's prediction had come to pass.

In the wee hours of July 23, 1967, just six days after authorities had quelled a major race riot in Newark, New Jersey, racial violence erupted seventy miles southeast of Flint in the city of Detroit. The fighting began after police officers arrested black patrons of an after-hours drinking establishment, or "blind pig," located at Twelfth Street and Clairmount Avenue. The violence spread quickly to black neighborhoods throughout the Motor City. In all, the Detroit riot of 1967 raged for five full days, claiming nearly fifty lives and resulting in millions of dollars in property damage, making it one of the most violent and costly racial conflagrations in American history.[98] But it was not simply a local disturbance. Within hours of the eruption in Detroit, the violence had spread to Flint's North End, where angry demonstrators gathered to vent their frustrations. On the evening of July 24, hundreds of protesters assembled at the corner of Leith and Saginaw Streets. During the night, small groups of rioters threw rocks at cars, looted commercial establishments, and firebombed at least seventeen stores and homes. The following day, Governor George Romney declared a state of emergency for all of Genesee County. Hoping to stem the spread of violence, Romney also ordered state troopers to monitor checkpoints at the Flint city limits, prohibited the sale of alcohol throughout the county, banned all nonofficers from carrying firearms, and ordered the closing of all North End gas stations, presumably to prevent any additional firebombing.[99] For their part, Mayor McCree, local ministers, and other prominent black community members maintained an around-the-clock presence in the North End, walking the streets, imploring people to return to their homes, and helping police officers track down suspects. By July 27 the North End disturbances had ended and a measure of calm had returned to the city.[100]

Still, the restoration of order did little to remedy the grievances of fair housing reformers. On July 27, just as city officials were declaring an end to the disturbances, a group of fifty activists, half of them white, attended a city commission meeting to demand a vote on open occupancy. This time, however, the group included downtown merchant Sydney Melet, owner of the Vogue clothing store; Saul Seigel of the Greater Flint Downtown Corporation; and Harold Stine, president of the Flint Board of Education. The outbreak of civil unrest had forced these and many other Flint leaders—many of whom had boycotted the 1966 hearings—to reevaluate the merits of a fair housing ordinance.[101] More unified than ever, members of the open housing coalition moved to force a vote on the issue.

As a concession to racial moderates on the city commission, the NAACP's open occupancy ordinance contained several loopholes that allowed for housing discrimination under certain circumstances. The ordinance would not apply to owners of two-unit apartments, for instance, if the deed holder occupied one of the dwelling units. Nor would the law cover religious institutions or homeowners who took in boarders. The bill did contain a controversial penalty clause for violators, though, which included punishments of up to ninety days in jail or a fine of up to three hundred dollars. Already openly hostile to any nondiscrimination legislation, Commissioners Carl Mason and William Polk used the penalty provision to persuade the commission's racial moderates to oppose the proposal. On August 14 the commissioners voted five to three to reject the ordinance.[102] Stung by the defeat, Mayor McCree obtained special permission to address the commission immediately following the vote. "Last November this city commission saw fit to make me mayor," McCree stated, his body shaking with emotion, "and that was fine all over the country, very wonderful. . . . Tonight, however, I've changed my mind. I'm not going to sit up here any longer and live an equal opportunity lie." In a move that surprised nearly everyone in the packed city hall chambers, McCree went on to announce his resignation as mayor.[103]

The defeat of the open occupancy bill and McCree's shocking resignation spawned a firestorm of protest and recrimination. Fearful that the decision would unleash another round of disorders in the North End, many civic leaders began looking for opportunities to compromise. Following the commission's vote, the editors of the *Flint Journal* and the Mott Foundation's governing board publicly endorsed open occupancy and implored McCree to return to his post.[104] On August 17 dozens of black elected officials and political appointees announced that they too would resign in protest. A day later, thirty-four supporters of open occupancy, most of them teenagers, organized a "sleep-in" demonstration in front of city hall. Led by youth activist Woody Etherly Jr., the protesters—some carrying signs reading "Our City Believes in Lily-White Neighborhoods"—vowed to camp out on the lawn of city hall until the commission reversed its position.[105] The protests culminated on August 20 with a massive Unity Rally in front of city hall, where Governor Romney and Attorney General Frank Kelley expressed their unequivocal support for the mayor and open occupancy.[106] The demonstration was the largest civil rights protest in the city's history and helped to shift the political tides decisively.[107]

The ongoing city hall sleep-in and the August 20 demonstration inspired Floyd McCree to return to the public arena. On August 28 McCree, despite suffering from exhaustion and a stomach ulcer, rescinded his resignation.[108] Upon reclaiming his office, the mayor opened a series of secret meetings with racially moderate commissioners in the hopes of gaining a favorable vote on a modified proposal. After a week of private negotiations, McCree had assembled a majority on the commission. In order to gain that majority, however, he agreed to several concessions that further weakened the ordinance. Namely, the new proposal maintained special protections for religious institutions while expanding exemptions to include landlords who lived in buildings with five or fewer rental units. The new bill also made it illegal and punishable to file false claims of discrimination.[109] The revised bill was, in every sense, a compromise measure.

On Monday October 30, after weeks of controversy and negotiation, the

FIGURE 6.4. Open housing demonstration, 1967. The youth activists featured here participated in a lengthy fair housing "sleep-in" demonstration on the lawn of the Flint Municipal Center during the summer of 1967. Their efforts led to the enactment of Flint's open occupancy law. Courtesy of the Alfred P. Sloan Museum, Flint, MI.

Flint City Commission, in a close five-four vote, approved the revised bill.[110] Supporters of fair housing had little time to celebrate, however. Within minutes of the law's passage, white opponents had launched a petition drive to force a referendum on the issue. Led by Gerald A. Spencer, a member of the John Birch Society, the Committee to Repeal Forced Housing Legislation united a broad coalition of homeowners and real estate professionals who believed that the new law violated the rights of property owners.[111] Joining the opposition to open occupancy were at least fifty members of a local Ku Klux Klan chapter, who organized several protests during the month of October, including a small parade in downtown Flint.[112] By November 24 Spencer and his supporters had collected nearly six thousand signatures, more than enough to trigger a public vote on open occupancy. Upon submitting the petitions to the city clerk, Spencer read a prepared statement to reporters that outlined his opposition to the new law. "The issue," Spencer noted, was whether "the government, right or wrong, has the power to tell a person how and to whom he must dispose of his property."[113] After certifying the signatures, Flint City Clerk Lloyd S. Hendon ordered a referendum election for February 20, 1968.[114]

In all, just over forty thousand citizens decided the fate of the open housing proposal. Leading up to the vote, only the most optimistic proponents of civil rights believed that open housing would survive a popular vote in Flint. Although similar elections had already taken place in several cities including Seattle; Berkeley, California; and Akron, Ohio, voters in the United States had yet to approve a law banning real estate discrimination.[115] In light of both the historical evidence and the fact that Flint's electorate was still overwhelmingly white, most observers believed that the referendum to overturn fair housing would pass easily. Yet in a tally that surprised even Mayor McCree, who went to bed on February 20 believing that his side had lost, Flint citizens defeated the referendum initiative by a razor-thin margin of thirty votes.[116] The victory meant that Flint voters were the first in the nation to endorse open housing legislation via popular vote.[117]

With the new law in place, a months-long celebration ensued. On February 23 UAW president Walter Reuther hailed Flint residents for building "one of the cornerstones upon which to build stable community relations."[118] According to Dale Kildee, a Flint representative in the state legislature and future member of Congress, the election demonstrated that "voters can be trusted to do what is right when they are well informed."[119] Just a year after the referendum, Mayor Donald R. Cronin, McCree's successor, declared that with the 1968 vote Flint had become "the most progressive city in the country."[120] Across the city, liberals and moderates championed the referendum victory as a culminating moment in the decades-long movement against legal, popular, and administrative Jim Crow.

In truth, though, the February vote was far less demonstrative than Cronin and other boosters assumed. In Wards 1, 3, 5, 7, and 9, which housed nearly all of the city's black residents, open housing won by sizable margins. Yet in the virtually all-white Second, Fourth, Sixth, and Eighth Wards, voters overwhelmingly rejected the new law. Citywide, in fact, voters approved McCree's compromise measure in only 46 of 122 precincts. Without heavy black turnout in Floral Park, Elm Park, Sugar Hill, Evergreen Valley, and of course the North End, opponents of the ordinance would have won an easy victory.[121] According to a *Flint Journal* editorial assessment, "The results demonstrated . . . that the Negro does indeed have power at the ballot box."[122] Just how much power black citizens possessed, however, was up for debate.

Rather than revitalizing the city and uniting its citizens around a shared commitment to fairness and equity, the 1968 open housing referendum highlighted the depth of the region's racial fissures. During the years following the vote, struggles over urban renewal and segregation continued to dominate local politics. As the city moved forward with plans to demolish St. John and Floral Park, activists discovered a variety of new opportunities to attack the area's color lines. Consistently, though, Flint's fair housing law failed to live up to the expectations of its framers.

Jim Crow in the Era of Civil Rights

On June 7, 1973, just five years after city leaders celebrated the open housing triumph, Olive Beasley of the Michigan Civil Rights Commission sent
an angrily worded memorandum to Richard Wilberg, Flint's urban renewal
director. Her correspondence included a bitter reflection on the relationship
between urban renewal, municipal policies, and ghetto formation. Para-
phrasing the federal government's 1968 Kerner Report—whose authors had
famously blamed "white society" and "white institutions" for maintaining
the inequalities that triggered the 1967 race riots—Beasley identified Flint's
municipal government as the primary gatekeeper for the city's ghettos: "The
City of Flint is deeply implicated in its ghettoes, City of Flint institutions created them; City institutions maintain them."[1] According to Beasley and other
critics of urban renewal, Flint's tightly packed black neighborhoods were
products of intentional government-sponsored segregation.

The words in Beasley's memorandum spoke directly to the experiences of
African Americans in St. John, Floral Park, and numerous other communities of color scattered throughout urban America.[2] Although the city's plan
for slum clearance initially drew support from civil rights activists, urban
renewal ultimately ossified Flint's residential color line. During the late 1960s
and 1970s, officials from Flint's Urban Renewal and Housing Department,
the Department of Community Development (DCD), and other government
agencies cleared three hundred acres of land in St. John, relocating most of its
three thousand black residents into segregated replacement housing. Much
the same occurred in Floral Park, where the city demolished a physically deteriorated yet vibrant black enclave to make way for public housing and a
freeway interchange. Wilberg and other officials hoped to replace the two
neighborhoods with new GM plants, downtown businesses, and a first-rate

urban freeway system. Ultimately, though, the renewal projects severely undermined the city's economy. Furthermore, by funneling displaced African Americans to recently integrated areas, urban renewal helped to trigger new hybrid forms of popular and administrative Jim Crow by spurring waves of white panic selling and property abandonment in the racially transitional neighborhoods bordering St. John and Floral Park. As many residents ultimately discovered, the "white flight" that ravaged the city during the 1960s and 1970s was as much a consequence of deliberate public policies as a reflection of individual fears and choices.[3]

A Fragile Coalition

In Flint and many other cities throughout the United States, corporate leaders, downtown retailers, and municipal policy makers thoroughly embraced the federal urban renewal program.[4] With its pledge of large federal subsidies for the assault on "urban blight," the United States Housing Act of 1954 received a hearty endorsement from GM executives, city commissioners, and other civic elites.[5] And so too did the Federal-Aid Highway Act of 1956, which promised to fund a nationwide network of high-speed freeways. The city's corporate and political establishments supported these federal programs because, above all else, they promised to bring GM's ambitious postwar growth strategy closer to fruition. In order to pursue its decentralized, highway-dependent vision of metropolitan capitalism, GM required open space for freeways and interchanges within Flint's densely packed urban core. Hoping to link their desires for new freeways and industrial facilities with broader civic concerns over urban blight, GM managers waged an aggressive campaign to level St. John and Floral Park.[6]

Nevertheless, support for the city's urban renewal and freeway projects crossed racial, class, and partisan divides. In fact, the city's vast coalition for redevelopment also included representatives of the Mott Foundation, the chamber of commerce, the UAW, the *Flint Journal*, developers and real estate brokers, the NAACP, the Urban League, and frustrated black residents from the North End.[7] By the early 1960s this broad yet fragile coalition had congealed around five major issues, many of which had national resonance. First, corporate spokespersons promised that the freeway would expedite the flow of industrial traffic, thereby increasing economic efficiency and reducing road congestion. Second, local UAW members hoped to use the St. John project to open up new land for long-awaited employee parking lots in the congested North End.[8] Third, Flint's interstate advocates, like their counterparts in other cities, argued that the freeway would spur downtown commerce and tourism by linking the city's cultural and business districts to the booming residential developments in the suburbs. According to Richard Tavis, executive director of the Greater Flint Downtown Corporation, without slum clearance and freeway construction, the city's downtown core faced "economic strangulation."[9] Fourth, trade unionists and cost-conscious municipal officials such as Thomas Kay, Flint's city manager, hoped that urban clearance would lure new jobs and tax-generating industries to St. John and other neighborhoods that contributed the least in local property taxes.[10] For Tavis, Kay, and others who subscribed to the gospel of economic growth, jump-starting the local economy and increasing the city's tax base were the key aims of urban redevelopment.[11] But not everyone in the Vehicle City viewed urban renewal through a macroeconomic lens. In fact, black community activists, racially liberal UAW members, and other opponents of segregation also supported demolition and redevelopment. Among this diverse group, many viewed clearance as an opportunity to combat Jim Crow, poor housing, and the North End's growing public health crisis.[12]

From the start, power relations within Flint's urban renewal coalition were profoundly asymmetrical. Though vocal in their demands for relocation, St. John residents were the least powerful proponents of redevelopment. Constituting less than 20 percent of the city's total 1960 population, African Americans, even those who viewed slum clearance with skepticism, had few options but to try to obtain racial fairness from inside the prorenewal alliance. By contrast, GM executives were a dominant force within the city's highway and renewal coalitions. However, they often exerted their influence quietly through their representatives in the Manufacturers Association of Flint, the Mott Foundation, and a powerful organization known as the Genesee Community Development Conference (GCDC). During the years preceding the first property acquisitions, leaders from the GCDC—a private housing and renewal agency funded by the Charles Stewart Mott Foundation—sponsored "diagnostic" surveys of St. John and other targeted neighborhoods and assisted city officials with the drafting of renewal plans. For their part, Mott Foundation representatives offered direct cash grants to the city's newly formed urban renewal department and helped to open "community relocation sub-stations" in eleven of the city's public schools.[13] Through their direct involvement in the creation and promotion of urban renewal plans, corporate officials took on familiar roles as state actors and, in the process, blurred the distinction between the public and private sectors. As time progressed and the city moved closer to implementing its renewal plan, the boundaries between GM and the government would become increasingly less significant.[14]

Inching toward Renewal

In the years following the Segoe plan's release, the I-475 freeway, the Floral Park interchange, and the St. John clearance project quickly became the focal points of a hotly contested urban redevelopment agenda. Although the Flint City Planning Commission officially adopted the Segoe proposal in 1961, the renewal program that ultimately ensued did not, in fact, reflect the master plan's recommendations. Shortly after the plan's release, officials from the Michigan State Highway Department challenged the proposed interchange and highway routes. Because Segoe had envisioned M-78 as a suburban-oriented southern bypass for east-west traffic through Flint, the master plan called for a Burton Township interchange with I-475, which would have spared Floral Park from complete clearance. However, MSHD officials favored a more disruptive, racially insensitive urban route for M-78 that included a large Floral Park interchange.[15] Highway planners justified the new location by pointing to Floral Park's low property values and substandard housing, its proximity to downtown and numerous GM plants, and the lower cost of the new interchange.[16] If implemented, MSHD planners added, the new interchange promised to further the city's urban renewal agenda by recapturing white migrants and tourists, luring new industry, revitalizing downtown, and, crucially, removing large numbers of poor black residents from the urban core.[17] Persuaded by the MSHD's logic, in February 1962 GM and the MAF formally endorsed the state's revised routing proposal.[18]

State highway commissioner John Mackie directed the city commission to move rapidly on the new plan. Specifically, Mackie proposed a thirty-day limit for city commissioners to approve the new routes and interchange location, promising to divert Flint's freeway funds elsewhere if local policy makers did not abide by his timetable. "If we do not receive concurrence within 30 days," he warned, "then the people who are now assigned to the Flint [freeway] study will be put to work on other urban freeway projects, and the funds allocated for the Flint freeway system will be used elsewhere."[19] Upon learning of Mackie's ultimatum and GM's support for the new plan, the city commission promptly endorsed the state's revised blueprints for I-475, M-78, and the Floral Park interchange. Not long after that, city and state officials began the lengthy property acquisition and relocation process in Floral Park.

Despite a few early signs of progress, things did not move as quickly in the North End. While MSHD planners were completing their interchange and freeway plans, the city commission voted, in July 1962, to adopt a three-thousand-acre urban renewal plan. Eager to move forward with the seven projects included in the plan, city commissioners wasted no time in selecting retired suburban developer Charles Richert to direct Flint's Urban Renewal and Housing Department. In February 1964 Richert and members of the commission submitted their General Neighborhood Renewal Plan (GNRP) to the federal Housing and Home Financing Agency. Several months later, city commissioners created the Flint Housing Commission (FHC), a public housing authority designed to accommodate low-income residents displaced by redevelopment.[20]

Upon receiving the GNRP, federal officials noted that Flint's urban renewal program was one of the largest ever attempted in a city of its size. Massive though it was, however, Richert's initial urban renewal plan made no mention of St. John and the North End. In yet another departure from the master plan, the GNRP covered only the central business district, Floral Park, and Central Park, a nearly all-white neighborhood just east of downtown.[21] This was surprising, of course, because GM bosses and city leaders had eyed the North End for redevelopment dating back to at least the early 1950s. In fact, according to the master plan the St. John project was the city's top redevelopment priority.[22] Both Richert, who died soon after his appointment, and his successor Leo Wilensky explained the discrepancy by pointing to a lack of available replacement housing for displaced residents as well as an array of other obstacles. With regard to replacement housing, federal law required city officials to present a "workable" plan for eradicating blight and rehousing displaced persons prior to obtaining renewal funds. In the Central Park and downtown districts, the city's plans called for only limited clearance and relocation, which meant that renewal could proceed immediately. For St. John, by contrast, Richert and Wilensky had planned for complete clearance and the relocation of over three thousand predominantly poor black citizens. With the Flint Housing Commission still in its formative stages and an extremely tight local real estate market, municipal leaders could not pursue St. John redevelopment until they had a feasible plan for housing displaced residents.[23] The release of the GNRP marked the first in a series of delays and defeats for North Enders seeking clean air and better housing.

The federal government's complicated urban renewal finance policies constituted an additional impediment to inner-city redevelopment. Under the terms of the 1954 Housing Act, the federal government was responsible for financing two-thirds of the cost of approved neighborhood development projects, leaving cities to fund the remaining one-third share. To cover their portion of the costs, municipalities could either pay in cash or employ "noncash credits." If city officials wished to apply these credits, as most did, they had to demonstrate municipal investment equal to the local share of redevelopment costs in the neighborhoods slated for urban renewal. According to federal guidelines, local investments in roads, schools, and other public improvements all qualified as noncash credits. In Central Park and downtown the city had invested substantially during the postwar decades, providing sufficient credits to finance the two urban renewal projects without additional cash outlays. However, local policy makers had never benefited St. John and other North End neighborhoods slated for renewal with such investments. Therefore the city had to raise its portion of the North End redevelopment costs in cash, which further delayed the project's commencement.[24]

The cash shortages that postponed the North End renewal program stemmed in part from a severe fiscal crisis at city hall. Like most other American cities, Flint relied heavily on property taxes to fund municipal government.[25] Yet after decades of procorporate governance, the city commission presided over some of the lowest municipal tax rates in the state.[26] Although local UAW leaders sponsored numerous postwar campaigns to increase taxes—particularly to support infrastructure development and higher wages for city workers—their efforts were largely unsuccessful.[27] Consequently, municipal agencies struggled to provide adequate services to Flint's growing population, even during periods of economic growth. The situation reached crisis stage in 1963, when city commissioners announced ten million dollars in spending cuts to fill a massive budget shortfall.[28] In response to the gathering fiscal storm, voters agreed to enact a 1 percent income tax for all Flint residents and a 0.5 percent earnings levy on all nonresidents working in the city. Beyond funding additional city services and a wage increase for municipal employees, revenue from the new assessments was to provide a seventeen-million-dollar capital improvement budget to fund urban renewal and freeway construction. However, in order to gain the public's support for the income levies, city commissioners also pledged to enact a three-mill reduction in local property taxes. Though the new income taxes helped to ease the immediate fiscal crisis, the reduction in property-based revenues severely undermined the city's attempt to raise money for the St. John redevelopment project.[29]

To be sure, federal financing strictures, municipal revenue shortfalls, and severe housing shortages all played major roles in the delay of the St. John project. Nevertheless, the colorblind dialogue on revenues and expenditures tended to mask a very real fear among city officials that white taxpayers— who still formed a clear majority in the city—would revolt against massive government expenditures on the North End, especially if they believed those funds might be used to disperse tightly compacted black neighborhoods. According to the Urban Renewal and Housing Department's annual report for 1965, "The City did not have the cash nor was it willing to spend large amounts of money on a program [in the North End] that is very complex and controversial."[30] Unable (and perhaps unwilling) to generate the funding and the public support necessary to commence with North End renewal, the city turned toward the private sector to gain support for the St. John redevelopment project.

In a bid to jump-start the stalled North End renewal program, General Motors, the Mott Foundation, and the MAF jointly commissioned members of the Urban Land Institute (ULI) to conduct a study of the city's urban development needs.[31] Released in 1967, the ULI report endorsed the construction of a north-south freeway as the centerpiece of a progrowth agenda that revolved around increased industrial output, commercial and tourist development near the riverfront downtown, and a massive land reclamation plan for the inner city. The authors of the report hoped to allay the fears of skeptical white residents by openly acknowledging that racial integration was not among the primary goals of North End renewal. In fact, during a 1967 hearing the ULI's Charles Fleetwood maintained that urban renewal had nothing to do with integration, remarking, "If you want a truly integrated nation, you have to raise the standards of these [black] people. That will be a long, laborious effort."[32]

Still, advocates of renewal had a difficult time gaining white support for North End redevelopment. In 1968, after nearly five years of failed efforts, the city commission moved yet again to secure public financing for urban renewal by launching a five-million-dollar bond initiative for the redevelopment of St. John and the neighboring enclave of Oak Park. In advance of the April 7, 1969, vote, urban renewal advocates formed a group called the Citizens' Committee in Support of the Oak Park Bond Issue.[33] Cochaired by Saul Seigel, executive director of the Greater Flint Downtown Corporation, and Edgar Holt, president of the Flint NAACP, this pro-bond group solidified the city's vast urban renewal coalition. Once united, supporters of renewal designed a publicity campaign that drew financial and political support from the MAF, the Mott Foundation, GM, the UAW, the Flint Chamber of Commerce, the *Flint Journal*, the Urban League, the NAACP, the executive committees of the Genesee County Democratic and Republican Parties, the Flint Board of Education, and the Greater Flint Downtown Corporation. Throughout the 1969 drive, these powerful supporters of redevelopment consistently reminded voters that the bond—which the city commission planned to repay through already existing municipal revenue streams—would generate no new taxes for residents. Moreover, renewers framed the bond initiative as a referendum on Flint's future: "A 'yes' vote . . . means progress, growth, development, prosperity. . . . A 'no' vote means a dismal future, turning our

backs on progress and opportunity—continued spread of blight, indiffer-
ence, apathy, and hopelessness."[34]

On election day, voters, by an overwhelming three-to-one margin, signaled their opposition to North End renewal by rejecting the bond.[35] Like the
New Flint and annexation defeats, these vote tallies pointed to the limits of
GM's political power. In reality, though, the measure never really had much
chance of passing, even with the support of the city's most powerful people
and institutions. Crucially, because the April vote was a special bond election
and not a general referendum, only property owners in the city of Flint were
eligible to participate.[36] This important stipulation effectively disenfranchised
approximately half of the city's black population and all of its renters, many
of whom were impoverished. But the issue also failed because it galvanized
thousands of property owners and working-class whites who feared higher
taxes, corporate-sponsored property seizures, and the specter of racial integration. One such individual was Saul Shur, a Flint Township resident who
owned property on the North End. Shur helped to organize the Committee
to Save Our Homes (CSOH), an ad hoc group that successfully mobilized
white property owners, many of them North End slum lords, to defeat the
proposal. Throughout the winter and spring of 1969, Shur and other CSOH
members described the referendum as a case of "David vs. Goliath," a popu-
list crusade against subsidies for GM and company rule. However, the CSOH
also tapped into a potent combination of working-class economic frustration
and white racial fear during the campaign. In an especially revealing appeal
that illustrated how race, class, and property ownership intersected in the
April elections, a CSOH leaflet asked, "2,764 families to be re-located. Where
will they go? Don't hurt poor people!"[37] Among African Americans trapped
in the North End, urban renewal seemed to offer a chance to overcome decades' worth of popular and administrative segregation. For thousands of
property-owning white voters who mobilized against the bond issue, North
End renewal was a nonstarter for that very same reason.

By the close of the 1960s, the St. John renewal project faced an uncertain future. After waiting ten years for property acquisition, North Enders
increasingly viewed the city's promises of relocation with skepticism.[38] And
they had many sound reasons for feeling that way. Originally Interstate 475
was to open in 1972, but between 1966 and 1974, the MSHD pushed back
the completion date on at least five different occasions. In December 1966
property acquisition for both M-78 and I-475 virtually halted after the federal
government announced a forty-seven-million-dollar cut to the state's transportation budget in order to fund the Vietnam War. One year later, the state
announced a three-year delay due to inflation and soaring construction costs.

Additionally, state and federal investigations into the environmental impacts of the roadway virtually halted North End property acquisition and construction between 1972 and 1976. By 1974, two years after the original completion deadline, the state had completed only the southern and northern portions of I-475, leaving open the 5.5-mile section of the freeway that was to serve downtown, Buick, and the St. John Industrial Park.[39] To many of those who remained stranded in St. John after over a decade of renewal promises, the city's inability to relocate residents seemed willful. James Sharp, who went on to become Flint's second African American mayor in 1983, excoriated city leaders for the delay, maintaining, "The city designated those urban renewal areas. And they did so, I believe, with full knowledge that they would never ever buy them out."[40] Already simmering, the contradictions between the politics of growth and the quest for fair housing came to a head during the battles over the delayed projects.

Without question, financial assistance from GM could have expedited the I-475 and St. John projects. Yet when pressed by city officials to donate cash for the clearance and relocation efforts, GM officials refused.[41] For some this was puzzling, especially since GM had spent decades advocating for the clearance of St. John. Initially representatives of the company looked to redevelopment as a means of acquiring eighty acres of North End land for factory expansions, employee parking lots, rail facilities, and a power plant. However, during the long delay GM purchasers were able to acquire most of the land they desired through privately orchestrated agreements with North End property owners. With its short-term property acquisition agenda largely complete, the company refused to bankroll the more expensive St. John and I-475 projects, preferring instead to fund the GCDC's private housing and renewal efforts while quietly pressing for further government action in St. John.[42] Lacking the necessary cash and public housing capacity, the city delayed St. John property acquisitions until 1970.[43]

During the long delay, GM executives and city commissioners presided over the almost complete destruction of the North End. In their 1963 report *Flint Faces the Future*, officials from the National Association of Real Estate Boards had warned city leaders that a premature announcement of clearance plans could trigger the spread of blight in neighborhoods designated for redevelopment. The key to mitigating neighborhood deterioration, NAREB representatives maintained, was to establish small project areas and minimize the time between the planning and implementation phases of urban renewal. "To designate more than a thousand acres in the heart of Flint for renewal treatment at a future date that is only vaguely defined serves to put the blight tag on a major section of the community," they argued. "The result

FIGURE 7.1. Aerial view of downtown Flint and freeway construction projects, 1972. An unknown photographer from the Michigan State Highway Department captured this image of downtown Flint (background) and the partially completed Interstate 475 (foreground) and Floral Park interchange (right) projects in 1972. The delayed completion of the city's freeway system frustrated tens of thousands of local residents, including many African Americans who were desperate to move away from the Floral Park and St. John neighborhoods. Once finished, Flint's new freeways and interchanges drew criticism for hardening the color line and hastening the process of deindustrialization. Courtesy of the *Flint Journal.*

of such action in other cities has been that owners defer maintenance, cancel plans for new construction, and put off improvements because of uncertainty over what will happen to their property."[44] Despite this and other warnings, however, local policy makers and administrators from the DCD did little to prevent the deterioration of the North End.

State Action and Ghetto Formation

The I-475, interchange, and St. John renewal projects never met the expectations of civil rights activists. In fact, at virtually every stage of their development, the city's revitalization initiatives intensified black poverty while generating new forms of administrative and popular segregation. Although the city had first announced its renewal plans in 1960, it was not until 1970 that elected officials produced the finances and the public housing necessary to gain federal approval for St. John Street redevelopment. Already well worn by 1960, the St. John Street of the early 1970s was among the most polluted and unlivable neighborhoods in the nation.[45]

Between 1965 and the early 1970s, officials from the urban renewal department and the MSHD assigned members of the Flint Board of Real Estate and the Society for Residential Appraisers in Flint, both all-white organizations, to conduct property appraisals throughout the city. In deference to federal guidelines, appraisers calculated reimbursements not based upon the actual relocation costs for families but rather by assigning "fair market values" to properties. Consequently, appraisers judged the overall condition of a neighborhood in determining the value of individual homes and businesses. The number of abandoned structures nearby, the quality of a neighborhood's schools, and other factors beyond the control of individual property owners thus played key roles in determining property reimbursements.[46] The policy of assigning fair market values to already condemned structures effectively penalized homeowners and shopkeepers—many of whom had invested substantially in their properties—for the larger processes of inner-city disinvestment.[47] Appraisers further handicapped local property owners by relying heavily on external appearances when assaying the market value of homes and businesses.[48] Already stretched financially by high-interest land contract payments—and largely excluded from the home improvement loan market—black property owners tended to invest in basic necessities and structural repairs at the expense of curbside appeal.[49] Yet appraisers offered little in return for such improvements. In 1965, for instance, homeowner Cecilia Ryan, a fifty-year-old seamstress from Floral Park, received an appraisal of just $6,500 for her residence despite the fact that she had purchased the home in 1954 for $6,900 and had invested thousands of dollars to make structural repairs and install new insulation.[50] As planner William Walsh noted in a 1967 study of North End housing, exterior inspections exaggerated the extent of structural decay, further depressing appraisal values: "We have a situation then in which properties in the area are declining in value because they are being negatively perceived by the market while virtually all of the people are making major expenditures to improve the homes."[51]

Municipal officials generally responded unsympathetically when black property owners challenged the appraisal program. After receiving petitions from a large delegation of Floral Park residents in the spring of 1966, City Manager Thomas Kay asserted, "Low property values for some properties are the result of injudicious purchases of property at inflated prices by the owners."[52] In February 1969 six hundred North Enders descended on city hall to protest urban renewal appraisals. During the demonstration, Fred Dent of the St. John Citizens for Improvement Association (SJCIA) addressed the inequities of fair market appraisals: "The majority [of North End residents] have kept their homes up despite the difficulty imposed by the damaging smoke from the Buick plant, and now they are being asked to start anew. If they must do so, they should receive enough money to buy another decent home in a convenient location."[53] While still expressing a preference for re-settlement, leaders from the SJCIA, the NAACP, local block clubs in Floral Park and St. John, and hundreds of ordinary citizens continued to fight from within the renewal coalition for more equitable relocation policies.

However, activists in Flint, like their counterparts in Detroit, New York, Atlantic City, New Jersey, and other places, consistently failed in their at-tempts to negotiate with city hall.[54] In truth, throughout the 1960s and 1970s municipal policies helped to exacerbate blight in Floral Park and St. John. Yet appraisers consistently overlooked the city's role in depreciating property values. Starting with the release of the 1960 master plan until the end of North End property acquisition in 1977, residents of St. John experienced an espe-cially harsh form of municipal disinvestment. Because the master plan had prescribed complete clearance for St. John, city commissioners sharply cur-tailed housing code enforcement, waste collection, and other basic services there beginning in the early 1960s.[55] City commission members formalized their St. John disinvestment policy in November 1965 by passing a unani-mous resolution instructing residents "not to spend large amounts of money in rehabilitation of their properties, but only those repairs necessary for or-dinary maintenance."[56] The policy resulted in rapid neighborhood deteriora-tion and infuriated thousands of St. John residents. "Rather than renewing the [North End] area," a Flint Urban League member asserted, "it appears that the city of Flint is stopping everything except decay."[57]

Municipal policies, corporate practices, and private disinvestment under-mined the entire North End housing market during the long era of urban renewal. In the years following the 1965 moratorium on building repairs, homeowners and shopkeepers in the North End reported receiving notifica-tion that insurance companies had simply canceled their policies.[58] For their part, local bankers responded to the commission's decree by refusing to fi-nance even basic repairs for homes in St. John. In effect, then, urban renewal triggered a second era of redlining in the North End. As one anonymous resident noted, "We are faced with a situation where you can borrow money for a Cadillac but you can't get it to fix up homes."[59] Just between 1972 and 1973, housing values in the city as a whole declined by over 10 percent, a drop fueled in large part by the collapse of the North End real estate market.[60] In a 1973 communication to US Senator Donald Riegle Jr., Olive Beasley described how government policies had devastated the North End: "Citizens have been waiting up to ten years, unable to make changes in their homes, get building permits or improve their living conditions. The promised extensive rehabilitation of homes, even just to bring them up to code, has been stopped. Property values do not remain constant. . . . Here," Beasley predicted, "the U.S. government actions over the past years and the current delay, will further reduce minority property values."[61] Fearful of investing even small sums of money in condemned properties, homeowners and shopkeepers in the St. John area had little choice but to allow their properties to deteriorate.

On December 22, 1970, the Flint Department of Community Development purchased its first property in St. John, a home at 1335 Rhode Island Avenue. After receiving a check for seven thousand dollars, the seventy-five-year-old homeowner, Beneva Smith, relocated to the Flint Housing Commission's Centerview Apartments.[62] By 1977, when the city concluded property acquisition in the neighborhood, hundreds of black families had followed Smith's path from home ownership to public housing. Many business owners simply closed for good following property acquisition. "I'm 52 years old," lamented Joe's Tavern owner Walter Christich, "and right at the stage where there's no sense in going into debt. My building is [owned] free and clear and I'm being put out."[63] The process of relocation was especially difficult for Christich and other North End property owners because appraisals there often matched those assigned to vacant parcels of land in white neighborhoods.[64] According to DCD reports from the early 1970s, property acquisition payments in St. John averaged less than $8,000 per home, yet citywide the median cost of private replacement housing in 1970 was $14,600.[65] As Congressman Riegle noted, the property acquisition processes made it all but impossible for St. John residents to secure adequate replacement housing: "Because of the undesirable conditions for residential living, the appraised value of these parcels is not very high. Particularly when it is necessary to convert this appraised value into safe and sanitary housing in some other location. It is virtually impossible."[66] For a large portion of residents in St. John and Floral Park, many of them homeowners, public housing was the only viable option.

After turning over their properties and receiving their payments, displaced families faced the difficult prospect of finding new housing. To assist them, the Department of Community Development, the state highway department, the Mott Foundation, and the GCDC jointly operated the city's Central Relocation Office.[67] Although all of Flint's displaced persons were eligible to seek assistance in finding new homes, very few whites ever visited the CRO. Instead, most whites displaced by the freeway relocated to segregated

private housing in suburban Flint, predominantly in the high-growth sections of southern Genesee County, where new subdivisions seemed to sprout
overnight.[68] For African Americans, on the other hand, the relocation process
was onerous.[69] Part of the difficulty stemmed from the fact that representatives of the CRO consistently ignored the issue of racial segregation, preferring to manage relocation "with the hope that the question of integrated
housing would best be solved without focusing on the question."[70]

An additional set of problems derived from the close collaboration between the MSHD, the CRO, and the local real estate board. From the outset,
representatives of the highway department and the relocation office categorically shunned the services of black real estate brokers in favor of white agents
from the segregated Flint Board of Real Estate.[71] With very few exceptions,
members of this exclusive board—who supplied the CRO with home listings from their multiple listing exchange—refused to show homes in white
areas to black realtors and purchasers. In a 1966 memorandum to the state
civil rights commission, relocation officials indicated that the Flint Board of
Real Estate provided listings only in "predominantly Negro or integrated
neighborhoods" for African Americans displaced by the freeway.[72] After a
long string of unsuccessful attempts to purchase replacement housing for
her family, Mrs. Robert R. Turpin of the Flint Human Rights Commission
complained, "Just in recent months . . . at least eight real estate agents have
denied us the privilege to buy, and most of these have refused to grant us even
a chance to inspect the home they advertised for sale."[73] Though made aware
of the situation by state civil rights commissioners, the Flint city attorney's
office did not enforce the municipal fair housing ordinance on behalf of displaced families. Moreover, officials from the US Department of Housing and
Urban Development (HUD) declined to offer protection under the federal
Fair Housing Act of 1968, which outlawed the refusal to sell, rent, or finance
dwellings based on race. In a remarkable forfeiture of authority, Chantland
Wysor, an official from HUD's Chicago office, informed the Michigan Civil
Rights Commission that his supervisors had reassigned civil rights officers
from Flint and nearby Saginaw to other higher-profile cities: "Flint-Saginaw
will have to struggle along without my help. . . . I would suggest that you
ignore everything I have said and do your eagle-eyed best to watch our programs yourself."[74]

Unable to create wealth through property ownership, more than half of
the displaced families in the city required public housing to meet their relocation needs. To accommodate the three thousand impoverished families
slated for relocation, city commissioners had first established the Flint Housing Commission in 1964.[75] Initially FHC members planned to construct five

hundred public housing units, all of them near the Floral Park interchange.[76] Citing the high cost of land in other areas, the FHC refused to consider developments in or near the all-white neighborhoods of the far west and south sides.[77] Yet after significant pressure from the MCRC and the NAACP, the commissioners eventually abandoned their original proposal.[78] Instead, they designed a plan for family housing that culminated in one project in Floral Park (Howard Estates), another in the North End (Aldridge Place), two additional complexes in isolated, largely unsettled areas on the urban fringe (the Atherton East Apartments and the River Park Apartments), and 183 units of scattered-site housing, almost all of which were within all-black or racially transitional neighborhoods. At the Floral Park and North End sites, the commissioners ensured segregation by locating public housing within heavily black urban neighborhoods, a common practice employed by housing authorities in Chicago, St. Louis, and elsewhere.[79] But with the two site locations on the urban fringe, FHC policies created islands of administrative segregation on the city's most remote borders.[80]

In 1977, when property acquisition concluded in St. John, the FHC conducted a census of public housing developments. During that year the ninety-six-unit Howard Estates project in Floral Park housed only one white family. In the River Park complex—built in a wooded, largely unsettled area near the city's northern border—FHC surveyors counted only eleven white families among the residents of its 151 units. At Atherton East—located in a sparsely inhabited district on Flint's southeast side—141 of 188 families were African American. To accommodate the seven hundred children who moved to Atherton East from the city's renewal zones, the Flint Board of Education opened a brand-new facility called the East Atherton School. Later renamed after Frank Manley, the founder of community education, the new building sat immediately adjacent to the public housing complex. Perhaps fittingly, Manley School was one of the most racially segregated and socially homogeneous educational institutions in the entire region.[81] When confronted by activists who objected to the site locations and racial composition of these and other new developments, urban renewal director Leo Wilensky argued, "If I am worried about whether segregation will re-occur in these projects, it will paralyze our present efforts to relocate." Without hesitation, Wilensky and other city officials vehemently defended Flint's rigid system of administrative segregation.[82]

Facing both long waiting lists for public housing and severe private housing shortages, activists from Floral Park and St. John could not mount an effective opposition to the FHC's site plans. By the mid-1970s even urban renewal proponents from the partially cleared St. John district, fearful of finding

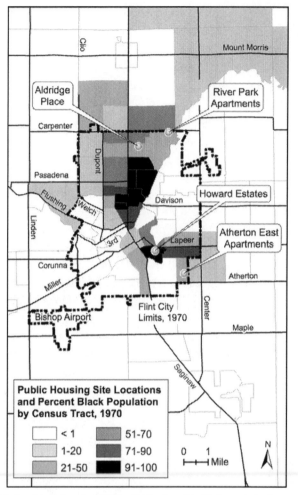

MAP 7.1. Public housing site locations and the distribution of the black population by census tract, Flint and vicinity, 1970. In order to accommodate the thousands of local families displaced by urban renewal during the 1960s and 1970s, members of the Flint Housing Commission oversaw the construction of four public housing complexes. Although many African Americans had initially supported neighborhood clearance and relocation, the site location policies of the FHC ultimately intensified administrative forms of school and residential segregation. *Sources*: Flint Housing Commission; Minnesota Population Center, *National Historical Geographic Information System: Version 2.0* (Minneapolis: University of Minnesota, 2011); Michigan Center for Geographic Information, *Michigan Geographic Framework: Genesee County* (Lansing: Michigan Center for Geographic Information, 2009).

no replacement housing at all, had turned against Flint's redevelopment program. In 1974 Howard Simpson, president of the St. John Council, lamented the reversal, noting, "We regret our antiprogressive posture in this attempt to save an area that, frankly, is not worth saving. But this is the only neighborhood we have."[83] In the end, all but a few black families in the path of the

FIGURE 7.2. Children playing in front of the Atherton East Apartments, 1967. Officials from the Flint Housing Commission ordered the construction of Atherton East and several other public housing complexes to accommodate displaced persons from Floral Park, St. John, and other urban renewal areas. The inauguration of Flint's public housing program marked a major expansion of administrative segregation. Courtesy of the *Flint Journal.*

freeway moved to neighborhoods surrounding Flint's demolished ghettos. Among the St. John residents displaced by urban renewal, home ownership rates dropped from approximately 50 percent to 15 percent.[84] Reflecting back on the history of slum clearance, fair housing sleep-in leader Woody Etherly Jr. expressed profound regret for his support of redevelopment: "That's one of my greatest disappointments in life . . . not having a true understanding of what the country was doing when they did the urban renewal or when they d[id] expressways."[85]

Despite initial support for slum clearance among civil rights activists, the I-475 and St. John renewal projects ultimately increased black poverty and residential segregation. The new social geographies that emerged because of urban renewal were not merely cases of de facto segregation, however. Rather, urban renewal produced a policy-driven web of administrative segregation that helped to make Flint one of the most racially divided, economically polarized, and spatially fragmented cities in the United States.[86] By the late 1970s Olive Beasley had come to realize that urban renewal was really a program of "Negro Removal"—an insidious plan to expand urban commercial, in-

FIGURE 7.3. Olive Beasley, 1972. After spending much of her adulthood working within various government agencies and civil rights organizations, in 1964 Beasley accepted a position with the Flint office
of the Michigan Civil Rights Commission. From then until her retirement in 1980, she served the commission by investigating civil rights complaints from Flint-area residents. Despite her affiliation with the
government, Beasley was an outspoken critic of administrative segregation. Courtesy of the Alfred P.
Sloan Museum, Flint, MI.

dustrial, and tourist economies by removing black residents from valuable
inner-city land. In 1976, three years after she had charged Flint's municipal
government with maintaining policy-driven ghettos, Beasley authored another blistering memorandum regarding redevelopment. This time, though,
Beasley indicted the entire federal renewal program, writing, "The whole history of urban renewal . . . has resulted in a history of black removal from
low income substandard housing in areas that are eventually redeveloped for
higher income residents."[87] In hindsight, Beasley's claim seems all but self-
evident. Yet as the Flint case suggests, it was not always apparent to civil rights
activists who initially supported redevelopment that urban renewal would

end up fortifying Jim Crow. Nor was it clear that the process of "Negro Removal" and resettlement would have a powerful ripple effect in other parts of the city and region. But that, in fact, is precisely what occurred.

Exodus

During the years preceding the redevelopment of St. John and Floral Park, local proponents of renewal echoed the claims of renewers nationwide in arguing that the demolition of "blighted" areas would help lure new residents and investors back to the city while stabilizing transitional neighborhoods. In reality, though, slum clearance only hastened the movement of whites to the suburbs.[88] As displaced persons from St. John and Floral Park searched for new homes in previously all-white sections of the city, neighborhood battles over race and real estate exploded throughout the Vehicle City and its innermost suburbs. Although whites initially fought to defend their segregated neighborhoods, most eventually left the city, leaving behind thousands of vacant and boarded-up properties.

As was the case in Chicago, Atlanta, and most other cities, racial succession in Flint often occurred in a linear, block-by-block progression.[89] On the south side of the city, displaced families from Floral Park tended to move along a southeasterly axis toward Elm Park, Sugar Hill, Evergreen Valley, and other nearby neighborhoods. In the North End, where racial transitions were more rapid and widespread, black population expansion during the 1960s and 1970s followed a northwesterly route toward Flint Park, Civic Park, Manley Village, Forest Park, and other neighborhoods west of Saginaw and Detroit Streets.[90] Upon occupying their new homes, black residents of previously all-white blocks often faced hostility from neighbors. In May 1967, for example, Mr. and Mrs. Charles Bell, an African American couple with several children, purchased a home at 3521 Sterling Street in a nearly all-white northwest side neighborhood close to Forest Park. Several days after moving into their home, the Bells discovered that unidentified hoodlums had spread toilet paper around their lawn and smeared butter and ketchup on their windows and doors. Throughout the late 1960s and 1970s, Beasley and other members of the Michigan Civil Rights Commission fielded scores of complaints regarding similar acts of racial harassment and popular segregation in northwest and southeast side neighborhoods.[91]

As black families poured out of St. John and Floral Park, local realtors launched a number of blockbusting campaigns within the city. Although panic peddling left few Flint neighborhoods untouched, the northwest side took the brunt of the assault. By the early 1970s fights over blockbusting had engulfed virtually every neighborhood in the city west of Saginaw Street. Noted one white resident who moved away from Flint Park in 1972, "I guess if I had to call it anything I'd say it felt like we were invaded."[92] Cognizant of the speed with which Evergreen Valley had transitioned to a black majority, white homeowners in Flint Park, Manley Village, and other northwest side neighborhoods vowed to resist panic peddlers if and when they arrived in their communities. Leading that effort were hundreds of northwest side residents who expressed their opposition to blockbusting by displaying signs reading "We're Staying" in their front yards.[93]

Other activists turned to block clubs and neighborhood associations. During the late 1960s and early 1970s, residents formed approximately 250 such organizations within the city.[94] Though they proliferated throughout Flint, these clubs were most common in the racially transitional neighborhoods of the north and northwest sides. Pitting "those who care against those who don't," as a reporter from the *Flint Journal* claimed, block clubs engaged in hundreds of neighborhood improvement and stabilization projects. Among their many campaigns, neighborhood associations and block clubs sponsored clean-up drives, home repair projects, and community watch patrols.[95] As diverse as the neighborhoods and residents they served, the agendas of block clubs reflected the grassroots concerns of residents. Some groups such as the Nosy Neighbors Block Club, which operated in the nearly all-white Glendale Hills neighborhood on Flint's west side, focused almost exclusively on crime and neighborhood watch activities.[96] Other groups were far more active and organized protests against slumlords, pressed city officials to improve municipal services, challenged builders to erect safer homes, and battled blockbusting realtors.[97]

Despite their best efforts, block club members were unable to stem the white exodus from racially integrated neighborhoods. On May 11, 1975, the *Flint Journal* published an article by Michael J. Riha describing the impact of blockbusting on local residents and neighborhoods. The article, which detailed African American homemaker Rosaline Brown's unsuccessful quest to find integrated housing on the northwest side, spoke directly to the many problems that black people throughout the nation encountered as they transgressed the color line.[98] In 1973 Brown, her two young daughters, Timeka and Danielle, and her husband Gary, a sheet metal worker, moved into a home on Greenlawn Drive in the nearly all-white Manley Village section of Flint's northwest side. Like other black pioneers, Brown hoped that her move to Manley Village would break down racial barriers. "I wanted to bring up my kids in an interracial setting," she stated. "I felt that the only way to improve this world is to get to know people as persons." The Brown family's experiment in integrated living proved to be brief, however. Just after they arrived in the neighborhood, realtors began a door-to-door campaign to convince white homeowners to sell their properties to black buyers. "They were knocking on doors, leaving their cards and telephoning," Brown recalled. By the spring of 1975, all but four of Brown's twenty-four white neighbors on Greenlawn Drive had sold their homes and moved away, many of them to segregated suburbs. "I could sit here and look up the street and see nothing but for sale signs," Brown reported.[99]

On May 15 the *Flint Journal* ran a follow-up story about popular segregation in Manley Village. This time, though, Riha approached the issue from the vantage point of Patricia Montpas, a white mother of four who hoped to sell her house. Montpas and her husband Michael, a GM factory worker, had lived in their Manley Village residence for ten years, purchasing their home in 1965 with a low mortgage interest rate of approximately 5 percent. Although by 1975 mortgage rates had nearly doubled to 9 percent, the Montpases still hoped to relocate. Like so many white Americans who departed big cities during the 1970s, Patricia Montpas cited increased crime, inferior schools, and declining property values as her primary reasons for wanting to move.[100] "It's just starting to be overwhelming," she declared, "where I'm afraid to let my kids out of my sight." Echoing Rosaline Brown, Montpas claimed that unscrupulous realtors, white and black, fueled panic selling in Manley Village. On one occasion Montpas met with a black realtor who advised her to sell her house immediately. "If you don't want to lose everything, you'd better get out as fast as you can," she remembered the realtor warning. "You know, the only thing moving in here is black trash." By exchanging a 5 percent home loan for a new mortgage at 9 percent interest, the Montpases paid a premium for their decision to sell. Yet the cost seemed well worth it to Patricia Montpas and thousands of other whites who left Flint during the 1960s and 1970s.[101]

Like millions of Americans, the Browns and the Montpases felt a compulsion to seek out new neighborhoods during the 1970s. By acquiring new homes, both Patricia Montpas and Rosaline Brown sought to purchase safety and security for their families, better schools, and a wealth-creating investment in real estate. But there were push factors at work too. In Montpas's view, Manley Village no longer felt like home because residential integration brought higher crime rates, frayed community ties, and deteriorating property values. To Brown, however, Manley Village was appealing precisely because of its racial diversity. While the local housing market afforded the Montpases the option of relocating to a more racially homogeneous neighborhood, the quick transformation of Manley Village left Rosaline Brown wondering whether she could ever purchase integration in the Vehicle City.

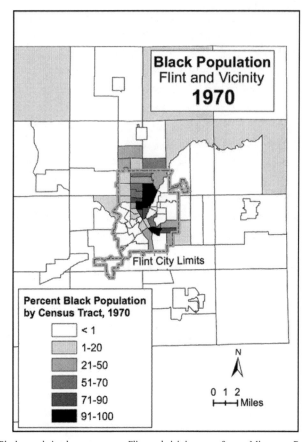

M A P 7 . 2 . Black population by census tract, Flint and vicinity, 1970. *Source*: Minnesota Population Center, *National Historical Geographic Information System: Version 2.0* (Minneapolis: University of Minnesota, 2011).

In the end, the 1970s campaigns in support of neighborhood racial stability were unsuccessful. Between 1970 and 1980, the number of whites in Flint declined from 138,065 to 89,470 while the city's black population increased from 54,237 to 66,164. By the close of the decade, the city's population was over 40 percent African American.[102] Over the same period, cities throughout the United States experienced similar demographic transformations as millions of urban whites decamped for the suburbs.[103] On Flint's northwest and southeast sides, the areas of the city that saw the largest black population increases, the demographic transformations were dramatic. Just between 1970 and 1975, the black population in census tract 42, which included the city's far northwest corner, increased from 13 to 71 percent. During those same years the black population of tract 2 on the city's northern border with Mt. Mor-

ris Township rose from 48 to 81 percent. Virtually every neighborhood in
the northwest side of the city posted similar figures. The white exodus was
equally intense on the southeast side, particularly in the neighborhoods
stretching east from the Floral Park interchange. In tract 45, for instance, the
African American proportion of the population jumped from 33 to 78 percent
during the first half of the 1970s.[104] Desperate to escape their changing neighborhoods, many white homeowners moved away before they could sell their
houses, causing a sharp rise in home vacancies and vandalism. At decade's
end, nearly 10 percent of Flint's homes sat unoccupied; in some areas of the
northwest side, vacancy rates surpassed 20 percent.[105] To Jack Litzenberg, director of community development programs in the city of Flint, it seemed as
though whites had simply abandoned the city: "While nearly everyone who

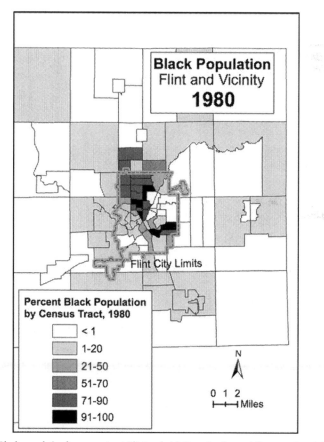

MAP 7.3. Black population by census tract, Flint and vicinity, 1980. *Source:* Minnesota Population Center, *National Historical Geographic Information System: Version 2.0* (Minneapolis: University of Minnesota, 2011).

is able flees the city, we are left with the poor, the elderly, blacks, and other minorities."[106] During the course of the decade, dozens of neighborhoods once off limits to African Americans quickly became predominantly black and poor. The For Sale signs, empty homes, boarded-up stores, and waist-high lawns that departing whites left behind in Flint and hundreds of other American cities were semiotic manifestations of one of the most significant human migrations in the nation's history.

The late twentieth-century battles over renewal, race, and residence that claimed nearly all of Rosaline Brown's neighbors in Manley Village angered citizens on both sides of the color line. As she looked out at the For Sale signs that had sprouted all around her home, Brown felt insulted that so many neighbors feared her presence. Equally upset, many whites in racially transitional neighborhoods were angry at what seemed to them like an invasion of African American home seekers. Back in July 1968, just as the struggles over desegregation were beginning to flare near her home in northwest Flint, an unnamed local resident had articulated the frustrations that many whites across the country felt as their neighborhoods changed. In a letter to Flint NAACP president Edgar Holt, she asked, "Why is it there is so much talk about the Ghetto? . . . Still the Colored people make it a point to flock into one neighborhood. They have a right to go any where they please—I am white and have no objection to a Colored neighbor (some of my nicest friends are Colored) but I do not like the idea of living in a 'Ghetto.' Must they all live in the same block?"[107] Clearly annoyed, the letter writer wondered why African Americans, in a city covered by so many fair housing statutes, continued to segregate themselves by race.

  With neither a name nor a return address posted on the correspondence, Holt never replied to the letter. But had he done so, he might have delved into the long, complex history of the residential color line in twentieth-century Flint. He could have pointed to the racially restrictive housing covenants that once excluded African Americans from white neighborhoods; the discriminatory practices of builders, lenders, realtors, and homeowners that limited choices for black home seekers; the public policies that helped to build ghettos first in St. John and Floral Park and later in the city's numerous public housing developments; or the racial violence that helped to enforce Jim Crow. Holt might also have reminded the letter writer that federal mortgage insurance policies, one of the primary engines of America's postwar economic boom, had effectively enshrined residential segregation as official government policy during the middle decades of the twentieth century. Holt's imaginary reply need not have been entirely historical, however. The veteran activist also could have pointed out that both popular and administrative forms of residential segregation such as blockbusting, racial steering, and even redlining persisted long after the demise of legal Jim Crow and the enactment of municipal, state, and federal open occupancy laws.[108]

Although the public housing complexes and vacant neighborhoods that dotted the city's landscape provided some of the most compelling evidence of the durability of the color line, there were also countless examples of inequality in evidence outside Flint's boundaries. In fact, one of the most striking instances of administrative segregation from the 1970s took place not in St. John or Floral Park but in the tiny northern suburb of Beecher. There, local and federal officials implemented a controversial "subsidized" housing program that hardened the color line and laid the groundwork for the subprime lending crisis of the early twenty-first century. Just as urban renewal policies in St. John triggered racial succession in northwest Flint, the concentration of affordable housing just to the north of Flint's municipal boundary set off a devastating wave of home foreclosures, blockbusting, and social conflict in Beecher. By the close of the 1970s, the intense contestation over low-income housing in the Beecher district had helped to push Flint's burgeoning urban crisis into the formerly all-white suburbs of Genesee County.

Suburban Crisis

In Flint, the state of Michigan, and the nation as a whole, 1968 was a momentous year for the fair housing movement. Two months after Flint's open occupancy referendum, President Lyndon Johnson signed the landmark Fair Housing Act of 1968, which prohibited many forms of residential discrimination. Shortly thereafter, Governor George Romney of Michigan followed suit, signing a tough open housing bill. By the summer of 1968, African Americans in Flint could cite three separate pieces of legislation that prohibited racial discrimination in housing.[1] Despite such legal protections, however, popular and administrative forms of segregation continued into the 1970s, 1980s, and beyond. The sheer inefficacy of fair housing legislation became readily apparent in the spring of 1976, when representatives of the Flint City Council, formerly named the Flint City Commission, held several well-publicized hearings on mortgage redlining.[2] Following the hearings, a specially appointed task force produced a report confirming that redlining and mortgage disinvestment, though no longer occurring in most of Flint's suburbs, continued to plague the North End and other predominantly black sections of the city. According to the report, only about 10 percent of the home mortgages issued in the Flint region during the 1970s were for properties located in the city. The remaining loans supported home purchases in outlying areas, particularly those in the booming all-white suburbs of southern Genesee County and other areas where voters had embraced the tenets of suburban capitalism. "After careful analysis," the report stated, "the Task Force is convinced that mortgage disinvestment is occurring in Flint."[3]

Missing from the report, however, was an acknowledgment of the significant changes in the mechanics of residential racial discrimination that had occurred since the fair housing revolution of 1968. Chiefly, the report failed to acknowledge the destructive wave of predatory mortgage lending that swept across the Flint metropolitan region and the nation at large during the late 1960s and early 1970s. Triggered by the implementation of an obscure federal housing program known as Section 235, this boom in predacious real estate practices illustrated that government housing "subsidies" could be just as discriminatory and injurious as redlining. At the same time, the battle over Section 235 highlighted important demographic transformations that were occurring in some of the formerly all-white suburbs of Flint and other American cities.

Section 235 was a novel government finance program for low-income home purchasers. It came into being just a few months after the passage of the Fair Housing Act as part of the federal HUD Act of 1968. Together, these two federal laws helped to stimulate the nation's flagging real estate industry while opening up many sectors of the private housing market to African Americans, the poor, and others with limited access to credit.[4] Between 1968 and 1973, when President Richard Nixon ordered a moratorium on all "subsidized" residential developments, Section 235 funds supported the sale of 465,972 new and used homes nationwide.[5] In Flint and the surrounding communities of Genesee County, nearly fifteen hundred poor and working-class families enthusiastically participated in the program.[6]

Popular support for the policy eroded quickly, however, as purchasers began discovering major flaws in their homes. Beyond these individual complaints, decisions regarding the locations of the new units provoked a torrent of discontent among homeowners and politicians, many of whom blamed Section 235 for creating federally subsidized ghettos in Flint's North End. By the early 1970s the battles over race and housing had spilled over Flint's northern border into the inner-ring suburb of Beecher, which received far more than its share of low-cost homes. There, Section 235 hardened administrative segregation, stimulated a rash of property abandonment and home foreclosures, and inflamed racial tensions. For a brief period during the early 1970s the strife over Section 235 turned Beecher into a focal point in the national debate over suburban desegregation. As part of that dialogue, local residents grappled with the implications of predatory lending and mass foreclosure in ways that foreshadowed the early twenty-first century outcry over subprime mortgages and "reverse redlining."

By the early 1970s the tiny Beecher district had come to resemble the predominantly black deindustrialized neighborhoods of Flint's inner city more than the all-white middle-class communities of the suburban cliché. Nevertheless, there were very clear limits on how far the color line and Beecher's gathering suburban crisis would spread. Indeed, among suburbs in Genesee

County the situation in Beecher proved to be somewhat exceptional. In the wake of the Section 235 saga, homeowners and other proponents of suburban capitalism in the remainder of Genesee County consolidated their independence from the city and worked to harden the color line through a combination of restrictive zoning and land use plans, municipal incorporation initiatives, and other strategies. During this era of suburban secession, public and private housing and development policies helped to insulate most of Flint's suburbs from the racial and economic diversification that was occurring in Beecher.

Working-Class Suburb

The so-called Beecher Metropolitan District occupies five square miles of land immediately north of the Flint city limits. Bounded by Carpenter Road, Dort Highway, Stanley Road, and Clio Road, this unincorporated utilities and school district is an urbanized suburb superimposed on portions of Mt. Morris and Genesee Townships. Because of Beecher's proximity to Flint, the two areas have often shared a similar historical trajectory. As Flint grew to prominence in the first half of the twentieth century, Beecher quickly became a popular destination for migrant workers seeking jobs at GM. Workers settled in the district for a variety of reasons. Many of them preferred Beecher to the city because it seemed to combine the best elements of urban and country living. Homeowners there could plant large gardens, raise chickens, and breathe moderately clean air while still commuting to their jobs in Flint in a matter of minutes. Moreover, because land in Beecher was cheaper than in the city, working-class families could buy lots and erect new homes there without exhausting their savings. The district's lenient building code even allowed cash-conscious residents to construct their own dwellings if they preferred.[7] Consequently, self-building was a common practice, with owner-built homes composing over half of Beecher's housing stock as late as 1960.[8]

By 1950 over seventy-five hundred people lived in the district, but that number grew quickly as officials built new roads and schools and established modern utilities. Policy makers from the FHA responded to those investments during the 1950s by insuring a number of home mortgages in Beecher, which brought additional growth to the area. The fact that government agents and lenders collaborated to exclude black buyers from participating in these federal mortgage programs only enhanced Beecher's desirability among most whites.[9] Even a major tornado strike that devastated a large section of the community in 1953 could not halt Beecher's postwar boom. Directors from GM did their part to fuel the district's rise by opening the massive Ternstedt parts plant on Coldwater Road in 1953. Overnight, GM's new facility helped to triple the community's tax base. The new plant also helped to push Beecher's population to nearly fifteen thousand.[10]

A combination of racially restrictive housing policies and popular segregation had kept Beecher virtually all white prior to the 1950s. However, an influx of black migrants during the late 1950s helped to make it one of the only racially integrated communities in suburban Flint. By 1960, 10 percent of Beecher's residents were African American. Even though racial violence flared in the district periodically when black families moved in, many African Americans seeking to leave the North End still looked to Beecher as a prime destination.[11] Much like their white counterparts in the district, African Americans selected Beecher because of its proximity to Flint and its reputation for affordable housing. Equally important, Beecher was one of the only communities outside of Flint where realtors—some of them acting as blockbusters, no doubt—willingly showed homes to African Americans.[12] This combination of forces helped to drive Beecher's black population to new heights in the years preceding the implementation of Section 235. Out of the twenty-five thousand mostly working-class residents who lived there in 1968, one-third were African American.

Virtually indistinguishable from the Flint neighborhoods that formed its southern border, Beecher was a suburb that looked much like the city. In the words of *Flint Journal* reporter Gene Merzejewski, it was "a chunk of transplanted city" in the otherwise lily-white suburbs of Genesee County.[13] Unlike its urban neighbor, though, the Beecher of the postwar decades seemed to be an unlikely breeding ground for social unrest. It withstood the "long hot summers" of the mid- and late 1960s without undergoing any major racial conflagrations. Furthermore, Beecher appeared to be a stable, integrated community, or at least a place where whites and blacks could coexist without violence.

Suburban Ghetto

Beneath that veneer of tranquillity, however, a crisis brewed as Beecher's racial demography shifted. The first clear signs of that crisis emerged as part of a revolt against low-income housing unleashed by the passage of the HUD Act of 1968. Section 235 of the HUD Act supported the construction and rehabilitation of low-cost housing and home purchases by poor and working-class buyers. The HUD law also provided funds for developers of affordable rental properties through an initiative known as Section 236. Under Section 235, members of Congress authorized the expenditure of $5.3 billion to enable the sales of up to two million affordable housing units. The program offered federal mortgage insurance to private lenders and developers who either rehabilitated existing housing units or constructed new low-cost homes to sell to poor buyers. The coverage offered through Section 235 was virtually identical to that provided through the selective credit mortgage insurance programs operated by the FHA and the Veterans Administration.[14] Unlike the recipients of such loans, however, Section 235 borrowers received special assistance from the federal government to pay their mortgages. But this support did not come in the form of cash grants and other demand-side subsidies to buyers. Instead, the federal government disbursed mortgage principal and interest payments directly to lenders and developers who backed and built low-cost housing. The funds for these corporate subsidies came from the revenues of the Government National Mortgage Association, popularly known as Ginnie Mae, a new public corporation created under the terms of the HUD Act. As part of its mission to expand the nation's affordable housing supply, Ginnie Mae purchased Section 235 mortgages from local lenders, reassembled them as mortgage-backed securities, and then sold them on the secondary mortgage market. By doing so, federal officials hoped to offset the cost of HUD's mortgage principal and interest payments while stimulating local real estate markets with an infusion of global capital.

Beyond members of the real estate industry and global investors, who stood to benefit immensely from the flow of new funds, the program targeted low-income Americans who were ineligible for either public housing or the selective credit loans backed by the FHA and the Veterans Administration.[15] Under Section 235, buyers whose incomes did not exceed 135 percent of the admission limit for public housing could obtain a new or used home with a thirty- or forty-year mortgage of up to $21,000 (or $24,000 in the case of large families needing bigger dwellings) with a down payment of $200 and fixed interest rates as low as 1 percent.[16] In real terms, this meant that a Section 235 buyer could purchase a $15,000 home for just $200 down and $48 per month, significantly less than the monthly payment of $97 required for a conventional mortgage of identical value. The architects of the HUD Act—a bipartisan group that included members of Lyndon Johnson's cabinet, Republican senator Charles H. Percy of Illinois, and representatives from the banking and building industries—hoped to use supply-side subsidies to revive the real estate market, reduce unemployment, and bring greater racial and spatial equity to federally sponsored housing initiatives. They intended to accomplish these goals by supporting the construction of affordable homes and apartments for people of color and the poor in both cities and suburbs.[17]

Local lenders, developers, and policy makers had their own aims in mind, however. For them Sections 235 and 236 arrived at an especially opportune moment, when rising interest rates and tightening mortgage markets had combined to depress residential construction and sales throughout the nation.[18] Eager to acquire any new business they could, lenders and builders across the United States began aggressively advertising the new program soon after its passage, often through widely publicized "real estate seminars."[19] In Genesee County, Citizens Bank, Deventeen Investors Company, Quality Construction, Whittier Building Company, D&B Golden, A&E Building Company, and other local firms began taking advantage of the new federal funds almost immediately, especially those offered under Section 235.[20] In the years leading up to President Nixon's 1973 moratorium, developers erected approximately fifteen hundred Section 235 homes in the Flint region—all but a few of them of new construction.[21] Nationwide, the HUD Act helped to fund the sales of nearly half a million housing units with a total mortgage value of $8.3 billion.[22] The new policy was so popular among builders that the phrase "Stay Alive with 235" became a popular mantra during the national housing slump of 1969–70.[23]

With urban renewal destroying thousands of inner-city homes at around the same time, municipal officials and representatives of the Flint Housing Commission also welcomed the sales of low-cost dwellings. Members of the FHC hoped that Section 235 would provide a path to home ownership for poor renters and in the process reduce long waiting lists for public housing. Although the FHC did not maintain a comprehensive record of move-outs, evidence suggests that many public housing tenants did in fact take advantage of Section 235. Just during 1971, for instance, seventy-two families moved out of the Atherton East complex into new Section 235 homes. Countywide, according to a 1974 study, one in four occupants of Section 235 housing had lived in public housing prior to acquiring their new homes.[24] Many of these former tenants initially supported the 235 program, believing that it would provide an opportunity for a fresh start in a new, trouble-free home.

In reality, though, quite the opposite occurred. The framers of Section 235 intended for the program to foster the "geographic dispersion" of affordable housing throughout metropolitan areas. To support that goal, officials from the FHA and HUD adopted an "informal guideline" to secure "widespread geographical distribution" of the units.[25] However, weak federal regulations on site locations and lax oversight from FHA and HUD administrators allowed local builders and policy makers to concentrate Section 235 homes in racially integrated neighborhoods.[26] One of these local actors was a man named Thomas Hutchinson, a Flint-based representative of the FHA who managed site requests for builders and lenders seeking to participate in the Section 235 program. Despite the government's guidelines on geographic dispersal, Hutchinson and his colleagues did not attempt to shift building activity toward all-white suburbs. Instead, they endorsed builders' requests to construct well over one thousand Section 235 homes in Flint and Beecher. According to a 1974 study conducted by the Genesee County Metropolitan Planning Commission (GCMPC), a countywide planning advisory board, 875 of the 1,387 Section 235 units in the region were located in Flint, most of them within the racially transitional neighborhoods of the northwest and southeast sides. The remaining 512 structures were all located in suburban Genesee County, but they too were concentrated in small geographic clusters. Of these suburban residences, 67 percent were located in Beecher.[27] For the nearly all-white suburb of Swartz Creek, by comparison, Hutchinson approved only six units of affordable housing. And Hutchinson authorized no low-income housing at all for the equally segregated areas of Flushing and Clayton Township. Despite containing only 5 percent of the county's overall population, Beecher received 25 percent of the units that Hutchinson approved in the metropolitan region, more than all other Flint suburbs combined.[28] This extraordinary concentration of Section 235 homes resulted in new forms of administrative segregation that in turn triggered widespread white panic selling and angry social conflicts that tore at Beecher's social fabric.

Disagreements over the geographic distribution of units were at the crux of many local disputes over Sections 235 and 236. From the start, members of the Urban League and other critics argued that the concentration of poor people and government-assisted housing would contribute to neighborhood blight, social conflict, and racial succession.[29] The case of West Beecher seemed to bear out such claims. Of the more than seven hundred Section 235 and 236 units that Hutchinson initially authorized for Beecher, 564 were located within the western half of the district. Overall, only 55 of the Section 235 homes in Beecher were located on "scattered" sites.[30] In several cases, builders simply carved out new dirt roads along which they erected dozens of architecturally indistinguishable homes. On Afaf Street, for instance, one developer constructed thirty very similar Section 235 homes, while West Genesee Street received a dozen virtually identical structures.[31] Likewise, the three Section 236 apartment complexes built in Beecher were all located within a three-block radius.[32] According to one survey, the "monotonous design" and "homogeneous occupancy" of Section 235 units in Beecher was creating "block ghettos" and a "pattern of development equivalent to traditional public housing projects."[33] Mt. Morris Township supervisor Donald J. Krapohl agreed, noting, "It's what I call federally-aided ghetto creation."[34]

Beyond fielding criticism over the concentration of architecturally monotorous units in West Beecher, local and federal officials received a large number of complaints regarding the abysmal quality of Section 235 homes. "Many of the complaints," a 1971 report revealed, "involved the lack of storm doors, screens, lawns, and in some instances, paved streets and roads."[35] Because federal regulators enforced the strict twenty-one-thousand- and twenty-four-thousand-dollar price ceilings for all 235 homes, builders often failed to install driveways, left lawns unseeded, and ignored major construction flaws in order to maximize profits. Even proponents of the 235 initiative such as Thomas Jean of the Genesee Community Development Conference acknowledged that builders erected "some very questionable housing under this program." "We had some units built, for instance, that crowded a four-bedroom home into well less than 1,000 square feet. We had some units put up with inadequate driveways, no sodding, and things that this community should not be very proud of."[36] Although Krapohl and other district and township officials loudly decried the defective new units, Beecher's weak zoning and building codes—which local policy makers had stripped down even further in the wake of the 1953 tornado to help replenish the area's housing stock—allowed rogue developers to work with virtual impunity.[37]

Beecher's experience with Section 235 and 236 housing was in many ways representative of what occurred in other parts of the United States. By the early 1970s complaints about the government's low-income housing ventures had surfaced across the country. In the towns of Elmwood, Missouri; Everett, Washington; and Huntsville, Alabama, for instance, congressional investigators discovered that builders were creating "instant slums" with Section 235 funds. Similarly, developers on the outskirts of Columbia, South Carolina, employed Section 235 appropriations to erect what one *New York Times* reporter referred to as "shantytowns." In Asbury Park, New Jersey, new homeowners complained of collapsing ceilings and floors, bathroom leaks, and homes without closets.[38] After discovering that a local builder had used rotting scrap wood to construct her home, Virginia Simpson, a new homeowner from Citrus Heights, California, wrote an angry letter to US Senator Alan Cranston in which she charged that Section 235 units were "worse than the shacks we were forced to live in while in the ghettos."[39] Although substandard developments sprang up in scores of communities from coast to coast, the worst abuses were concentrated in the metropolitan areas of Chicago, Philadelphia, Newark, Detroit, and Flint.[40]

As many critics noted, the preoccupancy home inspections required under the HUD Act should have prevented the sales of defective units. According to at least one local report, however, "The FHA Service Office in Flint has not considered it their responsibility to exercise quality control through

project review, appraisal/valuation, or additional inspections when dealing
with low and moderate income housing programs."[41] Instead, FHA representatives in Flint and other cities conducted only cursory "windshield" tours of
new and even used structures and seldom required builders to repair flawed
or unfinished dwellings.[42] Meanwhile, lenders in Flint and Beecher often
skipped inspections altogether under the assumption that newly built or re-
habilitated units would be structurally sound.[43]

Reality painted a much darker portrait, however. Like their counterparts
in other communities, residents of Beecher protested the defective new units.
"We were promised quality homes for low-income people, but all we received was an overpriced home of inferior quality," complained Mrs. Vernon
Dean, a resident of a 235 unit on Afaf Street.[44] Following a September 1972
visit to Beecher, Donald Riegle, then the district's congressional representative, informed local reporters that Flint's FHA office had received three hundred complaints regarding substandard 235 units. "All 300," Riegle declared,
"received little or no action by the Flint [FHA] office."[45] Unable to afford the
repairs to their units, many residents defaulted on their house payments and
forfeited their properties within months of occupancy. Nationwide, over a
quarter of Section 235 homeowners had defaulted on their payments by the
summer of 1972.[46] As homeowners abandoned their units, home foreclosure
and vacancy rates skyrocketed in communities such as Beecher. According
to a 1971 report on federally assisted housing in Genesee County, delinquencies and foreclosures on Section 235 mortgages "were running three to four
times" the average rates in the Flint metropolitan area.[47] By October 1971
there were two hundred abandoned homes in Beecher, a quarter of which
were less than a year old.[48]

Regardless of the city in question, investigators had little trouble confirming the claims made by critics of the Section 235 and 236 programs. In June
1971, members of the United States Commission on Civil Rights took the
rare step of denouncing another government agency in a report that stated,
"This agency [the FHA], traditionally attuned to serving the needs of white,
middle-class families, has . . . done little to develop affirmative procedures
and mechanisms to assure that the lower-income buyers are treated fairly."[49]
Shortly after that, HUD officials released the results of a national audit of Section 235 units. The study revealed that 25 percent of newly built Section 235
homes and 34.5 percent of existing structures covered under the program suffered from faulty construction. To the great dismay of HUD Secretary (and
former Michigan governor) George Romney, surveyors discovered overhead
lights without switches, cracked walls, inoperable heating units, roof leaks,
and a host of additional problems—some minor, others severe—in Section

FIGURE 8.1. An abandoned Section 235 home in Beecher, 2012. Local builders erected this home in Beecher during the early 1970s as part of the Section 235 low-income housing program. Critics of the initiative argued that it exploited poor people and led to racial segregation, social conflict, and property abandonment. Photograph by Andrew R. Highsmith, 2012.

235 homes. "We concluded," the auditors wrote, "that many of these [Section 235] houses contained easily observable but unreported deficiencies that seriously affected safety, health, and/or livability."[50] Equally troubling was their finding that Eugene A. Gulledge and other HUD and FHA leaders in Washington had specifically ordered local policy implementers to "relax the inspection requirements" for low-cost units in order to boost construction figures.[51] Just a few weeks later, federal officials published another damning report, this one on the Section 236 program. According to that audit, around one-third of the nation's Section 236 projects had "undesirable site characteristics," including locations near polluting industries and severe racial and economic "impaction."[52]

A 1974 survey of affordable housing sponsored by the Genesee County Metropolitan Planning Commission painted an even bleaker picture of the Section 235 program. According to this report, 91 percent of the Section 235 borrowers in Genesee County complained of problems with their homes. In spite of the complaints, however, auditors found that builders had repaired the homes of only 19 percent of the owners surveyed. Forced to absorb the repair costs themselves, many owners of Section 235 units succumbed to bankruptcy and lost their homes to foreclosure during the global economic con-tractions of the 1970s.[53] Between 1969 and 1973, foreclosure rates in the county increased by 184 percent, due in large part to the rash of mortgage defaults from Section 235 homeowners. By March 1974 more than one in four of the Section 235 homeowners in Genesee County had ceased making payments on their homes.[54] Because the HUD Act required the federal government to pay off the balance of the loan in all Section 235 foreclosure cases, local bankers and holders of mortgage-backed securities continued to profit financially even when homeowners defaulted on their payments. Consequently, investors remained supportive of the 235 initiative even in the face of the growing foreclosure crisis. As evidence of this fact, in March 1970, despite mounting criticism of the program, Howard L. Gay, a senior vice president of the Citizens Commercial and Savings Bank in Flint, wrote a letter to Congressman Riegle urging him to vote for the maximum appropriation for the 235 initiative, avowing, "This program is successful."[55]

On their own, the numerous reports documenting the shoddy construction and geographic concentration of the Section 235 units would have been sufficient to mar the image of the HUD Act. However, buried deep within many of these documents were charges of profiteering and usury that threatened to delegitimize the entire foundation of HUD's affordable housing program. The root of the problem was the government's laissez-faire approach to regulation. In their quest to "cut the red tape to pink ribbons," Gulledge and other federal administrators eviscerated the regulatory structures that might have protected individual citizens and local communities from unscrupulous developers.[56] Businesses participating in the Section 235 program thus committed acts of economic exploitation without fear of punishment.[57] Builders, of course, routinely cut corners in order to turn a profit. But the fleecing of poor buyers occurred at virtually every stage of the process. One of the most common schemes involved a practice known as "flipping," in which investors acquired ramshackle properties, completed only minor cosmetic repairs, and then resold them at inflated costs to unsuspecting 235 buyers. Fueling this process were legions of dishonest appraisers who were willing to approve thousands of substandard units for sale in exchange for a steady stream of new customers.[58] Lenders and realtors also acted as economic predators. Realtors routinely inflated their commissions in Section 235 transactions, in some instances pocketing up to 13 percent of a home's selling price. For their part, lenders often required Section 235 buyers to pay exorbitant "mortgage placement fees" as part of the purchase price of a home. According to HUD's 1971 national audit, placement fees frequently reached as high as 16 percent of a mortgage's overall value.[59] Although the campaign against predatory lending in the United States would take many more years to reach fruition, the

1970s boom in real estate profiteering that occurred under the auspices of the HUD Act clearly marked the beginning of a new era of nationalized reverse redlining in which federally insured debt served as a means of maintaining inequality.

Integration and Disintegration

Many white residents of Beecher and Mt. Morris Township resented the Section 235 and 236 programs, believing that they caused declining property values, "white flight," and school overcrowding. Seldom, however, did opponents of the low-cost units frame their opposition in openly racist terms. Instead, critics focused on economic issues—namely, that Section 235 families were poorer, paid less in property taxes, and had more school-age children than most other households in the community. Objections centered on the fact that approximately half of the purchasers of Section 235 homes in Beecher received public assistance, a figure more than quadruple the national average.[60] Detractors charged that the increase in poor families would necessitate greater public expenditures on social services and antipoverty programs. Regarding education, critics made a compelling case that the new housing generated large and costly enrollment increases in the district's schools. After remaining relatively stable during the years preceding the implementation of Section 235, the enrollment of the Beecher Public Schools rose by over one thousand students just between 1968 and 1971. According to Beecher's school superintendent Randall Coates, a large majority of the incoming students resided in the HUD units.[61] The enrollment increases forced principals throughout the district to hire more teachers and implement pupil "attendance shifts" to free up more classroom space.[62]

In addition to such economic considerations, racial fears played an important role in driving the resistance to the HUD units. Although only 50 percent of the occupants of 235 and 236 housing in Beecher were African American, the concentration of units there caused several schools and neighborhoods to "tip" toward black majorities. At Zink Elementary School, for example, the percentage of black pupils rose from 42 percent to 55 percent between 1969 and 1971.[63] During the four years preceding the passage of the HUD Act, the Beecher schools had maintained a stable racial mix of pupils: 70 percent white and 30 percent black. By the early 1970s, however, the combination of black in-migration and white departures had pushed several schools and neighborhoods past the racial tipping point. "We had one of the model integrated school systems in Michigan," Superintendent Coates complained, "and now it is being ruined."[64]

As the federally financed homes and apartments went up and hundreds of new African American families began arriving in Beecher, racial conflicts flared in schools and neighborhoods. The construction of 235 and 236 units also fueled widespread blockbusting across sections of Flint, Beecher, and Mt. Morris Township.[65] Describing the situation in Flint Park, a neighborhood just south of Beecher, white homeowner Ben Wisniewski blamed the 235 homes for intensifying panic selling: "They [Section 235 homes] were junk. . . . Within a very short time, the properties fell into disrepair. . . . If you lived next door to that you moved out."[66] As elsewhere, homeowners in Flint Park and Beecher responded to plans for 235 housing by working to protect their real estate investments. When that failed, many simply moved away. Others, however, turned to violence. To cite just one example, in July 1971 Clarence Hall, an African American who resided at G-1139 West Genesee Avenue in Beecher, reported that Ku Klux Klan members had burned a cross in front of his home. In a note left behind, the cross burners warned, "Niggers get out while you still can."[67]

Racially charged disputes also began erupting in the Beecher schools during this period.[68] The first major conflicts exploded in the spring of 1969, just as purchasers began to occupy the new Section 235 units. During the week of April 13, several brawls between white and black pupils broke out at Beecher High School. On April 16, school administrators had to summon police officers to quell a "major confrontation" between white and black students. According to Anthony Bell, a black student at Beecher High, African American pupils faced "constant provocation" from their white peers.[69] Angered by the situation, Superintendent Coates charged that the 235 program was responsible for Beecher's transformation from "integration to disintegration."[70]

In the aftermath of the spring disturbances at the high school, Coates and other district officials spent a great deal of time preparing for the 1970–71 academic year. To accommodate the approximately one thousand new students they expected to enroll in the fall, school board members and Superintendent Coates devised a complex plan for half-day sessions and sought out new classroom space in church basements.[71] The board also hired Paul Cabell, the first black administrator in the district, to serve as assistant principal at Beecher High School. Principal Robert Towns assigned Cabell the difficult task of handling student discipline and managing race relations. Ultimately the challenge of maintaining peace at Beecher High School would push Cabell past his emotional breaking point.

In February 1971, after receiving numerous complaints emanating from Genesee County, HUD suspended its 235 and 236 programs in the Beecher area pending the results of a complete investigation.[72] Released in May 1971,

HUD's investigative report confirmed that builders had saturated Beecher with more than its fair share of the contested units. The report's authors also acknowledged that many of the 235 and 236 units in the district were poorly constructed and architecturally monotonous and that the government's efforts were creating "suburban black ghettos."[73] Furthermore, the investigators determined that many other areas of Genesee County could have received affordable housing units "but HUD did not attempt to direct locations of building activities." Instead, Eugene Gulledge and his colleagues in Washington had given "100 percent responsibility" over the program to FHA district officers in Flint and other cities.[74] Beyond accepting responsibility for the poor oversight in Genesee County, HUD officials blamed Thomas Hutchinson and the FHA's Flint office for the concentration of units in Beecher.[75] "There was apparently very little, if any, discussion between the Flint [FHA] service office supervisory staff and builders and community interests regarding fund allocations, site allocations and community problems," the authors of the report charged.[76]

After reading the report, Secretary Romney admitted that poor federal oversight had contributed to the turmoil unfolding in Beecher and other communities. Romney also informed Superintendent Coates that he had "put a stop on additional funding of Section 235/236 units in the Beecher District."[77] While Romney's decision to cancel the 235 and 236 programs in Beecher came as welcome news to many homeowners and policy makers, blockbusting, property abandonment, and racial conflicts continued unabated. At Beecher High School, racial tensions erupted yet again during the 1971–72 academic year. By February 1972, as one journalist wrote, "[s]omething akin to anarchy prevailed at Beecher High. . . . Fires were set, windows were broken, youngsters were beaten and hundreds of students refused to attend class, choosing instead to loiter in the hallways." Ultimately it took two dozen police officers in full riot gear to end the February disturbances.[78]

Hired to curb the interracial acrimony that had plagued Beecher High School since the late 1960s, Paul Cabell worked diligently to serve as an intermediary between white and black students. Yet he drew harsh criticism from both sides of the color line while performing in that capacity. Militant black students referred to Cabell as an "Uncle Tom," maintaining that he favored whites and was insensitive to the grievances of Beecher's growing black minority. But Cabell also suffered the wrath of many white pupils, who believed that he was partial to black students. Cabell, *Life* magazine journalist Loudon Wainwright wrote, was "the man in the middle," squeezed on all sides by racial hatred. "No one was on his side," noted black pupil Donnie Odom."[79]

Cabell, a steadfast integrationist, initially embraced his appointment at

FIGURE 8.2. Paul Cabell, n.d., circa 1970. Cabell served as an assistant principal at Beecher High School from 1970 until his death in 1972. His job was to oversee student discipline and race relations. Ultimately Cabell killed himself in order to draw attention to the racial conflicts unfolding in Beecher. Courtesy of the Associated Press.

Beecher High School as an opportunity to improve race relations. Yet the seemingly ceaseless racial turmoil in local neighborhoods and at the school quickly destroyed his optimism. Unable to pacify the student body, Cabell became despondent and withdrew from his family and friends. On the morning of February 24, 1972, he sat down in his Flint home to write two letters—one for his wife Carlitta and another "For Beecher." In the two messages, Cabell admitted defeat in his quest to remedy the "unsolvable problem" of racism. He declared that the experience in Beecher had driven him to despair, forcing him "to totter precariously on the fine line between two cultures." "For what do I isolate myself in the middle, never right, always wrong?" he asked. In a direct appeal to racists and "hotheads" on both sides of the color line, Cabell decided to take drastic action.[80] After finishing the notes, the once resolute young administrator committed suicide by shooting himself in the head.

In the weeks and months following Cabell's shocking death, an eerie calm prevailed throughout Beecher. As members of the community grappled with the tragedy, the fighting and harassment quickly subsided. Yet in the end Cabell's death did little to stabilize the district. Between 1972 and 1977, the proportion of black students at Beecher High School doubled, rising to 70 percent.[81] And by the early 1980s blacks held a clear majority in both the schools and neighborhoods of Beecher. Meanwhile, the district's housing stock—hit hard by property abandonment, declining government services, and a new federalized system of predatory lending—emerged as one of the most dilapidated in the county.[82] Though Cabell did not die in vain, his passing failed to alter the district's fate.

Throughout the crisis in north suburban Flint, local and national observers speculated as to why Beecher, and not other area communities, had received so many units of "government housing." Most defenders of the 235 and 236 programs held that the market was the key factor driving location decisions. In support of his choice to approve so many units in Beecher, Hutchinson told a journalist from the *Washington Post*, "Our deal, as far as I can see, is this: is there a market?"[83] Clearly there was a market for affordable housing in Beecher. However, in light of that demand Hutchinson and his FHA colleagues gave builders almost complete freedom to select sites for 235 and 236 units. Moreover, Hutchinson allowed builders to self-regulate architectural and construction standards. Builders defended their actions by pointing out that the legal requirement that all federally financed homes have access to public utilities influenced their decisions to construct 235 and 236 units in Beecher.[84] Although unincorporated, Beecher did in fact meet two of the federal government's key financing requirements by providing residents with public water and sewer service.[85] Undoubtedly the guidelines regarding utilities placed clear limits on the potential locations for 235 and 236 housing projects. Yet Beecher was not the only suburban area that had access to such services. During the 1950s and 1960s, predominantly white suburbs such as Flushing, Swartz Creek, and Linden, which received few or no low-cost units, also built modern sewer and water systems.[86] How, then, could local officials account for the discrepancies? Hutchinson and other policy makers greeted those sorts of questions with a wall of silence.

Although builders and government administrators focused on the market forces that drove site decisions for 235 and 236 dwellings, other factors played a part in determining which suburbs received units. According to at least one investigation, builders concentrated their efforts in Beecher because "the area was already integrated, a water and sewer system existed, and vacant lots were relatively inexpensive." Furthermore, developers of affordable housing preferred Beecher to other areas of the county because of its minimal building regulations, which helped to keep construction costs below federally mandated price ceilings.[87] During a 1972 speech to county planners, township supervisor Krapohl admitted that Beecher had a "very inadequate" zoning ordinance and no controls whatsoever on subdivision development. "Many lots were plotted in the 1920s with the 40' and 60' frontage, unpaved streets

and very few sidewalks," he revealed. "With this situation existing, the builders with commitments from the local HUD office had a field day."[88]

Because Beecher was already integrated, provided public utilities, had
weak construction and zoning standards, and was affordable, builders effectively colonized the area with hundreds of substandard housing units. And in
their rush to stabilize the nation's real estate industry by increasing the flow of
supply-side subsidies, federal officials ignored even the most egregious local
practices. "This district has been the victim of what I call neo-colonialism,"
noted Ira Rutherford, a black official in the Beecher school district.[89] Just as
realtors and public housing administrators had done for many years, builders
and FHA administrators identified northwest Flint and Beecher as black population expansion zones. By funneling poor and African American residents
into Beecher and other such zones, federal housing administrators and other
participants in the Section 235 and 236 programs helped to create new forms
of administrative segregation while spreading the city's growing economic
crisis into suburban Genesee County.[90] However, the rejection of affordable
housing in most of the county's remaining jurisdictions meant that the color
line would travel only so far.

Exclusionary Suburbs

During the late 1960s and early 1970s, battles over site locations for Section 235
and 236 units and other types of "affordable housing" provoked intense conflicts in suburban Genesee County. As suburban homeowners and policy
makers watched and read about the discord occurring in Beecher, many
vowed to resist low-cost housing units in their communities. In support of
that goal, suburban officials enacted increasingly restrictive land use policies.
Although defenders of suburban capitalism and exclusivity seldom spoke
openly about the full range of considerations that drove them to tighten land
use controls, racial and class fears were never far from the surface of debates
over affordable housing.

Plans to build HUD housing in several of the all-white suburbs of Genesee
County triggered a wave of activism in the early 1970s. In April 1970, in one
of the first of these eruptions, the editors of the *Fenton Independent* angrily
denounced a plan to build Section 235 units in their city, warning that low-
income residents would create "a slum area."[91] Later that year 150 residents of
Davison Township organized against a developer's proposal to build a complex of low-cost townhomes.[92] Then in 1971, residents of Davison mobilized
after hearing a rumor that a builder planned to erect a Section 236 apartment
complex. Although the rumor turned out to be untrue, the Davison City

Council responded to the specter of affordable housing by implementing the most restrictive zoning and building ordinance in the county. Under normal circumstances, the Davison City Council voted on proposed ordinances only after a ten-day waiting period. In this case, however, Mayor Judson Davis ordered an immediate vote on a new building and zoning code, claiming that the rumored project constituted a "danger to the health, safety and welfare of the Davison community." City council members concurred with Davis's opinion and immediately enacted new zoning and building guidelines. Davison's 1971 code—designed specifically to restrict the construction of homes and rental units for low-income families—required developers to allocate at least 5,500 square feet of land for each three-bedroom apartment erected in the city. The law also prohibited developers from building any more than 8.5 apartment units on each acre of land. Moreover, to block the construction of freestanding Section 235 houses, the Davison ordinance mandated that all single-family homes have a minimum of 1,096 square feet of living space on the ground floor.[93] Together, these restrictions—all of them forms of administrative racial and class segregation—helped to banish virtually all low-income housing from Davison.

Not all Davison residents appreciated the city's exclusionary policies, however. Within months of the code's passage, young leftists had risen to challenge the city's leadership. In June 1972, student activists helped to elect an eighteen-year-old community organizer and future filmmaker named Michael Moore to the Davison Board of Education. At the time of the election, Moore was the youngest person ever to win elective office in the state of Michigan. Although many adults in the community vigorously opposed Moore's insurgent candidacy, newly enfranchised eighteen-year-olds—who had won the right to vote in 1971 with the passage of the Twenty-Sixth Amendment to the US Constitution—turned out in droves to support his campaign.[94]

Not long after taking office, the young activist drafted a resolution asking school board members to welcome African Americans and other minority groups to Davison. The proposal read, "Now, Therefore, be it resolved that the Davison Board of Education publicly invites and encourages all members of racial, ethnic, and religious minorities to come and reside in the Davison area and enhance the learning and living environment of the school district and help our students learn the true meaning of brotherhood." Overnight, the resolution elicited a firestorm of hostility from local taxpayers and school board members, many of whom viewed Moore's move as a crass publicity stunt. "The welcome signs are out and always have been," responded Charles N. Mitchell, an opponent of the proposal. "At the rate we taxpayers have had to build new schools to accommodate those who have found our city and school system so attractive I do not feel it is necessary to have anyone go on record as extending a welcoming hand to anyone." On December 3, 1973, Moore's colleagues formally and unanimously rejected the resolution. Moreover, within weeks of the December vote, Davison residents had launched a grassroots campaign to remove the young iconoclast from office. Though he ultimately survived the recall effort, Moore lost his subsequent reelection bid by a landslide. In Davison—as Moore learned the hard way— residents and officials reacted harshly to even the idea of integrated housing.[95]

Davison was far from atypical, however, as suburbanites from throughout the region and nation adopted similar approaches to maintaining administrative racial segregation and class homogeneity.[96] Shortly after the enactment of Davison's zoning and building code, members of the Grand Blanc Township Board passed their own exclusionary housing ordinance. Only slightly more lenient than the Davison law, the Grand Blanc Township code allowed a maximum of 9.3 apartments per acre. In 1973, board members in Vienna and Mundy Townships joined the movement against affordable housing by voting to raise the minimum size requirements for new homes. Over time, most other suburban governments in Genesee County followed suit by enacting their own restrictive zoning and building codes. Because the Section 235 and 236 programs required builders to abide by inflexible pricing guidelines, codes such as those passed in Davison and Grand Blanc Township effectively banned all HUD-backed housing units by raising the cost of construction above federal limits. Much the same occurred in the private sector, where would-be developers of mobile home parks, apartment complexes, and other types of low-cost housing quickly discovered that it was impossible to construct affordable residential units under the new codes.[97]

Beyond restrictive building and land use policies, some suburban officials looked to prohibitive utility and service regulations as a means of resisting low-cost housing and racial integration. In 1970 an anonymous elected official in Genesee County described how local utility departments in the suburbs routinely harassed developers of low-cost housing: "A contractor of ours will call up and say it's going to cost $5,000 to run water to five houses. I get on the phone and call township officials. After several minutes of conversation they finally get around to asking who is going to move into the houses. I tell them, 'He's a factory worker, has two children and by the way he's white.' Then the price of the water hookups goes down."[98] Conversely, when suburban policy makers suspected that African Americans were going to move into newly built units, they often increased utility charges or threatened to deny service altogether. Suburban officials who engaged in such practices hoped to resist efforts to sprinkle "fair share" housing throughout the out-county. Ultimately their results were quite impressive. According to a 1974 survey, fewer than two hundred of the fifteen hundred Section 235 units built in Genesee County were located outside of Beecher and Flint.[99]

Due to the proximity of the battles occurring in Beecher and north Flint, suburban opponents of affordable housing focused on blocking Section 235 and 236 units. In truth, however, most suburbanites loathed traditional public housing even more. The depth of this hostility became clear in December 1972, when Harold Black and Aaron Blumberg, on behalf of the Genesee County Metropolitan Planning Commission, a relatively powerless local advisory board, drafted a series of policy papers on the status of public housing in the out-county. After determining that Flint's suburbs had a disproportionately small share of the region's affordable residential units, Black and Blumberg proposed the creation of a countywide public housing commission to oversee the construction of fifteen thousand low-income apartments spread evenly across the metropolitan area.[100] On September 6, 1973, a subcommittee of the GCMPC endorsed the proposal. Mindful of the tumult occurring in Beecher, Thomas H. Haga, director of planning for the GCMPC, pledged that the proposed commission would not build any public housing units in the out-county without first gaining the approval of local officials. Still, Flint Township commissioner Russell Zaccaria and other suburban representatives on the Genesee County Board of Commissioners immediately denounced the idea. "This was told us [before] in Flint Township," Zaccaria claimed, "but the 236 program was shoved down our throats."[101] Hoping to preempt any action by the GCMPC, Zaccaria and a host of fellow commissioners moved quickly to block public housing from their communities. Board members in south suburban Grand Blanc Township were among the first to act. In March 1973, while GCMPC members were still debating the merits of the proposal, legislators in the township voted unanimously to supplement existing restrictions on affordable housing by establishing a new minimum lot size of twelve thousand square feet for single-family homes. Their goal was to ensure that no "subsidized" housing of any kind could go up in the township.[102]

The following year, members of the Grand Blanc Township Board continued their campaign of bureaucratic exclusion by voting to create their own public housing authority. The formation of the Grand Blanc Housing Commission did not stem from a desire to construct low-cost units, however. Rather, the decision simply established the township's legal authority vis-à-vis the county in all matters pertaining to public housing. According to Grand Blanc Township attorney Lyndon Lattie, who drafted the ordinance and several others like it, the establishment of the commission allowed lo-

cal officials to preserve "a maximum amount of local autonomy" in deciding whether to construct public housing units.[103] "Without a local commission," Lattie pointed out, "the county could determine when, where and
how public housing came into the township and the Township Board would
have no control over it." Within months, board members in Mundy, Gaines,
Davison, and Argentine Townships had followed suit by forming their own
public housing authorities.[104] Inspired by their counterparts in Grand Blanc
Township, suburban legislators voted to establish public housing commissions solely as a means of obtaining veto power over the proposed county
authority.

Throughout the 1972–74 controversy, GCMPC planners consistently
pointed out that they were working in an advisory capacity and lacked the authority to order the construction of public housing. They also reminded suburban critics that they would not propose building public housing units in
any community without first gaining consent from local officials. However,
suburban capitalists took little comfort in such reassurances. Instead, they
worked to forestall the "Beecherization" of their communities by preemp-
tively blocking public housing before local officeholders had even formed a
regional housing authority. Sensing the depth of the opposition to their plan,
GCMPC officials ultimately abandoned their efforts on behalf of fair share
public housing. The decision to drop the proposal ensured that all of the
county's public housing complexes—as well as almost all the region's affordable residential units—would remain in Flint or Beecher.[105]

Overall, suburbanites proved to be remarkably successful in resisting plans
for affordable housing in Genesee County. According to a 1980 survey, fourteen of the eighteen townships and seven of the twelve cities and villages in
the county contained no HUD units at all. Moreover, of the 1,735 "subsidized"
residential units that were located in Flint's suburbs, almost all were occupied
by elderly persons. Across the entire metropolitan region, there were just 391
HUD rentals specifically reserved for low-income families, and almost all of
these were in or near the Beecher area.[106] As was the case elsewhere in the
United States, restrictions on low-cost housing in suburban Genesee County
helped to sustain rigid patterns of racial and class segregation.[107]

In the end, suburban capitalists and their elected representatives fought to
defeat affordable residential developments because they accepted what forty
years of federal and local policies had taught them about property values,
community building, and quality of life. Specifically, they believed that low-
cost housing attracted low-rent individuals—the sort of people who were
purportedly destroying Flint and Beecher. For many suburban homeown-

ers, then, the struggle against affordable housing was part of a much broader campaign for complete secession from Flint and its problems. And that struggle did not end with the proscription of unwanted housing developments. Indeed, by the mid-1970s the battle lines in the campaign for suburban capitalism and independence had shifted toward the educational arena, where Flint's leaders contemplated an equally controversial plan to desegregate the region's schools.

The Battle over School Desegregation

Over two decades after the Supreme Court's landmark ruling in *Brown v. Board of Education*, federal officials discovered school segregation in Flint. On August 29, 1975, representatives from the US Department of Health, Education, and Welfare charged the Flint Board of Education with operating illegally segregated public schools in violation of Title VI of the 1964 Civil Rights Act. Outlined in a lengthy letter to Superintendent Peter L. Clancy, the complaint documented how school board members for decades had maintained a separate but unequal, dual education system through a combination of exclusionary employment rules, racially biased school location and pupil transfer policies, discriminatory construction and real estate practices, and deliberately gerrymandered student attendance boundaries.[1] Although board members vehemently denied the charges, federal reports left little room for doubt about the local government's pivotal role in maintaining administrative segregation.

In the area of student assignment, HEW officials painted a detailed portrait of intentional government-sponsored segregation in the heart of the ostensibly liberal urban North. Beginning in the mid-1930s, when small numbers of black families first moved into all-white neighborhoods bordering the North End and Floral Park, the Mott Foundation and the board of education had enforced a rigidly segregationist attendance boundary policy that occasionally kept white and black neighbors in separate schools. During the postwar era, as the city's African American community gradually expanded into previously restricted neighborhoods, board members repeatedly redrew attendance boundaries to maintain administrative segregation, often plucking black pupils from all-white facilities and returning them to Jim Crow schools. Ironically, the "neighborhood schools" to which black children returned were often outside of their residential districts. As the Second Great Migration intensified in the 1950s and 1960s, the board of education struggled to maintain its segregated schools in the face of large black enrollment increases and intense pressure from civil rights activists. Ultimately local organizers succeeded in winning open housing legislation, which many hoped would solve the problem of the educational color line. In reality, though, administratively driven school segregation continued in Flint throughout the civil rights era.[2]

Although representatives of civil rights organizations had long criticized the Flint Board of Education for operating unequal and segregated facilities, the widely held idea that the city's schools suffered from constitutionally innocent "de facto" segregation helped to delay the process of desegregation. As a matter of fact, it was not until 1975, several years after national public, legislative, and legal opinions had shifted decisively against the "busing" of pupils for the purpose of achieving racial balance in schools, that Flint activists filed their first of two unsuccessful federal desegregation lawsuits. Those suits, though of immense symbolic value in undermining the legitimacy of the long-standing de facto segregation myth, failed to sway school officials to alter their policies. In the absence of judicial remedies, activists pinned their hopes for desegregation on sustained local protest and federal enforcement of the 1964 Civil Rights Act. After years of inaction, HEW finally ordered the Flint Board of Education to desegregate its schools in late 1975, sparking a region-wide debate on race and community education. Stretching over a ten-year period in which the issue of busing exploded in the United States, Flint's lengthy desegregation crisis brought the issues of race, space, class, and neighborhood to the forefront of local political culture.

Shortly after HEW officials issued their desegregation order, leaders from the Mott Foundation abruptly canceled the community education program, leaving behind a massive funding shortfall and a rudderless educational power structure. Nevertheless, the school board continued to champion the neighborhood schools policies that had kept pupils segregated, in the end agreeing only to a weak desegregation plan that relied on magnet schools and other forms of voluntary desegregation. In a district that had already lost tens of thousands of white pupils to suburban and private school systems, Flint's voluntary program could not and did not achieve integration.

Justice Delayed

In the decade following the *Brown* verdict, with the nation's attention riveted on pupil desegregation in the South, activists in the North and West launched a number of campaigns against school segregation. As part of this effort, civil rights attorneys filed a flurry of school desegregation lawsuits in northern cities.[3] In Flint, however, the movement for school desegregation did not take hold until well into the 1970s. The reasons for the delay were manifold and stemmed from both national and local considerations. At the national level, the NAACP's emphasis on combating legal segregation in the South left would-be litigants in cities such as Flint with few resources for attacking their own Jim Crow schools. However, even when organizers targeted northern school districts, they faced obstacles from federal jurists. Although civil rights activists won an important 1961 federal desegregation case in suburban New Rochelle, New York, in which black plaintiffs claimed to have experienced de jure–style forms of legal and administrative segregation, a pair of 1963 rulings pertaining to Chicago and Gary, Indiana, effectively shielded the vast majority of the public schools of the North and West from judicial challenge. In his decision in *Bell v. School Board, City of Gary, Indiana,* Judge George N. Beamer of the Seventh Circuit Court of Appeals ruled that segregation in Gary's "neighborhood schools" derived not from intentional government policies but rather from de facto segregation in housing. By refusing to hear an appeal of the *Bell* decision, the members of the US Supreme Court sent a powerful message to prospective litigants in Flint and elsewhere about the regional barriers to school desegregation.[4]

At the local level, a number of factors converged to stall the assault on school segregation. Most importantly, many African Americans in the Vehicle City felt profoundly ambivalent about the issue. Because educational quality usually trumped racial integration as the central concern of black parents, local activists focused on the equalization of resources more than the issue of pupil segregation.[5] Furthermore, even among activists who were inclined to challenge the color line, the popular idea that housing discrimination was to blame for the city's "de facto" segregated schools helped to marginalize educational struggles. Still, proponents of civil rights remained outspoken in their increasingly harsh assessments of the city's community education program. In February 1964, members of a little-known ad hoc group called the Flint Citizens Information Committee (FCIC) drafted a blistering denunciation of the Mott Foundation: "The fact is the Mott people use this community school as a means of segregating the Negro and 'keeping them in places.' The schools are in neighborhoods that are segregated and the people living there must attend that school. . . . The Mott people really control all of Flint—the city, the schools, the banks and Flint Newspapers in general."[6]

The letter offers an accurate critique of the Mott Foundation's record of discrimination and political dominance. Yet the document also includes a contradictory statement on the causes of administrative Jim Crow that helps to explain why the city's school desegregation movement was so slow to develop. In the opening paragraph, the authors argue that the Mott Foundation used community schools "as a means of segregating the Negro and 'keeping them in places.'" In the next sentence, however, the letter implicates only "neighborhoods that are segregated," a passive-voice construction that exonerated school board members and Mott Foundation officials for the policies they created. The letter's grammatical structure is especially revealing because it indicates how the language of de facto segregation distorted analyses of the color line in Flint and many other American cities. Although most instances of Jim Crow in Flint's neighborhoods and schools derived in some fashion from government-sponsored education, housing, and urban development programs, proponents of the de facto framework blamed pupil segregation primarily on privately orchestrated residential discrimination. Several months after the release of the FCIC's letter, Robert Rawls, chair of the Flint NAACP's housing committee, offered an even more succinct articulation of this inaccurate assessment. According to Rawls, "The housing pattern directly influences the education of our children, creating segregation and, therefore, inferior schools."[7] For the thousands of African Americans who saw no link between integration and quality education, discussions about the nature of the color line were far less important than the quest for equal resources. Among integrationists such as Rawls, however, belief in the exculpatory narrative of de facto segregation helped to delay and undermine the attack on segregated schools.

The national discourse on de facto segregation reflected long-standing contradictions within both American liberalism and the northern civil rights movement that had first surfaced in the aftermath of the *Brown* verdict. In the wake of the court's ruling, black activists in Flint and other cities immediately grappled with its implications. Shortly after the decision, the leaders of the Flint Urban League published a commentary illustrating the friction between an acknowledgment of rigid segregation locally and the deeply ingrained belief that northern and western patterns of segregation were fundamentally different from those in the South. According to the editorial, the *Brown* ruling clearly did not pertain to Flint schools, even though they were starkly segregated: "On the surface the [*Brown*] decision would have little effect on Flint since its schools are already integrated. . . . Over the past few years certain grade schools have progressed closer and closer toward being all Negro. Is this because of segregation in the public schools? No. Will the recent decision on schools affect this situation? No. . . . Each public school is organized to serve its immediate neighborhood. As the direct result of neighborhood segregation, some Flint schools inevitably are segregated."[8] Despite much evidence to the contrary, the editorialists from the Urban League maintained that school officials were innocent of the charge of segregation.

Residents of Flint were not alone in attributing school segregation to housing discrimination. During the post-*Brown* era, the notion of de facto segregation gained a great deal of traction among national civil rights leaders. After visiting Flint in 1964, June Shagaloff, then an assistant to NAACP director Roy Wilkins, all but excused local education officials for the city's deeply entrenched patterns of segregation. "What is significant," she argued, "is the growing realization that even though problems of imbalance and de facto segregation may not be the creation of boards of education and administrators, they have a responsibility to try to correct them."[9] Figures such as Shagaloff offered these sorts of regional distinctions not to thwart the attack on administrative segregation but rather as a means of linking segregated schools and housing and assailing them both. In truth, though, the opposite occurred. Throughout the 1960s the NAACP's national leadership largely eschewed legal strategies in the North and West, preferring to win school desegregation outside the South through electoral gains, open housing legislation, and political pressure on boards of education.[10] In Flint the local NAACP followed the national pattern by forgoing legal action against the Mott Foundation and the board of education in favor of a political campaign for fair housing. Although this strategic decision ultimately led to the open occupancy victory of 1968, the focus on housing clearly inhibited the campaign for educational equity. Furthermore, the emphasis on de facto segregation allowed Mott Foundation officials and school board members to emerge from the civil rights battles of the 1960s largely unscathed. Shielded by the politics of de facto segregation, leaders from the Mott Foundation and the Flint Board of Education continued to operate racially gerrymandered community schools well into the 1970s.

Widespread belief in the concept of de facto segregation played a significant role in the persistence of Jim Crow in Flint and other cities. For opponents of desegregation, the de facto formulation provided discursive, political, and legal space to defend administrative segregation as an accidental outcome of "colorblind" neighborhood schools programs that segregated not by state action but by individual market-based housing decisions. In 1970 Flint's congressional representative, Donald Riegle, crystallized the false de jure–de facto and housing-schools dichotomies that helped to forestall desegregation: "In dealing with segregation, a distinction has been made between de jure and de facto segregation. The former means intentional segregation. . . . De facto segregation results from housing patterns."[11] Effectively, the language of de facto segregation allowed leaders such as Riegle and members of the board of education to denounce Jim Crow in the South while supporting it at home. The phrase "de facto segregation" thus became something of a rallying cry for opponents of desegregation during the school board's prolonged showdown with HEW.[12]

Battle Lines

Federal officials initiated school desegregation proceedings in Flint with a startling degree of reluctance. The process began in 1968, not long after HEW officers announced a plan to investigate pupil segregation in eighty-four school districts in the North and West. As part of that initiative, HEW's leaders dispatched a team of surveyors to the Vehicle City to determine "whether racial imbalance is the product of discrimination or of de facto segregation in housing."[13] During their three-year inquiry, investigators uncovered a long record of popular and administrative school segregation dating back to the mid-1930s. Hoping to capitalize on the publicity that accompanied HEW's northern campaign, in 1971 some members of the Flint NAACP began calling for a desegregation lawsuit against the board of education.[14] In the end, however, representatives of the cash-starved group opted to delay legal action until the conclusion of HEW's enforcement efforts and the resolution of several pending desegregation cases from other cities. Despite the threat of litigation, HEW's leaders proved to be less than enthusiastic in acting on the findings of their investigations. In fact, agency officials declined to pursue enforcement proceedings in Flint and forty-five other school districts until a 1975 federal lawsuit by the national NAACP forced the issue. In his ruling against the government, US District Court judge John J. Sirica determined that HEW had "failed to fulfill its duties and responsibilities" in the North and West through its non-enforcement of the Civil Rights Act of 1964, which mandated the cutoff of federal funds to school districts that maintained illegally segregated facilities.[15] Representatives of the agency responded to Sirica's ruling by releasing their long-awaited findings against the Flint Public Schools. The arrival of the HEW report in August 1975 marked the official commencement of school desegregation negotiations in Flint.

A full five years transpired between the conclusion of the HEW investigation and the onset of the desegregation process in Flint. During that long period, conflicts over school desegregation swept across the country. The controversy escalated in 1971, when members of the US Supreme Court, in the *Swann v. Charlotte-Mecklenburg Board of Education* case, unanimously sustained countywide busing as a remedy for legal and administrative school segregation in metropolitan Charlotte, North Carolina.[16] Two years later, in a lawsuit that held important implications for northern and western school systems, the Supreme Court ruled in its *Keyes v. School District No. 1* decision that the local board of education in Denver enforced a de jure form of segregation through the racial gerrymandering of attendance boundaries.[17] By finding a large school system outside of the South guilty of state-sponsored segregation, the justices chipped away at the legal basis of regional exceptionalism.

With its deeply entrenched patterns of segregation, Michigan quickly became a key battleground in the nation's school desegregation saga.[18] In February 1970 US District Court judge Damon Keith found officials in the school district of Pontiac, located just thirty miles southeast of Flint, guilty of intentional segregation.[19] To protest Keith's ruling, which ordered the busing of pupils to achieve racial integration, whites in neighboring Oakland County organized chapters of the National Action Group (NAG) and Save Our Schools (SOS).[20] Within weeks, the two groups had attracted thousands of white supporters in Genesee County and other areas of southeast and mid-Michigan. Although Flint had not yet been the subject of any school desegregation litigation, the Pontiac ruling inspired white parents and homeowners in the Vehicle City to form their own chapters of SOS and NAG. In September 1971 five hundred whites attended a raucous SOS rally at Flint's Zimmerman Elementary School to denounce desegregation.[21] These and other preemptive actions affirmed the depth of white resistance to school desegregation in Genesee County.

Prior to the 1970s no court had mandated "cross-district" metropolitan busing to remedy school segregation in the North or West. Judges had ordered busing in the Pontiac and Denver cases, but only within single school districts. In Flint, however, such an approach could not have succeeded in securing stable integration. Between 1950 and 1970, the proportion of black pupils in Flint's elementary schools increased steadily from 12.6 percent to 42 percent. By the time HEW issued its findings in 1975, that figure had risen to 50 percent, well beyond the "tipping point" at which whites typically fled big city schools. In all probability, therefore, a city-only desegregation plan would have simply hastened white departures from the Flint Public Schools and the city itself. As was the case in other cities with large black student populations, education officials in Flint could not achieve racial integration without implementing a metropolitan desegregation plan that included the nearly all-white school districts of the suburbs.[22] Because Genesee County included twenty-two jurisdictionally independent school districts, however, only the courts could order such a plan.[23] Yet before attorneys had filed such litigation, the United States Supreme Court effectively invalidated cross-district busing in the North.

In September 1971, in the *Bradley v. Milliken* case, US District Court judge Stephen Roth found that "governmental actions and inactions at all levels, federal, state, and local" had contributed to "de jure" modes of segregation in the schools of metropolitan Detroit.[24] To remedy what he viewed as a multi-jurisdictional problem, Roth ordered state officials to construct a massive desegregation plan that included over eighty school districts in metropolitan Detroit. White suburbanites revolted in opposition, however. In the aftermath of the *Bradley* decision, school desegregation emerged as the central political issue in southeast Michigan and perhaps the nation, fueling the rapid growth of NAG, SOS, and numerous other antibusing groups. Anger over busing and other issues pushed Michigan's white voters sharply to the right in 1972, providing former Alabama governor and renowned segregationist George C. Wallace with his margin of victory in the state's Democratic presidential primary. Notably, Wallace carried the UAW stronghold of Genesee County by a substantial majority in the 1972 Michigan primary election.[25]

The reaction to Roth's ruling in the judicial arena was equally polarized. In 1974, in a controversial decision that had profound implications for the desegregation process in Genesee County, the justices of the US Supreme Court narrowly overturned Roth's verdict. Though it endorsed busing within the Detroit city limits, the court's *Milliken v. Bradley* decision all but halted the quest for metropolitan desegregation in the North. Significantly, the justices issued their opinion a year prior to the release of HEW's findings on the Flint Public Schools. In the absence of metropolitan legal remedies, the battle over school desegregation in Flint would ultimately hinge on the resolution of the HEW enforcement proceedings and a last-ditch attempt from civil rights activists to win a favorable judicial ruling.[26]

Showdown

The federal government listed its charges against the Flint Public Schools under three broad categories: discriminatory hiring and staff assignment policies, unequal educational opportunities for students, and racially biased pupil assignment procedures.[27] Members of the Flint Board of Education responded to the allegations by issuing an open letter on November 3, 1975. Published in the *Flint Journal*, the letter signaled the board's refusal to accept blame for decades of administrative segregation and its continued embrace of neighborhood schools and community education. Though board members admitted to having a "preponderance of schools that are racially identifiable," they denied responsibility for the problem, arguing, "Segregation is not the result of deliberate policies or actions of the Flint Boards of Education of the last several decades." Instead, the letter writers blamed segregation on "housing patterns, economic factors and social mores once widely accepted."[28] Board members, emboldened by the growing hostility toward busing, listed seven restrictive criteria for the development of any school desegregation plan. Among the demands were two key conditions that indicated the board's basic opposition to all but voluntary forms of integration. First, the authors of the letter insisted, "[a]ny plan must have the involvement and support of the community." Additionally, members of the board stipulated that any desegregation plan "must reflect the integrity of the community education concept."[29] In recognition of the board's recalcitrant stand, HEW officers immediately forwarded the Flint case to their Office of General Counsel to begin enforcement proceedings.[30]

After issuing the open letter, Superintendent Clancy and the board of education held six public hearings throughout the city and solicited input regarding desegregation from at least fifty-four school-community advisory councils, nineteen civic organizations, and numerous parent groups. The feedback from community members included a vast array of recommendations that ranged from SOS's uncompromising defense of segregated neighborhood schools to proposals for extensive regionwide busing. In all, though, a clear majority of the presenters at the hearings favored voluntary desegregation over mandatory plans.[31] For their part, representatives of the Mott Foundation staked out a contradictory position as the crisis unfolded, simultaneously defending neighborhood schools and attacking both segregation and "forced desegregation."[32] In a familiar refrain, foundation personnel maintained, "Rigidly segregated housing patterns contribute to maintaining de facto segregation in Flint schools."[33] Superintendent Clancy, himself a former Mott Foundation intern, took a similar stance, embracing neighborhood schools and freedom-of-choice desegregation as essential to sustaining a vibrant program of community education. Meanwhile, Flint NAACP officials abandoned their earlier wait-and-see position and endorsed a metropolitan desegregation plan for the twenty-two school districts in Genesee County. With the group's coffers nearly empty, however, there were no funds to initiate legal proceedings.[34]

On March 10, 1976, the Flint Board of Education released its voluntary desegregation plan.[35] As expected, the plan preserved neighborhood schools for the overwhelming majority of the city's 21,607 elementary students. The proposal did include thirteen new magnet and specialty schools, but they were purely voluntary. At the junior and senior high school levels, the board's

FIGURE 9.1. Flint Public Schools superintendent Peter Clancy, 1974. While serving as district superintendent during the 1970s, Clancy spearheaded the opposition to school desegregation in Flint. In the
end, Clancy and other proponents of "neighborhood schools" succeeded in resisting federal demands for
mandatory pupil desegregation. Courtesy of the *Flint Journal.*

remedy was similarly flaccid. For those students and parents who chose segregated neighborhood schools over magnet schools and "open academies,"
the board would not impose desegregation.[36] Dismayed by the board's timid
proposal, Melvyn Brannon of the Flint Urban League remarked, "I think
the Flint Board of Education is being terribly, terribly optimistic in thinking that people will voluntarily take advantage of the options."[37] Equally un-
impressed, HEW announced its opposition to the Flint plan in May 1976,
proposing instead a limited system of mandatory school pairing for eight
adjoining elementary districts.[38] Despite the opposition, the board moved to
implement the voluntary plan in the fall of 1976.[39]

Although white parents and civic elites embraced the board's plan, it
drew widespread criticism from Flint's black community. Frustrated by the
board's unwillingness to force integration and HEW's reluctance to withdraw
over four million dollars in federal funding from the district for its noncom-

pliance, a group of black citizens ultimately bypassed the local NAACP in pursuit of legal action. On July 20, 1976, the parents of eight black students joined members of a group called the Black Teacher's Caucus in filing a metropolitan desegregation lawsuit against the Flint Board of Education, HEW, the state of Michigan, and twenty-one surrounding school districts in suburban Genesee County. Unlike the HEW officers, who hoped to implement a city-only remedy, the plaintiffs in *Holman v. School District of the City of Flint* sought to desegregate all of Genesee County's schools through a metropolitan desegregation plan.[40] Specifically, the complainants sought a de jure finding against each of the twenty-one suburban school systems in the county based on the racial gerrymandering of attendance zones, the long history of interdistrict collaborations, and regional pupil transfer agreements that allowed white children from the city to transfer to segregated suburban schools. By demonstrating that "acts of commission and omission" spawned administrative segregation throughout Genesee County, the plaintiffs sought to reverse judicial trends in the post-*Milliken* North.[41]

From the outset, the petitioners in *Holman* faced a difficult battle. To meet the restrictive judicial requirements established by the *Milliken* ruling, the plaintiffs had to demonstrate that pupil segregation resulted from the intentional and collaborative acts of officials from the Flint Board of Education and twenty-one independent suburban school districts. To support this claim, lead attorney William Waterman—who had successfully argued for school desegregation in Pontiac—presented a variety of exhibits featuring student demography as well as evidence that suburban pupils participated in Flint Public School ventures through cosponsored activities and programs such as Junior Achievement, the Youth Service Bureau, the Continuation School for Unwed Mothers, and the Genesee Vocational Educational Program. For their part, the defendants contested neither the existence of cross-district collaborations nor the presence of racial imbalances. Rather, defense attorneys argued that interdistrict racial imbalances were the outcomes of thousands of perfectly legal decisions made in the private housing market.[42] The defendants also maintained that if Waterman could not uncover specific examples of intentional board-sponsored segregation—and if the court could not trace those acts down to the level of the eight child plaintiffs—then the case had no merit.

On the issue of collaboration, the plaintiffs further contended that school superintendents in Flint and the suburbs routinely conspired to allow pupil transfers that took white students from racially transitional facilities in the city to all-white districts in surrounding areas. Here the charges centered on the collaborative relationship between the Flint Public Schools and the

Kearsley Community Schools, a predominantly working-class, all-white suburban district on the city's northeastern border. The partnership began in
1973, when officials from the tiny, perennially impoverished Kearsley system
implemented an open residency policy, which allowed cross-district pupil
transfers as a means of "filling empty chairs" while simultaneously generating
pupil tuition payments and increased funds from the Michigan State Board
of Education. According to Edward R. Hintz, Kearsley's school superintendent, the criteria for pupil transfers between Flint and Kearsley were exceedingly vague. "They would take our pupils if there was a special need," Hintz
stated, "like a youngster couldn't get along with his peers in our school, and
he wanted to go to Flint or Bentley or what have you because he thought he
would make it there, and graduate, or we had some pupils or vice versa, a
reciprocal type of thing."[43] Though the program was nominally colorblind,
Waterman claimed that Hintz and his subordinates accepted only white pupils into the district. Defense counselors countered by ridiculing the conspir-
acy claims: "Are Plaintiffs really alleging that 22 Boards of Education, 22 Su-
perintendents, the Governor of Michigan, the Attorney General of Michigan,
the State Board of Education, and the Superintendent of Public Instruction
got together and planned to discriminate against Plaintiffs?" "If so," they
continued, "it should not be asking too much for Plaintiffs to disclose what
hall was rented for that purpose and when the convention took place."[44] As
they had from the outset, defense lawyers argued that evidence of collaboration and segregation, no matter how compelling, was not enough to trigger
judicial intervention. Due to legal precedent, the defense maintained, Waterman and the plaintiffs had to prove that school officials in Flint, Kearsley, and
other districts intended to segregate students with their interdistrict transfer
arrangement.

Regarding property transfers, the plaintiffs argued that interdistrict land
and boundary agreements consistently brought black students to Flint while
relinquishing white pupils to suburban schools. Waterman supported that
claim by documenting property transfers from 1955, 1962, 1967, and 1969,
which "allowed approximately several thousand white students . . . living
in the Flint School District to attend schools in their [suburban] districts
in order to avoid attending schools with black students."[45] In response, the
defendants acknowledged both the property transfers and the enrollment
changes that resulted in segregation but denied that such outcomes stemmed
from any discriminatory or segregationist intent. Yet again, defense lawyers
countered Waterman's broad structural claims with a defense of de facto segregation that denied all but the most obvious, intentional, and exhaustively
documented acts of racial discrimination.

As the trial wore on, members of the Black Teacher's Caucus and Waterman's legal team struggled to prevail despite their limited financial and personnel resources. However, because the case involved so many separate school jurisdictions, legal research proved to be both time and cost prohibitive. Not wishing to deplete their meager financial resources on a polarizing lawsuit that had little chance of succeeding, leaders from the Flint NAACP donated only five hundred dollars to support the plaintiffs' efforts.[46] Further complicating matters, superintendents from the twenty-one suburban districts involved in the case consistently refused to release school board minutes and other important documents requested by the plaintiffs.[47] In addition, Stewart A. Newblatt, the presiding judge in the case, looked askance at many of Waterman's requests for information. The plaintiffs, Newblatt claimed on several occasions, were overburdening the suburban districts with too many requests for data.[48] Incongruously, though, Newblatt also repeatedly chastised Waterman and his colleagues for failing to produce specific documentation of de jure segregation.

Ultimately Newblatt determined that the plaintiffs had not proved their case. In his 1980 decision to dismiss the suit, the judge wrote, "[The plaintiffs have] repeatedly pled broad, non-specific allegations which encompass numerous theories of racial discrimination." "Without indicating a position on the validity of such claims," he added, "the court does not believe that such a shotgun approach should be tolerated, particularly when plaintiffs have been afforded time and opportunities to refine their claims."[49] Among those who had participated in the *Holman* proceedings, the rationale behind Newblatt's dismissal must have come as a surprise, for the same judge who had repeatedly resisted the plaintiffs' attempts at discovery ultimately threw out the case for a lack of specificity and evidence. Regardless, the decision marked the end of any viable plan for school desegregation in metropolitan Flint.

The Demise of Community Education

Shortly after the close of the 1976–77 academic year, as the *Holman* case languished in the judiciary and HEW officials and members of the board of education continued to wrangle over the terms of the city's desegregation plan, representatives of the Mott Foundation made an unexpected announcement that took the struggle over Flint's schools in a surprising new direction. On July 5, 1977, after a long private dialogue about funding priorities, foundation officials announced that they were ending their sponsorship of community education and severing all formal ties to the Flint Board of Education. "Except for administrative support grants," a foundation spokesperson said, "all present funding of FBE [Flint Board of Education] programs will be phased out within at most seven years."[50] Foundation representatives maintained that the school board could still solicit funds following the phaseout, but the announcement made clear that "FBE and its requests will be considered in the same manner as other organizations and their proposals." According to an April 1977 position paper authored by foundation trustees, the decision to cancel the forty-two-year relationship stemmed from a desire to transfer financial responsibility for community education to the taxpaying public. "If a program is assisted by a foundation long enough for its constituents to determine its value to them," the trustees claimed, "the program should in most cases pass to them for on-going funding." By cutting official ties between the foundation and the public school system, the trustees hoped to serve a larger purpose, though. The decision also reflected a desire among the foundation's leadership "to give FBE the freedom and encouragement to effect change in its program and to keep abreast of new challenges to an urban educational system."[51] The city's ongoing desegregation dispute constituted the greatest such challenge, of course, but the Mott Foundation appeared to want no part of it.[52] Indeed, after sustaining a decades-long partnership with the board of education, the Mott Foundation announced the end of its community education initiative at the precise moment when the Flint Public Schools faced its largest, most intractable crisis.

The July announcement sent shockwaves throughout the city. Financially, the decision constituted a devastating blow to the Flint Public Schools. Between 1935 and 1970, the foundation had donated nearly $1.5 million per year to the school system. By the mid-1970s that annual figure had soared to approximately $5 million.[53] With the 1977 policy shift, however, all of those funds vanished. Equally important, the decision stripped the city of its public education leadership during a period of extraordinary turmoil. Given the timing of the announcement, many residents wondered whether there was a connection between desegregation and the cancellation of community education. Did the foundation abandon the board of education in response to HEW's assault on neighborhood schools? Was the withdrawal of financial support an indication of the foundation's opposition to even limited forms of desegregation? Despite widespread demands for answers to such questions, Mott officials remained largely silent about their reasons for withdrawing support from the public schools. In fact, beyond explaining that the move would somehow empower city leaders and residents to solve their own problems, foundation personnel provided almost no additional information to the public regarding their decision. Clearly, though, segregation had been a defining feature of the community education program—so much so, in fact, that Mott officials may well have canceled the initiative to avoid complying with HEW's desegregation demands.

Although no concrete evidence supports a connection between the HEW proceedings and the end of community education, Charles Stewart Mott, Frank Manley, and other trustees of the foundation were, without question, staunch opponents of involuntary desegregation. During the two decades preceding his death in 1973, Mott was an uncomfortable witness to the civil rights revolution. In 1956, as southern resistance to the *Brown* ruling was congealing in the US Congress, Mott wrote warm letters of support to well-known segregationists such as Senator James Eastland of Mississippi, urging defiance. "I agree with all that you said, including 'State Rights' and Constitution," Mott informed the senator. "Certainly I agree with all that you said regarding the Negro Question."[54] Like Eastland, Mott opposed the *Brown* decision, viewing it as an unconstitutional assault on states' rights and racial gradualism.[55] As civil rights protests intensified during the 1960s, Mott moved even further to the right politically, and his defenses of segregation and "law and order" became much more pronounced. During the summer and fall of 1965, for instance, Mott exchanged cordial letters regarding community education and other matters with George Wallace, Alabama's openly segregationist governor.[56] Significantly, Mott authored his letters to Wallace just weeks after Alabama state troopers had assaulted peaceful voting rights protesters in the city of Selma. Mott avoided the harsh language and tactics of Wallace and other demagogues, but his personal commitment to Jim Crow ran deep, and his work in Flint's schools was an outgrowth of his belief that racial segregation was an essential prerequisite for community building and civic unity. Like Mott, Manley abhorred involuntary desegregation, once stating, "I am one of the kind that thinks that this busing is for the birds."[57] There is no question, then, that the two men most responsible for designing and overseeing Flint's system of community education were deeply committed to maintaining the color line. Anything beyond voluntary, gradual desegregation—and any proposal that undermined the city's prized neighborhood schools—threatened to destroy the delicate social fabric that Manley and Mott had spent much of their lives weaving.

Resolution

Despite their disappointment over the end of the Mott program, members of the school board and Superintendent Clancy remained committed to the principles of community education throughout the HEW proceedings. School board officials expressed that allegiance by continuing to embrace

their voluntary desegregation plan, even though it had failed to reduce pupil segregation. By 1978, 81.6 percent of Flint's elementary students still attended segregated schools, while thirty of the city's thirty-seven elementary
schools remained "racially identifiable." In May of that year, HEW officials
responded to the board's recalcitrance by instructing lawyers from the federal
Department of Justice to prepare legal action against the Flint school district.[58] Like their colleagues at HEW, investigators from the Justice Department had one overriding question in deciding whether to proceed, which
they communicated directly to the MCRC's Olive Beasley in 1978: "Did you
ever reach a determination as to whether the segregation [in the Flint Public
Schools] was of a de facto or de jure character?"[59]

Members of the board of education reacted to the threat of federal litigation with thinly veiled indignation. Buoyed by the local and national opposition to "forced desegregation," the board ignored HEW's threats of legal action and flatly refused to consider busing or any other involuntary approach.
The editors of the *Flint Journal* concurred with the board's decision: "It is
becoming apparent that extensive busing on the elementary level has failed to
reduce racial bias while being destructive of efforts to improve education."[60]
For their part, hundreds of white parents responded to the mounting conflict
over desegregation by placing their homes on the market and withdrawing
their children from the city's public schools. Between 1971 and 1980, the total
elementary school enrollment in the city declined by nearly 30 percent. Over
the same period, the proportion of black pupils in the Flint schools increased
to 59 percent.[61] With each passing year, the prospect of school desegregation
grew dimmer.

Beyond hastening suburbanization and galvanizing thousands of the region's suburban capitalist secessionists, the long duration of Flint's desegregation saga served to intensify white hostility toward the city's public school
system. In June 1978, white voters expressed their opposition to desegregation by overwhelmingly rejecting a thirty-four-mill school tax levy.[62] In an
election that saw voters split along racial lines, white critics of desegregation
joined the late 1970s tax revolt by rejecting the board's appeal for additional
funds. The vote was noteworthy because it represented the board's first tax
defeat since 1948.[63] The measure passed in the four wards that contained the
city's highest black concentrations, but it lost decisively in the five wards with
the largest white majorities. Even the UAW's local political action committee,
a perennial champion of public education, refused to endorse the school levy
of 1978.[64] Following the June defeat, the board of education announced that
the city's schools would close indefinitely unless voters approved a second
tax initiative on August 8, 1978. Fearing the complete shutdown of the city's

FIGURE 9.2. Flint-area residents protesting school segregation, 1979. In May 1979, members of the Flint
NAACP and other activists commemorated the twenty-fifth anniversary of the Supreme Court's *Brown
v. Board of Education* decision with a march through the city. During the event, speakers denounced the
Flint Board of Education's continued resistance to involuntary pupil desegregation. Courtesy of the Alfred P. Sloan Museum, Flint, MI.

school system, voters narrowly authorized the new levy by a slim margin of
two thousand votes. Once again, however, the votes split sharply along racial
lines. Under way since the 1950s, the city's slow march toward limited school
desegregation and a permanent black majority significantly eroded white
support for public education.[65]

The 1978 school tax election highlighted the fragility of the supposed postwar white consensus on neighborhood schools and public education. For decades white residents of Flint had expressed their overwhelming support for segregated public education by enthusiastically voting for new tax levies, participating in Mott Foundation programs, and, most important, enrolling their children in the city's schools. However, when federal intervention threatened to dismantle Flint's neighborhood schools, white taxpayers responded with acts of civic secession. In the end, the popular revolt against desegregation reverberated upward to shape the final resolution of the HEW crisis. Hoping to regain white taxpayers' support for the public schools, the Flint Board of Education resumed negotiations with HEW in the fall of 1978. As it had been since 1975, the key issue in the dispute was the board's refusal to consider busing and other involuntary forms of desegregation. Specifically, the school board declined to design a backup busing plan to implement in the event that magnet schools failed to reduce segregation. Lacking the authority to mandate systemic desegregation, HEW officials continued to withhold funding from the board of education while modifying their insistence on a pupil transportation plan.

By the close of the 1970s, only a small minority of Genesee County residents endorsed school desegregation. In fact, the issue was controversial even among civil rights activists, many of whom favored equal funding over desegregation as the surest path to educational equity. Ultimately HEW negotiators responded to the city's demographic shifts, the board's recalcitrance, and public opposition to busing by backing away from their insistence on mandatory desegregation, choosing instead to speed the resolution of the conflict by endorsing the expansion of Flint's magnet program. Much to the dismay of activists from the NAACP and the Black Teacher's Caucus, who were still holding out hope for a cross-district metropolitan remedy, representatives of the board of education and HEW signed a crisis-ending consent decree in the spring of 1980 that closely resembled the board's initial voluntary plan.[66] In exchange for the school board's promise to expand the elementary school magnet program from thirteen to eighteen schools, HEW agreed to drop its enforcement proceedings and submit a binding promise that it would not pursue any further legal action regarding desegregation.[67] Federal officials promised to fund the expansion of the magnet program by releasing $2.35 million to the Flint Public Schools that they had been holding in escrow pending the resolution of the negotiations. Although the accord allowed the federal government to establish enrollment goals to measure the magnet program's success, the consent decree contained no provision for involuntary busing. In effect, then, the signing of the 1980 agreement signaled the triumph of the neighborhood schools that for fifty years had formed the heart of a segregated system of community education.[68]

Reactions to the consent decree were polarized. As expected, the conservative editors of the *Flint Journal* cheered the end of the conflict, concluding, "The Flint Settlement, whatever its merits and deficiencies, is based upon a realistic appraisal of what could be accomplished at this time."[69] Meanwhile, many black critics denounced the 1980 agreement as a step backward for racial equality. For several days picketers from the NAACP and the Black Teacher's Caucus marched in front of the Flint Board of Education headquarters carrying signs that read, "The Board of Education and Segregation Are Still Shacking Together."[70] One of the protesters, Claire McClinton, condemned the resolution: "It is a segregation bid. The students have been sold down the river."[71] According to the agreement, which marked the formal end of the federal government's enforcement proceedings, the school board claimed neither legal nor historical responsibility for segregation in the Flint Public Schools: "The district denies legal liability in this matter and maintains that it has never and does not now operate its public schools in a racially discriminatory manner."[72] With the turn of a phrase and the stroke of a pen, the consent decree thus bestowed innocence upon the Mott Foundation and the Flint Board of Education by expunging their fifty-year record of deliberate administrative segregation. To black citizens such as Minnie Simpson, whose white neighbors attended different "neighborhood schools," or Wesley King, the child who had studied in a coat closet, the ultimate resolution of the desegregation standoff must have seemed utterly surreal.

In the years following the conclusion of Flint's desegregation crisis, the board of education implemented a voluntary, city-only magnet schools program that had little chance of succeeding. Just one year after the signing of the desegregation pact, principals of magnet schools reported their inability to meet interracial enrollment goals.[73] In a move that only exacerbated the situation, the federal government in 1981 cut nearly one million dollars from the board of education's desegregation budget.[74] In spite of a publicity campaign designed to promote the benefits of a "magnetized" education, white families consistently chose to remain in their neighborhood schools, enroll in private academies, or leave the city altogether.[75] Between 1980 and 1992, the proportion of white pupils in the Flint Public Schools dropped sharply from 52.5 to 29.6 percent, while the overall student enrollment plummeted to less than twenty thousand. As was the case in many other urban districts nationwide, the city's public schools underwent a rapid transition from a titular phase of desegregation to almost complete resegregation.[76] In 1992 African American school board member Paul Newman conceded defeat in a somber assessment of the desegregation process: "Our ability to integrate the school system as hoped is all but gone."[77] Though it provided opportunities for interracial contact among small numbers of children, the board's magnet schools program never overcame the twin legacies of administrative and popular segregation that had shaped the community schools movement since its inception. In the end, voluntary desegregation proved to be no match for decades of involuntary segregation.

The demise of Flint's community education program and the resolution of the desegregation crisis signaled the beginning of a new era of urban renewal in the Vehicle City. Beginning back in the 1930s, men such as Frank Manley and Charles Stewart Mott had sought to revitalize the city and its people by investing in public education. In the words of Mott's close friend Joseph Anderson, "He [Mott] was convinced that if we had a fine educational system, everything would follow."[78] The foundation's urban redevelopment strategy changed dramatically, however, in the era of civil rights and mass suburbanization. As Flint inched toward a seemingly permanent black majority and schools that were only nominally desegregated, foundation officials jettisoned their education-based approach to urban development in favor of a new bricks-and-mortar revitalization campaign designed to lure businesses, shoppers, and white suburbanites back to the city. Inherent in that policy shift was yet another ethos of urban renewal—one that emphasized place over people.[79]

"The Fall of Flint"

There was nothing foreordained about the deindustrialization of Flint. In the decades following World War II, GM executives and many other local employers began opening new plants and retail outlets outside Flint's municipal boundaries. However, in many instances these new suburban developments were not products of capital flight. On the contrary, GM and other area firms built most of their suburban facilities just beyond the city limits—close enough so that Flint residents could easily commute to work. Moreover, most of Flint's major business leaders openly supported annexation and other efforts to enlarge the city's boundaries. As an extension of their metropolitan capitalist strategy, members of the Vehicle City's progrowth coalition threw their support behind the 1958 New Flint proposal for regional government. When that plan foundered on the shoals of a burgeoning suburban capitalist independence movement, the city's business and political leaders looked to a combination of strip annexation and "slum clearance" in order to stave off economic decline. By the beginning of the 1970s, municipal officials had strip annexed a number of valuable suburban businesses, cleared hundreds of acres of residential land for redevelopment, and initiated construction on a new urban freeway and interchange, all in their attempts to lure white homeowners, tourists, jobs, and investors back to Flint.

Although annexation and redevelopment helped to sustain Flint's vibrant economy during the postwar decades, the 1970s marked a crucial turning point in the city's economic history. Collectively, the energy crises and economic slumps of that decade brought the city's postwar boom—and the nation's, for that matter—to a sudden halt. In response to the economic slowdowns and the ongoing loss of market share, GM implemented an austerity program that devastated Detroit, Pontiac, and many of its other plant cities. Between 1971 and 1991, GM cut its workforce in the United States by 50 percent. In previous decades corporate officials had shielded the Flint area from such painful job losses, even during major economic slowdowns. By the 1970s, however, GM's commitment to the city had waned. After failing to implement their metropolitan capitalist growth plan, company leaders gradually lost faith in their ability to control local politics. As that confidence declined, so too did their commitment to producing vehicles in Flint. Whereas the executives of the postwar period had seen Flint as a business-friendly company town, their counterparts in the 1970s and 1980s viewed the city as an inhospitable and unprofitable place due to its strong trade unions, high wages, obsolete infrastructure, and increasingly anticorporate political climate. Flint thus bore the brunt of GM's late twentieth-century restructuring campaign. Notwithstanding the numerous tax abatements and other incentives that GM received to keep jobs in the city, the corporation slashed nearly forty thousand local positions during the 1970s and 1980s while reinvesting heavily in new automated facilities in the so-called Sunbelt and abroad. These long-distance capital migrations, which created spatial mismatches that even car-owning commuters could not overcome, marked the onset of a new era of mass deindustrialization. The dismantling of GM's local manufacturing operations also led to a devastating human exodus from Flint. Between 1974 and 1982, the Vehicle City's population declined by a staggering 20 percent while mortgage defaults, business closures, and property abandonment surged to all-time highs.[1]

Nevertheless, Flint's municipal leaders continued in their efforts to revivify the city. The city's urban renewal program initially revolved around the St. John and Floral Park redevelopment projects. By the 1980s, however, the locus of revitalization had shifted to the downtown business district. As the industrial crisis intensified, city council members and downtown boosters designed a series of proposals to remake Flint into a center for research, education, commerce, and tourism. These efforts resulted in a flurry of new developments that included an automotive theme park, a shopping center, a high-rise office building, a waterfront "festival marketplace," a luxury hotel, and a branch campus of a major research university, all located in or near the heart of downtown. Within a few years, though, most of these developments had fallen into bankruptcy. At century's end, they stood as powerful reminders of the city's injurious experiment with urban renewal.

Crisis

The flight of jobs and industry from the Vehicle City and other industrial centers in the United States reached unprecedented proportions during the

1970s.[2] The decade began on a sour note with a grueling nationwide UAW strike against GM that ended in November 1970. The walkout ultimately resolved little, and it cost local workers millions of dollars in lost wages and pushed the city's unemployment rate to nearly 50 percent.[3] The bitter strike also put additional strain on the relationship between company leaders and local residents. That same year, downtown retail sales began dropping precipitously after the opening of the Genesee Valley Shopping Center in Flint Township. Long dissatisfied with the urban retail climate, executives from Sears-Roebuck, J. C. Penney, Hudson's, and Montgomery Ward promptly relocated their downtown stores to the new mall on Miller Road. Shortly thereafter, Smith-Bridgman, the Vogue, Woolworth, and several other downtown stores closed permanently. At around the same time, the famed Durant Hotel on Saginaw Street also shut its doors, another casualty of the downtown's long, painful decline. Between 1967 and 1972, sales receipts in the downtown area dropped by a startling 32 percent.[4]

In an attempt to recapture some of the city's dwindling tax base, municipal officials and GM executives turned yet again to the metropolitan capitalist strategy of annexation. This time, however, city leaders had their sights set on the industrialized suburb of Burton Township, home to a portion of GM's AC Spark Plug complex and the Eastland Mall. As annexation rumors swirled, township residents organized an aggressive municipal incorporation

FIGURE 10.1. The shuttered Vogue clothing store in downtown Flint, 1986. The late twentieth-century collapse of Flint's industrial economy had a devastating impact on downtown retail sales. The crash forced the owners of the Vogue, whose boarded-over storefront appears here, and many other businesses to close their doors permanently. Courtesy of the *Flint Journal*.

drive. Their campaign came to a head on April 13, 1971, when suburban vot-ers, over the objections of GM's leaders, overwhelmingly chose incorpora-tion. The defeat served as a reminder to local executives that their dreams of a metropolitan capitalist empire in Genesee County would go unfulfilled. Equally important, the incorporation of Burton left the city of Flint in an extremely vulnerable economic position. Though three of the four townships surrounding the city remained unincorporated in the aftermath of the 1971 election, suburban capitalists in those areas had made it clear that they were staunchly opposed to surrendering any land to the city—even if that meant choosing incorporation. According to Patrick Martin, the city's coordinator of economic development, the failure to annex meant that Flint no longer possessed the land necessary to lure new investment. "If any large manufac-turer wanted 100 acres," he claimed, "we couldn't begin to accommodate it." Without room to grow, the city faced a dark future.[5]

Although the incorporation of Burton proved to be a major blow to Flint's fortunes, the real economic shock arrived two years later, in 1973, cour-tesy of a worldwide energy crisis. Throughout most of that year, strong GM sales led to a robust economy in metropolitan Flint. Overall, the company's dealers sold 6,512,000 vehicles during the year, breaking all previous records. The increased sales helped to drive GM's local workforce to 73,900, a four-year high. All of that ended on October 17, however, when members of the Organization of Petroleum Exporting Countries (OPEC) announced an oil embargo against the United States and other nations that supported Israel in the Yom Kippur War. Within months, the OPEC action had resulted in depleted oil supplies and soaring fuel prices throughout the nation, which in turn drove consumer interest in smaller, energy-efficient Japanese cars to historic highs. With only a few compact vehicles in their lineups, American automakers suffered severe sales declines during the crisis. Between 1973 and 1975, domestic car and truck sales dropped from 12.6 million to 9.3 million units.[6] At GM sales decreased by 23 percent in 1974 alone. By 1975, purchases of GM's cars and trucks had fallen to a thirty-year low. In their 1975 annual report to stockholders, the company's directors painted a bleak picture of the business climate: "Inventories of unsold cars were alarmingly high, produc-tion was sharply curtailed, two of every five hourly automobile workers were laid off, consumer confidence was never lower, and in showrooms across the country shoppers were few—and buyers were even fewer."[7]

The national economic recession of 1973–75 hit Flint, Detroit, Chicago, and other manufacturing centers with tremendous force.[8] In December 1973 Buick general manager George Elges reported that his division would lay off 5,675 employees, including the entire second shift of assembly workers.[9] Two

TABLE 10.1. Average number of General Motors employees in the Flint metropolitan area by year, 1946–85

Year	Avg. no. of GM employees	Year	Avg. no. of GM employees
1946	42,600	1966	76,700
1947	50,900	1967	72,600
1948	48,700	1968	74,900
1949	54,900	1969	76,700
1950	63,700	1970	60,500
1951	57,800	1971	69,500
1952	53,400	1972	68,900
1953	67,300	1973	73,900
1954	72,900	1974	63,200
1955	82,200	1975	59,600
1956	74,400	1976	67,300
1957	70,400	1977	73,600
1958	56,600	1978	76,900
1959	59,300	1979	76,000
1960	65,500	1980	62,600
1961	60,900	1981	65,700
1962	66,500	1982	55,400
1963	68,100	1983	58,000
1964	67,300	1984	61,900
1965	75,900	1985	62,600

Source: Thomas R. Hammer, *Evaluation of Development Potentials for Metropolitan Flint, Michigan* (Evanston, IL: Northwestern University, 1986), 2.

months later, GM announced an additional round of layoffs affecting over fifteen thousand workers at the Buick and Fisher Body 1 plants.[10] By 1975 GM's area workforce had dropped to just over fifty-nine thousand.[11] Within the city joblessness reached a peak in 1975, when Flint's unemployment rate ranged between 15 and 20 percent.[12] Among young African Americans, however, the unemployment rate during that year approached 50 percent.[13] "It is painfully obvious," civil rights activist Wylie Rogers noted, "that our community sits squarely in the teeth of the storm."[14]

Legislators in Lansing and other state capitals responded to the economic crisis of the early 1970s by enacting a large number of procorporate tax reforms.[15] In 1974, members of the Michigan legislature passed Public Act 198, a bill designed to stimulate investment in depressed urban areas. The new law, intended to burnish the state's reputation with skittish corporate investors, granted municipal officials the authority to abate up to 50 percent of a company's property taxes for as many as twelve years on newly constructed industrial facilities. For the rehabilitation of "obsolete" plants, employers could receive property tax abatements of up to 100 percent.[16] The law had an immediate impact on local economies in Michigan. Between 1974 and 1983, municipal governments statewide granted 3,228 industrial tax abatements worth an average of $2.2 million per firm. Locally, lawyers from GM filed abatement requests for each of their plants. Desperate to stimulate new investment, members of the Flint City Council approved all of the company's proposals, as did officials in Swartz Creek, Genesee Township, and the new city of Burton. By 1983 the Flint City Council had authorized fourteen abatements for the automaker on over $400 million worth of local properties. The value of these abatements totaled $2,937 for every resident of Flint when calculated on a per capita basis. When asked about the wisdom of the abatements, former city manager Thomas Kay, by then a spokesperson for the Manufacturers Association of Flint, declared without reservation, "GM is Flint's best citizen and deserves every break they can legally get."[17]

Although most city council members accepted Kay's logic and saw the abatements as a way to reestablish the company-town bond, many Flint residents vehemently disagreed with the tax cuts. Critics of the abatements argued that corporate tax breaks did little to create new jobs and further impoverished the city during an already difficult economic period. In 1973 the city of Flint received over forty-four million dollars in revenue. By 1978, however, municipal receipts had fallen to slightly under forty million dollars.[18] Appalled by the city's declining fiscal fortunes, activist and corporate watchdog Michael Moore urged the city council to rethink its abatement policy, arguing, "GM has absolutely no plan to not only create jobs here in the City, they plan to eliminate jobs. Why do you keep giving them little presents as they continue to pull jobs out of the City?"[19] "This old friend [GM]," added corporate critic Barry Wolf, "has used the city the way a pimp uses a whore."[20]

Michigan's controversial abatement law required the recipients of tax breaks to invest in their local economies by erecting new buildings, rehabilitating old ones, or purchasing new equipment. Yet, in a surprising omission, the law did not stipulate that these employers had to create new jobs. In fact, as Moore and others often asserted, the wave of abatements issued during the 1970s and 1980s actually coincided with a net loss of nearly fifteen thousand local positions at GM. Together, the potent combination of GM layoffs, tax abatements, federal spending cuts, and runaway inflation created a severe fiscal crisis at city hall. Between 1980 and 1982, Mayor James Rutherford laid off six hundred municipal employees in an attempt to fill a seventeen-million-dollar budget hole.[21] He also authorized drastic cuts to city services, including a switch from weekly to bimonthly trash collection. As garbage piled up all over town, Flint's rat population exploded—so much so, in fact, that in 1986 officials from the Genesee County Health Department revealed that there were more rodents than people living in the city.[22]

Despite numerous signs that the abatements were not producing the desired results, Mayor Rutherford and others at city hall remained committed to a supply-side approach to municipal economics, arguing that the tax cuts at the very least would preserve existing jobs. He and other supporters of the abatements hoped to send a powerful message to corporate leaders that local officials, even if they were unable to deliver a growth-friendly regional government, could offer benefits to industry that rivaled the lavish subsidies available in the states of the South and West. Without explicitly saying so, Rutherford and his allies wished to reassure company leaders that Flint was still a company town. Phillip Hoffman, a manager of GM's local fiscal operations, responded favorably to Rutherford's overtures, maintaining that the tax cuts were essential because they helped executives determine where to build future plants. "If there is a tax incentive available," Hoffman told council members in January 1979, "that will enter into the annual operation costs of any new project and would be added to the bottom line."[23] Among Rutherford and his supporters on the city council, statements such as Hoffman's served as a reminder that the abatements were necessary. To GM's critics, though, Hoffman's words sounded like coercion.

Ultimately corporate tax breaks did little to halt the outward flow of jobs from Flint and other struggling cities.[24] After suffering devastating market share losses in 1974, GM sales and profits rebounded dramatically between 1975 and the first half of 1979. In 1978 the company's worldwide sales increased to 9.5 million units while its local employment roll rose to seventy-eight thousand.[25] Flush with optimism, Buick's general manager George Elges offered a bold statement in 1977 on Flint's place in the automaker's future: "Without question, Flint . . . now fits into the overall corporation pattern in a more solid and secure manner than it ever did before."[26] Elges was no prophet, however. Just two years later, the combined effect of high interest rates, inflation, and a second worldwide energy crisis helped to drive the nation's economy into yet another grueling recession. Although the crash hit Flint especially hard, no region of the United States managed to escape its wrath. Between 1978 and 1982, the country's largest employers—that is, companies such as GM, which employed more than one hundred workers—eliminated nine hundred thousand jobs per year. In all, one in fourteen Americans lost their job as a result of the plant and business closures of the late 1970s and early 1980s.[27] General Motors proved to be no exception. In 1980 alone GM's sales declined by 17 percent and the corporation lost $762.5 million, its first annual loss since 1921. By the end of the year, executives in Detroit had cut the company's domestic workforce from 468,000 to 376,000.[28] Locally, GM's 1980 layoffs pushed the unemployment rate to over 16 percent, the fourth highest in the United States.[29] The 1980 austerity measures were just the beginning, though. The following year, Roger B. Smith, GM's chief executive, vowed to cut "megajobs" and make the company a "leaner, tougher, and better corporation."[30] In his quest to make GM tougher, Smith invested over eighty billion dollars worldwide on new automated factories, product redesign, and plant modernization. To make the company leaner, though, he ordered a devastating series of layoffs, plant closures, and tax protests.[31]

Company v. Town

The economic turbulence of the 1970s helped to give rise to a powerful antitax movement in the United States. Drawing widespread support from conservative-leaning white property owners, many of them suburbanites, this revolt first gained national attention in 1978, when voters in the state of California passed Proposition 13, a controversial law that cut and capped property taxes. In the wake of the California election, voters from across the nation began demanding similar reforms. However, states such as Michigan proved to be less fertile terrain for the grassroots antitax insurgency. Several months after the passage of Proposition 13, conservative activists in Michigan placed three separate tax limitation proposals on the statewide ballot. Yet voters decisively rejected two of the three measures, including the "Tisch Amendment," a plan similar to Proposition 13. Michiganders did endorse Proposal E, the "Headlee Amendment," by a slim margin, but this law was significantly less severe than the tax rollbacks enacted in California and other states.[32]

Unlike their peers in California, antitax protesters in Michigan unanimously viewed the 1978 election as a defeat. Two years later, Republicans mounted a second campaign for tax reform by placing Proposal D, another version of Tisch, on the ballot. Once again, though, voters repudiated the initiative. Undeterred by these setbacks, conservatives continued to fight for California-style tax relief, but Michiganders resoundingly rejected these proposals in virtually every instance. Between 1970 and 1992, Michigan citizens voted down eight of the nine antitax propositions that appeared on the statewide ballot. In 1994, antitax crusaders won an important victory with the passage of Proposal A, which limited property assessments and increased the highly regressive state sales tax. Nevertheless, Proposal A, like Headlee, did not provide the sorts of tax rollbacks that most conservative political activists endorsed.[33]

The antitax movement in Michigan never gained the same traction among grassroots activists that it did in California and other states. To be sure, thousands of white homeowners from the Flint area and elsewhere actively supported the tax protests of the late twentieth century. However, these activists usually found themselves in the minority, even during the height of the 1970s rebellions. Moreover, most Michiganders, even those who embraced the Tisch Amendment and Proposal A, viewed taxes with a profound degree of ambivalence. As evidenced by the polarizing 1978 school levy elections in Flint, tens of thousands of white homeowners in the Vehicle City loathed paying taxes to fund government programs that allegedly offered disproportionate aid to African Americans and the poor. Moreover, many ordinary whites deeply resented what they saw as impersonal bureaucracies operating at city halls, in Lansing, and in the nation's capital. But many more happily supported local taxes for new schools, roads, sewers, and other improvements, especially those located close to home. Consequently, the antitax coalitions of the 1970s and 1980s often disbanded within weeks of statewide elections as supporters with opposing local interests and viewpoints went their separate ways. Indeed, six months after the 1980 Tisch defeat, *Flint Journal* reporter Bob Sherefkin noted that the tax protest movement was "losing steam" in the Flint area.[34]

This was not entirely true, however. In Michigan and other states, GM executives helped to lead a powerful corporate campaign against taxes that flourished even as grassroots antitax protests were faltering.[35] In 1985 lawyers from GM filed a series of appeals against the company's property tax assessments in more than two dozen cities nationwide. The challenges in Michigan, all filed with the state tax tribunal, included a request for a $178 million reduction in property appraisals for plants in Flint, Genesee and Grand Blanc Townships, Burton, and Grand Blanc. Just in the city of Flint, where the appeal cited grossly inflated assessments dating back to 1982, GM sought to have its property valuations reduced from $303 million to $170 million.[36] At stake was over $30 million in contested tax payments and, many feared, the last remnants of Flint's tattered reputation as a company town. Like GM's abatement requests, the property tax appeal generated a hostile response from community activists, left-leaning public officials, and trade unionists. Even the editorial board of the *Flint Journal*, traditionally a staunch supporter of local manufacturers, publicly opposed the company's position. Led by Michael Moore and a local UAW activist named Mike Westfall, the campaign against GM's tax appeal gained support from prominent liberal activists across the country such as consumer critic Ralph Nader and actor Edward Asner. Mayor James Sharp, Flint's first popularly elected black chief executive, similarly vowed to fight GM, stating, "We are not going to balance GM's budget on the backs of our school children."[37]

Notwithstanding the breadth of opposition to the corporate tax revolt, the battle against GM's appeal proved to be a much more costly and difficult endeavor than local activists ever imagined. During the first year of the dispute, city leaders spent $1.5 million just to cover legal fees and property appraisal expenses. As the costs mounted and the city's economic crisis deepened, municipal officials began negotiating to resolve the controversy. In May 1988 Flint city attorney Richard Figura announced an agreement on GM's appeal for the Fisher Body 1 plant. The settlement included a tax refund of $426,667 for GM to be shouldered by the city of Flint, Genesee County, the Flint Community Schools, the Genesee Intermediate School District, and Mott Community College. Speaking to reluctant city council members just prior to their vote on the proposal, Figura acknowledged that the payout would provoke much bitterness among Flint's many unemployed workers. Yet he also pointed out that the city could not afford a sustained dispute with GM. With the city's tax base already ravaged by home foreclosures, layoffs, tax abatements, and impending plant shutdowns, all but the most strident of GM's opponents supported the measure. On May 25, 1988, city council members voted by a seven-to-two margin to approve the deal.[38] Several years later, municipal officials reached a final accord with GM on the remaining plants. The settlement called for the city to refund $34 million and provide the corporation with a 30 percent reduction in property tax assessments. For their part, company officials agreed to forgive all interest on the city's debts and coordinate a long-term payment plan. In truth, though, neither side left the negotiating table feeling satisfied. Throughout the city, residents and policy makers felt as though they had been defeated in an unwinnable and unfair war of attrition. "GM wore us down," state senator Joe Conroy admitted. "We couldn't afford to fight them anymore." At GM, by contrast, corporate decision makers interpreted the dispute as yet another sign that Flint's once-friendly business climate had vanished.[39]

The property tax dispute with GM coincided with a period of great trial for American workers, especially those in the auto industry.[40] As Michael Moore and other activists predicted, Roger Smith's restructuring campaign brought a new wave of layoffs and plant closures to metropolitan Flint. Locally, GM's disinvestment program reached a peak in 1986 and 1987. In December 1986 Flint's Chevrolet V-6 engine plant closed permanently. Five months later, GM shut down one assembly line at the Chevrolet truck and bus facility on Van Slyke Road. Meanwhile, thousands of workers from virtually all of the region's facilities received layoff notices. The combined effects of factory closures and layoffs reduced the automaker's Genesee County workforce from nearly eighty thousand in 1978 to just forty-eight thousand in 1987. With so few industrial facilities left in the city, members of the Manufacturers Association of Flint voted to disband in 1987. Throughout this period of retrenchment, the Vehicle City's unemployment rate routinely exceeded 25 percent.[41] One of the newly unemployed autoworkers was a woman known publicly only as Beth P., who suffered a nervous breakdown after losing her job at the V-8 plant. In an emotional statement provided to Mike Westfall, she revealed, "I feel anger, depression, uselessness, uncertainty, let down and I am not at ease with myself. I'm losing my house and everything I have always dreamed for. It tears me up to know that I can't do anything but sit and watch everything go down the drain."[42]

The 1990 federal census revealed a city in deep economic crisis. That year, the Vehicle City's unemployment rate of 18.3 percent was one of the highest registered in the United States. With so few people working, the per capita income for residents of Flint dipped to just $10,415, substantially lower than the statewide figure of nearly $15,000. The city's overall poverty rate of 30.8 percent was nearly three times as high as that in the state of Michigan as a whole. As was the case throughout the United States, the economic devastation of the 1970s and 1980s had a disproportionately harsh effect on children, people of color, and women. Among children residing in the city of Flint, the poverty rate in 1990 exceeded 44 percent. Within the black community the poverty rate was 36.4 percent, nearly double that for whites. While not as wide as the racial gap, the economic disparity between men and women was also striking. According to census records, women workers living in Flint earned median incomes that were only two-thirds as high as those of their male counterparts. Likewise, female poverty rates in Flint and across the state of Michigan were nearly five percentage points higher than men's were. The persistent economic gulfs between men and women in Flint and across the United States led many scholars and social commentators to conclude that the 1970s and 1980s witnessed a widespread "feminization of poverty." This was undoubtedly the case in Flint.[43]

The city's urban renewal and freeway construction projects were supposed to prevent the sort of economic shocks that beset Flint and its people in the 1970s and 1980s. In actuality, though, the economic growth that boosters promised would accompany I-475, the Floral Park interchange, and the St. John Industrial Park projects never materialized. At a total cost of more than one hundred million dollars, the freeway created few new jobs and generated only short-term private investment in Flint's manufacturing and commercial sectors. Moreover, as in other cities, the freeway and interchange

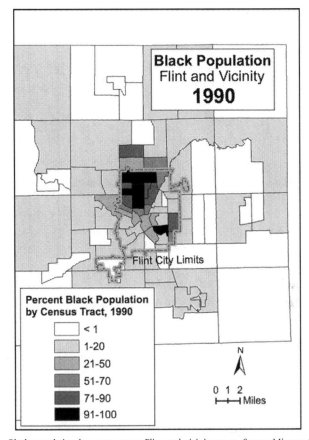

MAP 10.1. Black population by census tract, Flint and vicinity, 1990. *Source*: Minnesota Population Center, *National Historical Geographic Information System: Version 2.0* (Minneapolis: University of Minnesota, 2011).

projects removed hundreds of acres of taxable properties from city rolls and facilitated the suburban migrations of both corporate capital and white taxpayers.[44] For those who remained in the neighborhoods bordering St. John and Floral Park, the freeway and interchange undermined commerce, impeded pedestrian traffic, and exerted a negative influence on long-established community ties. "The expressway has eliminated a lot of homes, and it has eliminated a lot of the community spirit," observed Angela Sawyer.[45] Equally important, the transportation projects—which resulted in the closure of no fewer than eighty-five east-west roads connecting the black enclaves of the North End with the predominantly white east side—sharply inhibited travel between neighborhoods and ossified the city's already stark patterns of administrative segregation.[46] Nowhere was this clearer than in the far north side of the city, where the completion of I-475 created a new St. John–style island

neighborhood of three hundred homes, almost all inhabited by black families, sandwiched between the freeway and a forbidding web of railroad lines.[47]

The freeway and urban renewal projects did enable several new business
openings in and around the North End, but the resulting job creation was
minimal. Aside from that, not all of the new enterprises benefited the region
and its people. One such venture was the Genesee Power Station, an eighty-
million-dollar waste incinerator and electric power–generating facility that
opened in 1992 in an area just north of the demolished St. John neighborhood.
Hoping to make a profit from the mountains of demolition waste created an-
nually by urban renewal and other activities in the Flint region, the operators
of the facility designed it to run entirely on wood fuel. The burning of the
demolition wood also created a substantial amount of electric power, which
the operators of the plant sold to Consumers Power Company, the Flint area's
primary energy provider. Much of the wood burned at the Genesee Power
Station was coated in lead-based paint, however, which resulted in the release
of dangerous toxins into the air. Although local residents and activists were
quick to point out the many dangers associated with lead exposure—which
include high blood pressure, kidney and brain damage, premature birth, and
learning disabilities—officials from the Michigan Department of Environmental Quality, unwilling to block a job-creating venture in the midst of a
long-term economic slump, ultimately approved the plant's operations. Undeterred, local residents joined with members of the Flint NAACP and an ad
hoc group called United for Action in filing suit to close the plant. In the end,
though, the courts allowed the plant to remain open. For the citizens of the
North End who had suffered through decades of popular and administrative
Jim Crow, endured foundry pollution and the city's disinvestment campaign,
and then lost their homes to urban renewal only to be slowly poisoned by the
burning of some of those same toxic structures, the irony and indignity of
the incinerator's existence must have been impossible to escape.[48]

The completion of the North End renewal projects also resulted in the
opening of the St. John Industrial Park, a $30.5 million project that plan-
ners predicted would create thirty-six hundred new jobs.[49] Yet in the end the
industrial park produced only five hundred new positions. Making matters
worse, the disappointing job-creation numbers at the industrial park sent
North End real estate prices tumbling. For the leaders of Buick, the completion of I-475 in 1981 brought improved deliveries from the Fisher Body supply
plant. Those deliveries ended when GM shuttered the Fisher Body 1 facility
in 1987, however.[50] Ironically, the same freeway that GM workers and plant
managers had demanded for decades actually hastened the demise of the historic Fisher plant, birthplace of the 1936–37 sit-down strike.[51]

Company officials sealed the fate of Fisher Body 1 in the late 1970s when they began developing a "just-in-time" production strategy modeled after the *kanban* system used in the Japanese auto industry. Proponents of the *kanban* method believed that manufacturing efficiency depended upon smaller work-forces, increased automation, decreased production time, and, crucially, spa-tially integrated manufacturing and assembly plants.[52] In order to reduce the high storage costs, damaged inventory, and production errors that resulted from stockpiling parts, GM transformed its massive North End facility and a portion of the St. John Industrial Park into a behemoth known as Buick City, a highly automated and centralized manufacturing and assembly complex that opened in 1985.[53] Unlike the company's earlier metropolitan capitalist strategy, which entailed a centrifugal movement of capital, the just-in-time model depended on a recentralization of production. Consequently, after the opening of the new North End complex, GM shifted the stamping and body-welding operations from the Fisher 1 plant on Flint's far south side to Buick City. Ultimately the opening of Buick City—an event made possible only by the completion of I-475—rendered Fisher Body 1 and its thirty-two hundred workers redundant. In 1987 GM closed the south side facility permanently.[54]

In the absence of stockpiled supplies, Buick City's productivity hinged on a constant stream of components delivered via freeways. Initially corporate designers had envisioned a North End complex that resembled the Ford Mo-tor Company's massive River Rouge facility, "a totally integrated plant with steel entering one door and a car coming out the other end."[55] With that goal in mind, Buick City officials sought to lure as many suppliers as possible to the vicinity of the St. John Industrial Park. Though several supply firms did, in fact, purchase land adjacent to Buick City, they created few new jobs. In an ironic twist of fate, the freeway itself repeatedly undermined GM's plans to recentralize production on the North End. Citing the efficiency of Flint's interstate highway system, all but a handful of Buick's suppliers declined relo-cation proposals from GM and the city of Flint. Why relocate to Flint, many of these business owners asked, when the freeways made the city so acces-sible? Although General Motors invested three hundred million dollars in the Buick City project, the new plant was never as profitable as executives had hoped it would be. In 1986, just one year after the complex opened, GM laid off thirteen hundred Buick City workers.[56]

Desperate to find work, nearly thirty thousand residents moved out of Flint between 1982 and 1987, many of them to southern and western cities such as Atlanta, Houston, and Phoenix. Just between 1985 and 1990, the city's population dipped by 8.6 percent.[57] As the migration unfolded, enterprising local bookstore owners began selling copies of the *Houston Post* and other southern and western newspapers with thick and enticing classified sections. Citywide, booksellers sold over one thousand copies of southern and western newspapers every day during the 1980s crisis. "The plain truth of the matter," asserted unemployed *Houston Post* reader Marcus Cleveland, "is that there's work down there and none up here. Period."[58] Over time, the mass migration from Flint to points south created numerous difficulties for the proprietors of local moving companies, who could not keep their vans and trucks in the city. "All our equipment is down south," complained U-Haul employee Jerry Clark. "I don't have a third of the equipment that I had a year ago."[59] Unable to turn a profit, business owners throughout the city responded to the exodus by closing their shops in record numbers.[60]

The economic malaise of the 1970s and 1980s brought Flint to the verge of bankruptcy. The fact that Mayor Rutherford had to cut garbage collection services to twice a month was only one indication of how dire the financial situation had become. Throughout the city, whole neighborhoods began to look as though they had been abandoned, leading Neal R. Peirce, a journalist from the *Detroit News*, to refer to Flint as "Plywood City," a place where "there are so few people about that you might think the neutron bomb had hit."[61] During a 1980 interview with a writer from the *Detroit News*, one unemployed worker explained why the city seemed so empty: "Flint is folding up. There's nothing here, no opportunity."[62] Several years later, a young journalist named Alex Kotlowitz encountered even grimmer evidence that the city had fallen on hard times. Tacked in front of one North End home he found a handmade sign that read, "Night Crawlers $.60 a dozen. Open 24 hours."[63] Convinced that the city was doomed, in 1984 Michael Moore went so far as to write an obituary for Flint.[64]

Flint's "Great Leap Forward"

Moore's death notice proved to be pure hyperbole. In fact, even during the worst of the 1970s and 1980s economic meltdowns, Flint's civic boosters and elected officials refused to succumb to despair. Although the Floral Park interchange and Interstate 475 undermined the city's industrial economy, the new roadways did help to make Flint's downtown more accessible to tourists and other nonresidents. Yet the city still lacked the consumer outlets and cultural attractions necessary to lure new visitors and investors. To rectify the problem, municipal leaders launched yet another phase of urban renewal. Unlike the earlier campaigns, however, which entailed removing black residents from their neighborhoods, this next stage of revitalization centered on an effort to bring new people to the city's depressed downtown core.

Discussions about a comprehensive downtown renewal campaign began in earnest in the early 1970s. In April 1970, twenty-nine of the city's leading industrialists, retailers, bankers, and developers met in a northern Michigan resort area near Tippy Dam for a three-day conference on Flint's future.[65] By all accounts, the conclave was a major success. According to *Flint Journal* reporter Lawrence Gustin, "By the time the meeting broke up, the twenty-nine men were in an enthusiastic mood to solve problems." The mood was so euphoric that Harding Mott, vice president of the Mott Foundation, later termed it "the spirit of Tippy Dam." Upon their return to Flint, conference participants began holding weekly meetings to trade ideas on downtown redevelopment. In October 1971 they formed the Flint Area Conference, Inc. (FACI), "a nonprofit organization of civic and business leaders working with public officials in an effort to meet unfilled physical and economic needs of the community." The group's board of directors included local leaders such as Mott; J. L. Gillie, an executive from Consumers Power Company; Alfred W. Hewitt, a chief officer of the Michigan National Bank; former mayor and MAF leader Oz Kelly; and each of GM's local plant managers. By design, it seems, FACI's board members excluded representatives from labor and civil rights groups. As Gustin explained, "Since FACI will not be primarily devoted to solving social problems, no effort is being made to include all segments of the community."[66]

Like proponents of downtown renewal in Philadelphia, Boston, and other cities, the leaders of FACI believed that a revitalized business district would have a beneficial effect on the entire region.[67] According to FACI member and AC Spark Plug official Joseph Anderson, a vibrant downtown would generate a "renaissance of spirit, pride, and progress in the Genesee Metropolitan Area."[68] In pursuit of that goal, the new group hired several planning firms to create *Centric 80: A Revitalization Strategy for Flint*. Released in February 1972, the report included an ambitious plan to bring people and new attractions back to Saginaw Street and the rest of downtown Flint. Among the recommendations included in the report were a new downtown campus for the University of Michigan–Flint (UM-Flint), an automotive hall of fame near the Flint River, and other "people generating" facilities and activities on the riverfront.[69]

Although Harding Mott and other FACI board members welcomed the report's proposals, the economic recession of 1973–75 forced the group to shelve its downtown plans temporarily. Throughout the crisis, members of FACI continued to meet privately, though, and added new developments to their wish list. By the close of 1975, FACI, the Mott Foundation, and other civic groups had created a sweeping revitalization plan for the central business district that surpassed the modest recommendations included in *Centric 80*. In December representatives from the Mott Foundation and FACI publicly announced their support for AutoWorld, a downtown automotive museum and theme park; Riverfront Center, which included a high-rise hotel, shopping complex, and convention center; a Flint River flood-control and beautification program; and the Doyle project, a mixed-income residential development just north of the central business district. To help launch the enterprising plan, the Mott Foundation pledged $4.25 million in start-up grants, while the mayor and city council members offered their full support.[70]

The first obvious signs of downtown redevelopment surfaced in 1977, when UM-Flint unveiled its first campus building just south of the Flint River.[71] In 1979 Riverbank Park, the flood-control and beautification project, opened adjacent to the new campus.[72] At around the same time, the Windmill Place shopping center opened near downtown, and the Mott Foundation began purchasing land just north of the river for the AutoWorld museum and amusement park. Shortly after that, the thirty-million-dollar Hyatt Regency Hotel, also on the waterfront, welcomed its first customers.[73] To persuade visitors and shoppers to patronize these new developments, FACI leaders sponsored the "Buy Local" and "Flint Image Improvement" media campaigns.[74] Eager to do their part, writers from the *Flint Journal* championed the new developments, proclaiming that Flint's "Great Leap Forward" was officially under way. Supporters of the downtown projects even created a rap music anthem, "The Flint Booster," whose lyrics trumpeted the new and improved central business district:

> AutoWorld and Buick City
> Will help Flint gain vitality,
>
> Mayor Rutherford is turning Flint around
> And things are happening downtown,
>
> Flint is getting back in the race
> We have a nice new Windmill Place,
>
> U of M Flint's downtown Campus
> Has really helped all of us,
>
> We all can really have a lark
> At our splendiferous Riverbank Park
>
> The ripple effect from Hyatt Regency
> Is spreading to every community.[75]

Throughout this period of frenetic development, builders erected over a half-dozen major new attractions near the downtown waterfront.[76] Each of the projects was part of a broad strategy to lure new residents, shoppers, tourists, jobs, and investors back to the shrunken city. The real showpieces of the downtown campaign, however, were the Hyatt Regency Hotel, AutoWorld, and the Water Street Pavilion, a festival marketplace designed by acclaimed developer James Rouse.[77] AutoWorld was the linchpin of the overall program. After nearly two decades of planning, the seventy-three-million-dollar theme park opened with great fanfare in July 1984. The project, like most of the others downtown, had drawn financial support from a variety of public and private sources, including $31 million from the Mott Foundation, $36.5 million in public funds, $4 million from local donors, and nearly $9 million from Capital Income Properties, an East Coast investment firm. For their part, GM officials contributed $1 million to support the effort. In the lead-up to the grand opening, civic boosters raved that AutoWorld would be the largest enclosed amusement park in the world. In truth, this development—operated by the Six Flags Corporation—was equal parts theme park, museum, and shopping center. Unlike most shopping malls, though, it required an admission fee of $8.95.[78]

Beneath AutoWorld's studded dome—which bore a striking resemblance to a stegosaur's back—were a variety of attractions including a carousel, shops and restaurants, a carnival ride through "The Humorous History of Automobility," a movie theater, and a mock "Assembly Line of the Future" operated by robots. The park also contained a huge replica of downtown Flint, designed to resemble the area as it looked in 1900. This idealized version of the old downtown featured a climate-controlled main street, a river, comfortable benches, soothing music, and five hundred thousand dollars' worth of tropical foliage. The designers of AutoWorld hoped that the faux main street and other nostalgia-inducing features of the park would allow local visitors to escape the drudgery of life in Flint.[79] "Visit AutoWorld," one advertisement urged, "and leave the real world behind." For many area residents, the downtown simulacrum was the highlight of the AutoWorld experience because it brought back memories of "Old Flint," the city that existed before the plant closures, population loss, and perhaps desegregation. Joanne Ladd, a Burton resident who toured the complex, commented on how AutoWorld made her wistful for the Saginaw Street of her childhood: "The old time 'flavor' I remember of downtown Flint was there. It could only be topped if there had been a Kresge's and that candy store where you could get a phosphate in a soda glass." To some, though, AutoWorld served only as a reminder of how unattractive the real downtown had become. "Mostly," another visitor said ruefully, "I just remember thinking that being in 'Old Flint' was a lot nicer than being in the present Flint."[80]

In addition to "Old Flint," AutoWorld contained numerous exhibits designed to celebrate the history of the American automobile industry. Some of these, however, were emphatically racist. Several weeks after the opening, a group of Asian Americans objected to a display at the park that derided Japanese people and automobiles. The poster in question depicted "a car with caricatured Oriental features—buck teeth and slits for eyes . . . dive-bombing an aircraft carrier labeled 'Detroit' against a rising sun background." After receiving complaints from members of Citizens for Justice, an Asian American activist group, Six Flags spokesperson Kathy Schoch defended the artwork, claiming that it was part of a "satirical display." Surprisingly, she also suggested that tourists could construe plenty of other things in the complex as insulting. To illustrate her point, Schoch pointed out another poster, this one featuring an image of "a used-car salesman in a loud plaid coat with a large nose that might be taken as a Jew." "The pictures," she argued, "poke fun at all kinds of people—rednecks, housewives, and Texans. . . . We haven't had any comments or complaints about it." Clearly, though, AutoWorld's exhibits were deeply offensive to many visitors.[81]

Many others simply viewed the park as boring and overpriced. Prior to the July ribbon cutting, Mott Foundation president William S. White commissioned several studies to determine how many people would visit the facility. According to these reports, the development was supposed to draw around one million paying customers per year. Ever the optimist, White was confident that this goal was attainable. Even more than that, he assumed that throngs of tourists and locals would enjoy the amusement park, eat a meal downtown, and shop at the Water Street Pavilion. Some guests, he and others predicted, would even conclude their visit with an overnight stay at the Hyatt Regency Hotel.[82] These ideas turned out to be fanciful, though. By the fall of 1984, the crowds at AutoWorld had already started to dwindle, with the park appearing empty most days. For employees, most of whom earned the minimum wage, work at the facility was often excruciatingly dull. "I worked at 'The Humorous History of Automobility,'" a former employee recalled. "I had to take the job because jobs were scarce and I wanted to save up enough money to get out of Flint. I snorted a lot of ground-up No-Doz during those days, just because the repetitiveness of the job put me to sleep." "It is boring as hell," UM-Flint professor Neil O. Leighton fumed. "Even my sixteen- and twelve-year-old sons found it boring." According to one survey, 30 percent

FIGURE 10.2. A crowd of people waiting to enter AutoWorld, 1984. Hailed as one of the largest indoor theme parks in the United States, AutoWorld, with its distinctive dome and spiked roof, opened in July 1984 amidst great fanfare. Although civic boosters believed that AutoWorld would drive the revitalization of downtown Flint, it quickly closed due to poor attendance. By the end of the 1980s, AutoWorld and many of the other downtown developments erected as part of Flint's "Great Leap Forward" had succumbed to bankruptcy. Courtesy of the *Flint Journal*.

of AutoWorld's visitors indicated that they would not recommend it to their friends.[83]

When it became apparent that AutoWorld would not attract the customers it needed to remain profitable, the park's financiers moved quickly to close it down. In December 1984, just six months after the grand opening, officials from Six Flags announced that the park would be open only on weekends through the remainder of the winter. A month later, investors closed the facility altogether.[84] The building reopened for several brief periods over the next few years, but poor attendance continued to frighten away investors. In 1994 AutoWorld shut its doors permanently. Several years later, workers from the Best Wrecking Company in Detroit blew up the facility in an event downtown boosters predictably dubbed "Boom for Growth."[85]

The AutoWorld fiasco severely undermined the city's entire downtown renewal effort. Within a year of AutoWorld's initial closing, at least six of the stores in the new festival marketplace had gone out of business. Although the managers of the Hyatt Regency aggressively recruited conventioneers and other travelers, the hotel remained largely vacant on most nights. By

century's end, the Hyatt, Water Street Pavilion, and Windmill Place developments had all fallen into bankruptcy.[86] Significantly, the UM-Flint campus and a new office building owned by the state of Michigan survived the
downtown's collapse. However, these two commuter-oriented facilities failed
to attract enough people to support downtown businesses. Part of the difficulty stemmed from the fact that tourists and suburbanites alike clung to
the notion that downtown was dangerous in spite of overwhelming evidence
suggesting that it was the safest area in the city. Even among those who studied and worked downtown, though, most preferred to do their shopping
and dining elsewhere. This was in part a reflection of the university's housing policies and the architecture of the campus, which discouraged trips to
downtown stores. Because the new school had no on-campus dormitories,
most students commuted to class by car. After arriving downtown, students
and faculty parked their vehicles in one of several garages attached to classroom buildings by enclosed "skyways." Likewise, workers in the state's new
office building could avoid the street by using aboveground walkways that
connected their offices to parking garages. These overhead passages made it
possible for students, faculty, campus visitors, and government workers to
spend the day downtown without ever setting foot on Saginaw Street. And
many, in fact, did just that.[87]

Scores of downtown restaurants and shops closed during the prolonged
economic crisis of the late twentieth century. Even supposedly recession-
proof fast food establishments found it virtually impossible to survive in
Flint's depressed downtown core. In a 1986 report, researcher Thomas R.
Hammer noted that none of the major fast food companies had any fran-
chises left in downtown Flint. "In central Flint," he said, "you simply cannot
buy a Big Mac, Whopper, or Double with Cheese."[88] Willing to try almost
anything to turn the situation around, city workers and business owners attempted to mask the downtown's decline by painting faux storefronts onto
abandoned buildings. This act of desperation fooled very few people, though.
In 1986, as unemployment surged past 25 percent, *Money* magazine released
its "Best Places to Live in America" issue. Much to the chagrin of civic boosters, Flint came in last place.[89]

Roger and Me

The local battles over plant closures, property taxes, and urban renewal
helped to spawn a new generation of activists during the 1970s and 1980s.
The most noteworthy of these new organizers was Michael Moore. After losing his seat on the Davison school board, Moore moved to the Vehicle City,

where he founded the *Flint Voice*, eventually renamed the *Michigan Voice*, an alternative weekly newspaper. The newspaper attracted a talented pool of writers including Alex Kotlowitz and Ben "Rivethead" Hamper, a local auto-worker.[90] Moore and his colleagues won a wide following for their muckraking essays on unemployment, urban development, plant closures, and other local topics. However, Moore yearned for an even larger audience than his newspaper could generate. During the bitter campaign to defeat GM's tax appeal, he began considering a foray into filmmaking. In 1987 Moore revealed to a small number of friends that he was going to begin work on a film about Flint and General Motors, which he planned to title either "Dance Band on the *Titanic*" or "My Hometown." Later renamed *Roger and Me*, Moore's film debuted in theaters in 1989. Much to the surprise of his former nemeses in Davison, the movie quickly became one of the highest-grossing documentary films in American history.[91]

Roger and Me follows Moore across the country as he attempts to meet with his foil, GM chief Roger Smith. Throughout the film Moore hilariously stalks his elusive foe at shareholders' meetings, GM's corporate headquarters, a yacht club, and other locations—all in a desperate but unsuccessful attempt to convince the executive to come to Flint to witness the devastation caused by the company's plant closures. Interwoven throughout these comedic situations are poignant scenes of abandoned homes, tenant evictions, and interviews with local residents who reflect upon Flint's declining fortunes and the depressing fates of unemployed autoworkers. Like other articulations of the Rust Belt decline narrative, the storyline of *Roger and Me* turns on a series of juxtapositions between wealth and poverty, hope and despair, past and present. At several key moments in the film Moore uses cutaway shots and other cinematographic techniques to contrast scenes of postwar prosperity and abundance with the gloom of postindustrial Flint. During a 1990 taping of the *Phil Donahue Show* conducted on location in Flint, Moore elaborated on the past-present juxtaposition at the heart of the film: "Phil, you know, the American Dream used to be that if you worked hard, and the company prospered, you prospered. Now it's you work hard, the company prospers, you lose your job."[92] At its core, Moore's story is a tragic tale about corporate abandonment and the demise of the American Dream.

Despite its shoestring budget of two hundred thousand dollars, *Roger and Me* attracted millions of viewers and won extraordinary praise from film critics. Viewers in Flint greeted the film with a mixed response, however. During the taping of the *Phil Donahue* segment, local activist Herbert Cleaves spoke out in support of the film: "The reality is, what he [Moore] did, he put Flint on the map for everybody in the world to see the disgraceful kinds of things

FIGURE 10.3. Filmmaker Michael Moore at the world premiere of *Roger and Me*, 1989. After losing his seat on the Davison Board of Education in the mid-1970s, Moore moved to Flint, where he published an underground newspaper and began work on *Roger and Me*, the first of his many movies. One of the most acclaimed documentary films of all time, *Roger and Me* is a humorous but poignant exploration of Flint's experiences with deindustrialization. Courtesy of the *Detroit News* Archives.

that's happened to the people in this town. . . . I'm glad he did it."[93] Cleaves and others hoped that the publicity generated by the film would spur federal and state officials to intervene on the city's behalf. The film's many critics, though, feared that Moore had permanently sullied the city's reputation and further alienated GM's leaders. After viewing the controversial movie, Woodrow Stanley, the city's third black mayor, denounced Moore as a "pimp" and charged him with irrevocably damaging Flint's public image.[94] If Flint's problems derived primarily from negative beliefs, as Stanley and other civic boosters maintained, then *Roger and Me* surely dealt the city a knockout blow.

Roger Smith and his colleagues met great resistance from Michael Moore and other Flint activists who opposed GM's austerity measures. Yet even the company's staunchest critics could not sidestep the fact that decades of corporate disinvestment had left Flint's urban plants increasingly obsolete.[95] By the same token, the growing hostility between the company and local residents only underscored executives' negative perceptions of Flint's business climate. The deindustrialization of Flint thus continued unabated during the 1990s.

In May 1990, GM representatives announced that they were going to close Plant 2A, formerly Fisher Body 2, at the Chevy in the Hole complex. Two years later, the company revealed plans to shut down twenty-one factories nationwide, including the Chevrolet V-8 engine facility on Van Slyke Road.[96]

Hoping to forestall any additional plant closures and resurrect Flint's historic image as a business-friendly town, in June 1996, local leaders created the Billy Durant Automotive Commission, a blue-ribbon jobs panel consisting of representatives from the Genesee County Board of Commissioners, the Flint Board of Education, the UAW, Consumers Energy Company, the Flint-Genesee Economic Development Conference, the Flint Cultural Center Corporation, the Mott Foundation, and the Genesee Area Focus Council. In 1997, members of the commission drafted a "living agreement" that included formal pledges to GM in the areas of labor, education, health, government, and culture. On behalf of organized labor, UAW representatives promised that unionists would work to eliminate obstacles to productivity, support new automated technologies, and allow GM to negotiate local work rules on a plant-by-plant basis. School board officials vowed to create better factory laborers by formulating "school-to-work" partnerships, developing career training tracks at local high schools, and designing a curriculum that "puts all classroom learning in a workplace problem solving context." Commission members also agreed to inaugurate a countywide wellness campaign in order to reduce GM's rising health care costs. For their part, government officials promised to continue the tax abatement program, provide new water and sewer improvements, and implement an aggressive road maintenance plan. The covenant concluded with an open-ended commitment to assist the automaker as it adapted to meet the challenges of the twenty-first century: "The Greater Flint community stands ready to act as a full partner with General Motors by evolving institutions with the flexibility to respond to unanticipated trends and events, as well as articulated needs."[97] With its impressive package of pledges and incentives, the Durant commission's living agreement was an unbridled expression of civic love that harked back to the city's former identity as a company town.

Ultimately, though, members of the Durant Commission were unable to alter the city's fate. In November 1999 GM's directors did the unthinkable and shuttered Buick City. To many locals the decision came as a shock, not least because GM had only recently invested $475 million to modernize and automate the historic facility. Moreover, just a few years prior to the closure the independent market research firm J. D. Power and Associates had ranked the North End facility as the highest-quality automobile plant in North America.[98] Nevertheless, during that same period, rising demands for minivans, trucks, and sport utility vehicles severely eroded Buick's market share. Between 1980 and 1999, the number of Buicks sold annually in the United States declined sharply from 720,858 to 445,611.[99] In their attempt to "rightsize" production, GM executives closed Buick City and shifted its operations to newly constructed plants in the Michigan communities of Hamtramck and Orion Township.[100]

Shortly after Buick City closed, GM executives announced plans to raze the empty plants that had once employed more than twenty thousand local workers. On the fences surrounding the 235-acre complex, workers from MCM Management Corporation, an industrial services firm hired to level the plants, attached signs that read "Demolition Means Progress." Like the postwar slogans in support of North End redevelopment, the signs implied that the destruction of the old plants would pave the way for a brighter future for the city and its people. From their segregated public housing units just miles from the deserted Buick complex, former residents of Floral Park and St. John Street must have felt an overwhelming sense of déjà vu.

FIGURE 10.4. A sign posted in front of GM's demolished Buick City facility, 2005. Displayed prominently in front of shuttered automobile plants across the Vehicle City, signs reading "Demolition Means Progress" suggested that the destruction of outdated factories would facilitate Flint's revitalization. More broadly, though, the phrase is an apposite metaphor for the long history of urban renewal in the United States. Photograph by Andrew R. Highsmith, 2005.

For many observers, GM's decision to close Buick City marked a major historical turning point between the end of the industrial era and the rise of the Rust Belt. In reality, however, the North End plant closures were the culmination of GM's decades-long departure from Flint. Regardless, the loss of Buick City devastated many local residents. Several years after the complex closed, journalists from the *Detroit Free Press* adopted the phrase "The Fall of Flint" to describe the city's fate.[101] Though it offended many, the expression was apt because dating back to 1904, when Buick's founders chose Flint as their manufacturing headquarters, the Vehicle City had staked its future on the success of GM. During a 1953 "Flint Community Salute to Buick," chamber of commerce members celebrated that bond between town and company, proclaiming, "Buick and Flint . . . Flint and Buick. These are inseparable."[102] For the tens of thousands of citizens who grew up imbibing those words, the closure of Buick City created a grave identity crisis. As journalist Warren Cohen from *U.S. News and World Report* magazine asked, "How can a town nicknamed 'Buick City' not make Buicks?"[103] Like chameleons, the post–Buick City and its residents would have to find new identities—and new jobs, tens of thousands of them—to thrive in the hypercompetitive globalized marketplace of the twenty-first century.

Epilogue

"America Is a Thousand Flints"

General Motors' late twentieth-century retreat from Flint devastated all but the most sanguine of the Vehicle City's denizens. Not long after learning about the closure of Buick City, Michael Moore reaffirmed his earlier prediction that the city of Flint would soon cease to exist. "My fear," he admitted in 1999, "is that Flint is going to be one of the first ghost towns of the 21st century."[1] As economic conditions deteriorated following the turn of the century, apocalyptic visions of the city's future began circulating widely, perhaps most visibly on T-shirts, mugs, buttons, and caps featuring the ironic slogan "Flint: Last One Out, Turn Out the Lights." In a 2001 song about Flint titled "Buick City Complex," band members from the Old 97's added to the catastrophic din by contemplating what life would be like in an empty, deindustrialized city: "They're tearing the Buick City Complex down / I think we're the only people left in town / Where are you gonna move, where are you gonna move?"[2] Several years later, in a column that appeared in the *Uncommon Sense* newspaper, local satirist Mad Mac asked, "When Flint finally closes next year, what will people in the future think about all the expressways, bridges, ramps, railroads, and the airport that all intersect here . . . where there isn't anything?"[3] At the dawn of the twenty-first century, jobs and hope appeared to be in equally short supply in the Vehicle City.

Still, many residents remained confident about the city's fate. Among those who believed that Flint's best days were ahead, many viewed the plant closures and population losses as a blessing of sorts—a chance to obtain a brighter postindustrial future for a city that had bound its future to America's ailing automobile industry for far too long.[4] One such individual was Gary Ford of the chamber of commerce, who insisted that Flint was poised for a comeback: "Flint is like a Phoenix waiting to rise out of the ashes. The only thing working against us is our name."[5] Even as the city's economic crisis appeared to take on an air of permanence, Ford and other residents vowed to fight on for a new and better Flint. In the end, however, optimism and resilience proved to be little match for the large-scale structural adjustments that continued to buffet Flint in the twenty-first century.

Decline and Renewal

In spite of Moore's 1999 prediction, the city of Flint did not become a ghost town. Moreover, local business owners and politicians, much like their counterparts in other cities, refused to surrender in their perennial quest to bring homeowners, development, and jobs back to the shrinking urban core. Instead, advocates of renewal launched a sustained campaign to improve the city's tarnished reputation.[6] Although representatives of the chamber of commerce and other civic groups had long argued that Flint's distressed economy stemmed directly from its poor public image, embarrassment over *Roger and Me* and the city's growing renown for crime, unemployment, and labor militancy injected a new sense of urgency into the movement to refurbish the area's reputation. According to Mott Foundation researcher Thomas Hammer, a proponent of the public relations–centered approach, "A big part of the economic development problem is a need to undo adverse perceptions about Flint and its environs."[7]

In their attempts to polish Flint's image, area boosters drew from a much longer tradition of urban and regional development dating back at least to the middle decades of the twentieth century. Like their postwar predecessors in the growth-hungry states of the South and West, the Vehicle City's twenty-first-century boosters aggressively publicized Flint's probusiness economic climate, pointing to the city's two-tier wage structure, low taxes, severely weakened unions, surplus labor, and company town heritage.[8] In 2004 a *Flint Journal* editorial trumpeted the fact that wages in the Vehicle City had fallen below those in Detroit and other distressed urban centers. The editors, hoping to use "our well-known economic misfortune as an advantage," urged everyone in the city to "tell small manufacturers and other businesses how much this community has changed, that job-seekers no longer consider high wages a birthright, and moving here won't price them out of the market."[9] Along similar lines, municipal officials waged a successful turn-of-the-century battle to have Genesee County included in "Automation Alley," a cooperative business, technology, and public relations venture bringing local governments and corporations together to combat southeast and mid-Michigan's reputation for labor discord and antigrowth policies. Today large green signs announcing that Genesee County is an Automation Alley community greet motorists traveling on Interstates 69 and 75 as they enter the Flint metropolitan area.[10]

Politicians augmented these public relations efforts by enacting new policies to support job growth. As part of that endeavor, members of the Flint City Council created numerous low-tax "renaissance zones" in the inner city and designed new grant and loan programs to lure investment while continuing to offer long-term tax abatements and other incentives to encourage economic expansion. State and federal officials offered block grants, tax-exempt industrial development bonds, loan guarantee programs, and "brownfield" redevelopment tax credits to make Flint and its vacant industrial sites more appealing to potential investors.[11] Ultimately, though, the public relations campaigns and supply-side stimulus programs produced only mixed results. In 2001 area boosters celebrated when GM officials launched a multimillion-dollar campaign to upgrade the Flint Metal Center on Bristol Road. The following year, the automaker opened a massive new 1,165,212-square-foot engine plant adjacent to the Van Slyke facility. Eventually workers in the new factory would produce engines for the Chevrolet Volt, GM's hybrid electric car. Then in 2004 executives announced plans to spend $148 million to open new production lines at the truck and bus complex on Van Slyke Road.[12] The new and updated plants, all of them highly automated, helped to fill the economic void left behind by the closures of Buick City, Chevy in the Hole, and other local facilities.

Yet they created few new jobs, especially for city residents. As expected, GM adhered to its existing agreements with the UAW and awarded virtually all the new positions—two thousand in all—to unemployed autoworkers from Buick City and other shuttered facilities. Since most of the recalled workers were suburban residents, very few city dwellers benefited directly from GM's new investments.[13] Moreover, because city council members and Mayor Donald J. Williamson had offered millions of dollars in tax abatements and infrastructure improvements to secure GM's investments, the new facilities came at a significant cost to the city. In an attempt to dispel false optimism about the new plants, Second Ward councilman Ed Taylor warned, "Nobody in Flint is going to get a new job. . . . Nobody [who works at the new and updated factories] is going to buy a house [in Flint]. They aren't going to shop here."[14] Much to the chagrin of local residents, Taylor's predictions turned out to be prescient.

Despite GM's new investments, the deindustrialization of Flint continued apace in the new millennium. In the fall of 2004, a crew of workers tore down Plant 4 in the mostly empty Chevy in the Hole complex on Flint's west side.

As many locals noted, Plant 4 was the historic factory from which black autoworker Roscoe Van Zandt and his UAW comrades triumphantly marched at the conclusion of the Flint sit-down strike. Just prior to the demolition, thieves added insult to injury by stealing the two historical markers honoring the sit-downers from the front of the plant.[15] The following November, GM's leaders fended off bankruptcy rumors by unveiling an aggressive plan to close plants and eliminate thirty thousand manufacturing jobs in the United States. Among the factories slated for closure was the Flint Engine North facility, whose workers built V-6 engines near the demolished St. John Street neighborhood.[16] In 2006 production ended at the AC Spark Plug plant on Dort Highway, by then operated by Delphi Automotive Systems, a GM spinoff company. After hearing the news, Dale Mark, a GM retiree living in Genesee Township, expressed shock that the sprawling complex had closed: "I never thought it would. I thought they'd always need the things we made." Within two years the old AC factory, like so many others before it, had fallen to the wrecking ball.[17]

Notwithstanding the new and updated GM plants and other important success stories, the progrowth policies and pronouncements of the early twenty-first century generated only minor economic development in the city of Flint and its struggling inner-ring suburbs. Although the county's southernmost communities—especially those bordering the booming high-tech corridor in neighboring Oakland County—benefited substantially from the Automation Alley program and a robust turn-of-the-century real estate market, Flint and close-in northern suburbs such as Beecher, Genesee Township, and Mt. Morris Township continued to hemorrhage jobs, residents, and hope in the new millennium.[18] By 2006 additional layoffs and plant closures had brought GM's local workforce down to approximately fifteen thousand. With so few jobs available, Flint's population plummeted to 113,000. As residents fled the devastation, the city's housing vacancy rate surged to nearly 20 percent.[19] And things would only get worse.

In 2008 residents of the Vehicle City joined people throughout the world in honoring the hundredth anniversary of GM's founding.[20] Unlike the 1950s, though, when revelers at the Golden Carnival bragged that the company town of Flint was the envy of the world, GM's centennial celebration took place under dark clouds of uncertainty. Between December 2007 and the fall of 2008, a nationwide spike in subprime mortgage delinquencies and home foreclosures triggered a precipitous drop in both real estate values and the broader securities market, which in turn spawned a sharp decline in stock prices. A severe international economic recession ensued. By November 2008 the stock market had plunged and home and vehicle sales had all but halted

FIGURE E.1. An abandoned home in Flint, 2007. As Flint's population plummeted during the late twentieth and early twenty-first centuries, abandoned homes such as this one began appearing in large numbers in neighborhoods throughout the city. In their efforts to combat blight and crime, area officials allocated millions of dollars per year to tearing down vacant and dilapidated buildings. Photograph by Andrew R. Highsmith, 2007.

nationwide.[21] Seven months later, GM executives shocked millions of Americans by filing for bankruptcy.[22] Although federal officials ultimately approved nearly thirty billion dollars in loans to rescue the moribund auto industry, GM's collapse and the so-called Great Recession of 2007–9 wreaked havoc in Flint and other cities.[23] Across the country, GM plant cities as diverse as Pontiac, Michigan; Doraville, Georgia; Wilmington, Delaware; and Lordstown, Ohio, suffered significant job losses as the beleaguered automaker downsized its global operations.[24] And Flint, of course, was no exception. In the wake of GM's painful restructuring effort, the company's overall employee count in metropolitan Flint dropped to just 6,434.[25]

The collapse of the subprime housing market and the recession that followed cast a pall over Flint and its increasingly bleak suburbs. Already overstretched by a combination of underemployment, high taxes, and usurious adjustable rate loans, thousands of local residents began defaulting on their mortgages and property taxes when the real estate bubble burst. In a matter of a few years, predatory lending—just as it had in Beecher during the

1970s—transformed untold numbers of vibrant middle- and working-class communities into desolate landscapes of poverty. By 2009, officials from the Genesee County Land Bank, a public receiver for abandoned and tax-foreclosed properties, had repossessed nine thousand homes, businesses, and lots just within Flint, a figure that accounted for 14 percent of the city's overall surface area.[26] Meanwhile, in the private lending market, mortgage foreclosures were surging to Depression-era levels, routinely exceeding one thousand homes per month. Just during September 2010, for instance, nearly 1 percent of the region's two hundred thousand housing units slipped into foreclosure.[27] The boarded-over homes and businesses left in the wake of the subprime disaster served as poignant reminders that credit and debt can be as unjust and exploitative as redlining and other exclusionary lending practices. In the end, the toxic trio of deindustrialization, tax delinquency, and predatory lending caused more forced relocations of local residents than any other event in the region's history, including the St. John and Floral Park renewal projects.[28] The crash also pummeled the local housing market into a state of near collapse, with home vacancy rates surging to well over 20 percent and average sale prices for homes in Flint dropping to just seventeen thousand dollars—significantly cheaper than the Buick sedans that GM workers once assembled in the North End.[29]

When federal census takers arrived in Genesee County in 2010, they beheld a region of Dickensian extremes. Inside the city, officials counted just 102,434 residents, about half the number of people who lived in Flint during its postwar heyday. Not surprisingly, African Americans held a substantial majority, accounting for over 56 percent of the population. Composing just 37 percent of the total populace, whites were a clear minority in the Vehicle City. Although the number of Latinos had risen significantly since the turn of the century, they still accounted for less than 4 percent of the city's overall population. Together, Asian Americans, Pacific Islanders, and Native Americans constituted just 1 percent of the population of Flint. For all of its similarities to other cities, Flint's status as a predominantly biracial black-white city continued to set it apart from Los Angeles, New York, Houston, and other American metropolises with large multiracial populations.[30]

Economically, the census painted an austere portrait of life in Flint. The city's per capita income stood at just $14,910. With an unemployment rate of nearly 30 percent and over a third of its people living below the poverty level, Flint ranked yet again as one of the poorest cities in the United States. Among the impoverished, women continued to outnumber men, accounting for 54 percent of the city's residents living below the federal poverty level. To the

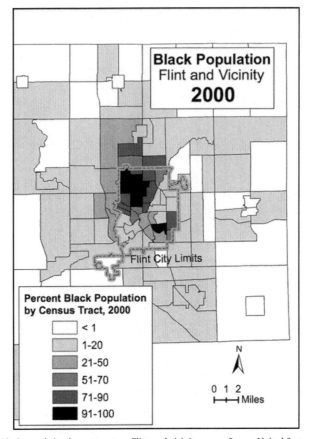

Black Population
Flint and Vicinity
2000

Flint City Limits

**Percent Black Population
by Census Tract, 2000**

- < 1
- 1-20
- 21-50
- 51-70
- 71-90
- 91-100

N

0 1 2
Miles

MAP E.1. Black population by census tract, Flint and vicinity, 2000. *Source*: United States Department of Commerce, Bureau of the Census, *2000 Census of Population and Housing*, https://www.census.gov/prod/www/decennial.html.

surprise of almost no one familiar with Flint's past, the 2010 census revealed a starkly divided city in deep economic crisis that was poorer, more isolated from its neighbors, and smaller than at any time in its recent history.[31]

Of those who departed Flint in the new millennium, some sought out new opportunities in the fast-growing states of the South and West. However, many others took up residence in the suburban and exurban areas of Genesee County. Although the percentage of African Americans living outside of Flint rose sharply in the new century, the suburbs of Genesee County remained overwhelmingly white. In 2010 there were 323,356 people living in the twenty-seven municipalities, villages, and townships surrounding the Vehicle City. Nearly 90 percent of these suburban dwellers were white, while just 9.5 percent were African American. Moreover, of the 31,013 African Americans living in suburban Genesee County in 2010, over 70 percent resided in the inner-ring communities of Mt. Morris Township (including Beecher), Flint Township, and Grand Blanc Township. By contrast, the remaining two dozen communities in suburban Genesee County, with a combined population of 232,418, housed fewer than ten thousand black people. The racial divide was equally stark in the exurban and rural fringes of the Flint area. For instance, in the Bavarian-themed tourist city of Frankenmuth, home to 4,816 white people (and the world's largest Christmas store), government workers counted just twenty-seven African Americans. Similarly, the western exurb of Durand, with nearly thirty-five hundred residents, contained only nineteen African Americans. Although courts and legislative bodies had long since overturned the statutory infrastructure of legal Jim Crow, the persistence of

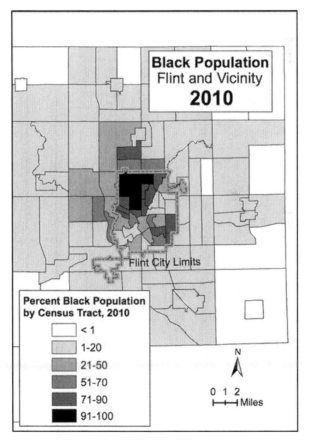

MAP E.2. Black population by census tract, Flint and vicinity, 2010. *Source:* United States Department of Commerce, Bureau of the Census, *2010 Census of Population and Housing*, https://www.census.gov/prod/www/decennial.html. Map by Gordon Thompson.

popular and administrative forms of segregation ensured that the color line remained an ever-present feature of the Flint region's twenty-first-century landscape.[32]

Still, evidence from Genesee County and the nation as a whole clearly pointed to increasing levels of suburban diversity. Although most of the suburbs and exurbs of metropolitan Flint remained highly segregated by race in the new century, their class dynamics were much more complex. For example, in 2010 the exurban hamlet of Owosso, once an all-white "sundown town," was home to only 117 African Americans who comprised a minuscule 0.8 percent of the population. However, a clear majority of the town's 14,540 whites were either poor or working class. In fact, the city's median household income of $35,850 was approximately $15,000 less than the national average.[33] And Owosso was hardly unique. For every elite community such as Goodrich—where median household incomes of $74,955 surpassed US averages by nearly $25,000—there were suburbs such as Montrose, Beecher, and Swartz Creek, where residents earned income at or below national norms.[34] Between the extremes of Goodrich and Flint, an increasing number of the area's formerly middle-class suburbs were dotted with bank-owned homes, boarded-up storefronts, and seemingly endless unemployment lines, all of them testaments to the nation's deep housing and jobs crises.[35]

Although signs of deprivation and division were evident throughout Genesee County in the new century, nowhere were they as visible as in the Flint Public Schools. In 2011, 81 percent of Flint's public school children were eligible for free or reduced-price lunches.[36] Poverty, however, was but one of the obstacles that the Vehicle City's children faced in the new millennium. Flint's schools were also profoundly segregated along racial lines. Between 1992 and 2005, the number of segregated public schools in the city—defined as those with an African American enrollment of at least 80 percent—jumped from nineteen to twenty-eight. By 2014 the Flint area's schools ranked second only to Detroit's in statewide levels of racial and class segregation. That year, which happened to be the sixtieth anniversary of the *Brown v. Board of Education* decision, many of Flint's elementary and junior high schools were more segregated than they were on the eve of the Supreme Court's 1954 ruling.[37] The Flint case was far from an isolated example, however, as segregation and income stratification increased sharply in school districts throughout the nation during the 1990s and 2000s. In what was perhaps the starkest manifestation of school resegregation, the early twenty-first century coincided with the powerful national resurgence of "apartheid schools"—facilities with a minority enrollment of between 99 and 100 percent.[38]

With so many dire conditions prevailing in the city's schools and neighborhoods, local boosters found precious few reasons to be hopeful about Flint's future. What optimism there was derived in large part from a variety of high-profile business openings and building projects in Flint's longsuffering downtown business district. The surge in downtown development first became obvious shortly after the turn of the century, when local business owners and investors formed Uptown Developments, a limited liability company devoted to acquiring, renovating, and marketing real estate in Flint's downtown core. Making use of generous financial support from the Charles Stewart Mott Foundation and a variety of both private and government sources, Uptown's growth-minded officials quickly acquired a dozen derelict buildings and converted them to mixed-use residential, commercial, and retail developments available for sale and lease. Shortly thereafter, new investors began lining up to take advantage of the downtown area's depressed real estate market and the many subsidies available for entrepreneurs. By 2014, representatives from Uptown Developments and other local investors had transformed the First National Bank tower and the long-vacant Durant Hotel into mixed-use residential and commercial structures; erected a new twenty-five-thousand-square-foot facility to house the Wade Trim engineering firm; converted several vacant buildings into a headquarters for the Rowe Professional Services Company and a luxury loft project; reopened the former Hyatt Hotel as a dormitory for UM-Flint students; transformed the shuttered headquarters of the *Flint Journal* into a new office and teaching space for the Michigan State University College of Human Medicine; demolished the decrepit Genesee Towers skyscraper, a major downtown eyesore; and relocated the Flint Farmers' Market to the central business district. As new residents, workers, and visitors began arriving downtown to make use of the new developments, a number of coffee shops, bars, restaurants, and specialty stores sprouted up on and around Saginaw Street.[39] Meanwhile, popular events and initiatives such as the Flint Film Festival, the Back to the Bricks Cruise and Car Show, the Flint Jazz Festival, the Buckham Alley Fest, the Flint Juneteenth Celebration, the Flint Art Walk, the Crim Festival of Races, and the Flint Public Art Project added immeasurably to the downtown's growing appeal. Together, the new shops, activities, and people helped to improve the downtown's sullied public image while generating new revenues for local businesses and the city.[40] According to Chris Everson, a longtime resident of downtown, the turn-of-the-century changes were dramatic: "I moved downtown in 1994, when it was desolate. Now, there are people out every night, hanging out and going for walks downtown. It's fun to see downtown coming alive."[41]

Although it was virtually impossible to ignore the downtown area's twenty-first-century makeover, the new developments along Saginaw Street

FIGURE E.2. Back to the Bricks Cruise and Car Show, 2012. A number of new housing developments, business openings, and popular events such as this one brought fresh life to downtown Flint in the new millennium. However, most of the city's neighborhoods continued to suffer from poverty, segregation, and disinvestment. Courtesy of the Flint and Genesee Chamber of Commerce.

were not a panacea for the city. In fact, the new lofts, wine bars, and restaurants raised the specter of downtown's gentrification for the first time since the failed Great Leap Forward. This was especially clear in the case of Uptown Developments' Berridge Hotel loft reconversion project. Prior to 2008 the Berridge's one hundred residential units—derisively dubbed a "flophouse" by Uptown investors and other downtown boosters—served as the housing of last resort for a significant number of ex-convicts, the poor, and other vulnerable and transient community members. Part of the Berridge's appeal was that impoverished and peripatetic renters could obtain basic rooms for as little as twenty dollars per night without making any long-term housing commitments. However, all of that ended with Uptown's acquisition and rehabilitation of the facility. Although the new Berridge Place loft project created dozens of new construction jobs and seventeen new postmodern loft spaces to add to the downtown's growing list of middle-class residential units, the erstwhile tenants of the Berridge had few, if any, comparable options available in the downtown area. So as the six-million-dollar renovation project got under way, residents of the hotel quietly vacated the central business district for points unknown. Long a haven for individuals down on their luck, downtown Flint, like other American city centers, became increasingly inhospitable to the poor in the new millennium.[42]

In addition to masking the growing problem of gentrification, boosters' almost singular emphasis on downtown's successes obscured the ongoing crises in the remainder of Flint. Despite GM's quick recovery from bankruptcy and the steady growth of Flint's academic, life sciences, and health care sectors, the city's stubborn unemployment rate, which typically ranged between 15 and 20 percent, remained one of the highest in the nation.[43] With so few jobs available for local residents, city officials faced massive budget deficits in the new century that routinely exceeded ten million dollars per year.[44] In response to the deficits and the threat of a second state takeover of the municipal government, members of the city council and Mayor Dayne Walling, a progrowth moderate elected in 2009, exercised some of their few remaining options by privatizing numerous city services and enacting a series of steep cuts to the police and fire departments.[45] As local firehouses began shutting down and police department layoffs multiplied, many of the city's nearly ten thousand vacant homes began to go up in flames. During the five years preceding the cuts, city officials had recorded an annual average of two hundred cases of residential arson. In 2010, 2011, and 2012, however, the number of criminal house fires in the city surged to over three hundred per year, making Flint the nation's undisputed arson capital. Unable to keep up with the conflagrations raging across the city, in 2012 firefighters adopted the controversial policy of letting abandoned buildings burn to the ground.[46]

Over the same period, Flint gained still more notoriety for being the most violent city in the country. With at least sixty-six homicides recorded in 2011, Flint's per capita murder rate was higher than those in Newark, New Orleans, Chicago, Los Angeles, and even the war-torn city of Baghdad, Iraq. The Vehicle City, according to writer Charlie LeDuff, had morphed into Murdertown, USA.[47] What is more, even though the cuts to the public safety budgets hit hard across the city, they were not enough to prevent Flint's descent into bankruptcy and state receivership. In November 2011 Republican governor Rick Snyder issued a formal declaration of financial emergency for the city of Flint—the second such pronouncement in a decade—and appointed a new chief executive to manage the municipal budget. In the wake of Snyder's proclamation, Flint's residents and officials suffered perhaps the ultimate indignity of losing the right to self-govern.[48] Flint was not the only city to suffer such a fate, however. Nearly a year and a half later, in March 2013, Snyder appointed an emergency manager to run the heavily African American city of Detroit, Michigan's largest municipality. By 2014, unelected managers led twelve of the state's most financially troubled cities as well as six struggling school districts. As critics noted, these autocratically run municipalities and school systems—which included the cities of Flint, Detroit,

Pontiac, and Benton Harbor—contained approximately half of the state's black population.[49]

On the ground, in the city's many distressed neighborhoods, signs of the long-term economic crisis were ubiquitous. According to Bill Vlasic and Brett Canton of the *Detroit Free Press*, much of Flint appeared abandoned: "Parts of the city are eerily empty, with vast stretches of vacant land standing in silent testimony to the factories that were once there."[50] At Buick City— which once employed over twenty-five thousand autoworkers—visitors in 2014 could find a vast postindustrial wasteland covered with broken slabs of asphalt, piles of discarded tires and mattresses, and weeds as tall as trees. On the city's east and west sides, the unobstructed vistas and waist-high prairie grasses that had replaced the old AC Spark Plug plants and Chevy in the Hole looked like Kansas dipped in concrete. In the North End, where St. John Street residents once lived, hipsters and urban explorers in search of industrial detritus and other forms of "ruin porn" could wander across vast tracts of vacant and sparsely populated land in virtual isolation.[51] For pessimists such as Vlasic, Canton, and others fixated on Flint's deeply rooted economic crisis, these and other signs of decline were the city's defining characteristics.

In truth, however, the American obsession with urban decline and abandonment has always concealed a more ambivalent and complicated reality in cities such as Flint.[52] Of course visitors in the new millennium could find a plethora of industrial ruins scattered throughout the region. The empty plants and vacant lots were poignant reminders that GM had clearly abandoned its postwar metropolitan capitalist growth strategy in favor of a more austere and truly globalized production map in which former company towns such as Flint figured much less prominently. Nevertheless, GM remained a significant presence in the Flint area long after the era of corporate abandonment had supposedly destroyed the city. As evidence of that fact, in December 2013, GM officials surprised many area pessimists by announcing plans to spend six hundred million dollars on a new paint plant and updates for their assembly facility on Van Slyke and Bristol Roads.[53] Although the new and updated facilities created only a small number of new jobs and failed to return Flint to prosperity, GM's investments were incompatible with the notion of corporate abandonment that had become synonymous with the Vehicle City.[54]

By the same token, the tens of thousands of whites who remained in Flint decades after it had become a majority-black city fit very uneasily within the popular narrative of white flight. Not to mention that civic boosters, elected leaders, and neighborhood activists refused to relent in their struggles to revitalize the city. Even during periods of austerity and economic retrenchment, area leaders continued to work for Flint's renewal and searched, though often in vain, to find reasons for optimism. Empty houses and shuttered businesses aside, Flint had not been abandoned.

In the wake of the subprime crash, a new crop of demolition-minded renewers coalesced around Genesee County treasurer Daniel Kildee, the founder of the Genesee County Land Bank (and future member of the US House of Representatives), who argued that Flint's path to salvation lay in "planned shrinkage." For Kildee and other members of the so-called shrinking cities movement, a popular international phenomenon, population decline was more of an opportunity than a problem because it presented new chances to demolish abandoned neighborhoods, scale back costly urban infrastructure, and improve municipal services for urban residents. The key to Flint's future, Kildee and his supporters argued, lay in embracing the possibilities of smaller, greener, and denser cities rather than fighting decline. In pursuit of that aim, Land Bank officials acquired no fewer than 10,721 tax-foreclosed properties between 2002 and 2011, nearly half of which ultimately became eligible for demolition.[55] In Beecher, north Flint, Kearsley Park, and other struggling areas, the Land Bank's aggressive residential demolition program quickly transformed row after row of boarded-up abandoned houses into uninhabited open spaces. Although residents often disagreed in their assessment of the Land Bank's work, the growth of the shrinking cities movement clearly illustrates the undying and chameleonic nature of urban renewal in the United States.[56] Regardless of the period in question or the crisis at hand, proponents of urban revitalization from every political stripe and every social station have always maintained a strong presence in Flint and other cities. And the new millennium proved to be no exception.

In the end, the most remarkable fact about Flint's past is not that white people, employers, and investors have turned their backs on the city, though many surely have. Rather, it is the ways in which government officials, civic leaders, and ordinary citizens have fortified social inequalities in their attempts to revitalize the city and region. Whether the campaigns for revitalization began within corporate boardrooms, the Mott Foundation headquarters, the offices of the Federal Housing Administration, city hall, suburban coffee klatches, neighborhood association meetings, churches, union halls, or the world-renowned Torch Bar in downtown's Buckham Alley, the transhistorical quest to renew cities such as Flint points to the limitations of the Rust Belt and urban decline narratives. As evidenced by Flint's experience, the story of metropolitan America in the twentieth century and beyond is at heart the story of the near constant campaign for revitalization and its often harsh consequences. Indeed, if nothing else, an examination of Flint's painful past and present confirms that the driving forces in modern metropolitan history have always been reinvention and perseverance more than declension and abandonment.

For those willing to see beyond the ruins, the signs of urban perseverance in twenty-first-century Flint were every bit as visible as the abandoned homes and factories that had come to define the city. Of course, they were easy to spot in the newly opened shops, bars, and restaurants sitting beneath the decorative metal arches lining Saginaw Street in the downtown business district. But they were also evident in deeply segregated and impoverished inner-city neighborhoods such as Carriage Town, where small numbers of young people, artists, entrepreneurs, and professionals, many of them gay and lesbian, began rehabilitating vacant Victorian homes, opening new shops, and reestablishing a long-lost sense of community solidarity.[57] Although Carriage Town's new residents could easily be confused with gentrifiers or investors looking to profit from the real estate bust, in reality many of them have spent years, decades even, building a multiracial and economically diverse community near the heart of the city. One such individual is Ken Van Wagoner, a longtime Carriage Town resident and, since 2000, the proprietor of the Good Beans Café, one of the city's most popular and gay-friendly meeting places. Another is Leanne Barkus, who moved to Carriage Town from the nearby college town of Ann Arbor in 2000. According to Barkus, a grant development specialist at Mott Community College, residents of Carriage Town succeeded in creating a powerful sense of community that simply did not exist in many other big-city neighborhoods. "I knew everyone within a month of moving in. If I need a jump for my car in the winter, I've got 20 people I can call. Four neighbors have my house keys. Whoever thought you could have that in Flint?" "We're very passionate about Carriage Town," she added, "and we don't give up."[58]

Several miles northeast of Carriage Town, near the old St. John neighborhood, signs of urban pluck as well as an emerging ethos of postindustrial survivalism were manifest in a massive garden on East Piper Avenue tended by the indefatigable Harry Ryan. In 2005, after seeing his street and neighborhood fall victim to numerous home foreclosures, acts of vandalism, and a series of fires, Ryan, a retired African American autoworker and musician, approached city officials with a modest plan to clear a few vacant lots and plant a garden. Upon obtaining approval, Ryan and several friends began cleaning up debris, pulling weeds, mowing long-neglected lawns, and planting rows of broccoli, turnips, collard greens, and even fruit trees. Over time, their vegetable garden and orchard grew to cover ten contiguous lots where a long row of vacant homes once sat. When asked to explain why he had devoted so much time and money to the project, Ryan, ever modest, simply quipped, "It needs to be done." Like millions of other urban Americans confronted with economic catastrophe and social division, Ryan refused to surrender in his fight for a better city.[59]

To be clear, individuals such as Ken Van Wagoner, Leanne Barkus, and Harry Ryan have experienced very limited success in their fight to revitalize Flint and other struggling urban centers. The new shops, gardens, and neighborhood associations that sprouted in Flint, Detroit, and other deindustrialized cities in the new millennium could not begin to replace the millions of manufacturing jobs that had departed urban America over the years. Nor could they effectively counteract the array of public policies that had left cities such as Flint so deeply segregated and politically polarized. In spite of the best efforts of urban renewers, life in twenty-first-century Flint remained extremely bleak for tens of thousands of city residents.

And yet the ongoing efforts to improve the city still demand attention— not least because they stand in sharp contrast to the apocalyptic caricatures that have become staples in the popular discourse on the urban crisis.[60] On their own, gardens and coffee houses could not save Flint, but they did become important sources of fresh produce for needy residents as well as vibrant and safe public spaces for community members. Moreover, they stood out as unmistakable signs of urban tenacity and determination—indications that residents of the Vehicle City and others like it continued to struggle and endure despite the structural calamities engulfing them.

Of all the signs and symbols of human perseverance that surfaced in the postcrash incarnation of the Vehicle City, perhaps none was as striking as the eye-catching roof over Sharon Dickinson's home on Jane Avenue. At some point in the winter of 2011, Dickinson decided to cover a large section of her roof with an enormous plastic banner advertising the Daewoo Leganza, an "affordable luxury" automobile. The sign features a tall, blonde female model on a white sand beach staring longingly at a Daewoo Leganza sedan. Dickinson's intent was not to market automobiles or make a social statement, however. She was simply trying to keep her home dry. Like numerous other homeowners on Flint's hard-luck east side, Dickinson could not afford to repair the leaky roof covering her deteriorated shotgun-style bungalow. So she improvised by purchasing the ninety-dollar sign and carefully placing it over the entire east side of her roof. The new covering, which quickly became a conversation piece among neighborhood residents, immediately stopped the leaks and kept the house much warmer during the long winter months. "That tarp was a blessing," she noted. "It saved me. It saved my roof."[61]

At first blush, Sharon Dickinson's home—like the hundreds of thousands of dilapidated and abandoned buildings that dot urban landscapes across the

FIGURE E.3. A home on Jane Avenue in Flint, 2012. In 2011 Sharon Dickinson purchased the large advertising banner shown here in order to cover the many holes in her roof. The tarp quickly gained notoriety as a symbol of both urban poverty and perseverance. Photograph by Joel Rash, 2012.

United States—seems to provide prima facie evidence only of the depth of the nation's urban crisis. However, the experiences of Dickinson and other impoverished but resourceful urban dwellers cannot be summed up by the simple concept of declension. On the contrary, telling the full story of metropolitan America in the twentieth century and beyond means grappling with both stories embedded in the powerful image of Sharon Dickinson's home. It means analyzing the structural forces that resulted in the home's decay, of course. But it also means acknowledging the resolve, care, and ingenuity that resulted in the makeshift roof on top.

There can be little doubt that mass suburbanization, deindustrialization, and deeply rooted forms of structural racism have hit cities such as Flint with a special ferocity. Yet the city and its people have not disappeared; nor have they vanished from Detroit, Philadelphia, New Orleans, Oakland, and other cities that once formed the backbone of America's economy. Quite the opposite, in fact. As the economic crises of the late twentieth and early twenty-first centuries battered American cities, urban renewers from all walks of life and all political affiliations only redoubled their efforts. And this, in short, is why even the most nuanced portraits of urban death and declension remain problematic. Though they are full of vibrant and sentient beings with unique and important stories, cities themselves are inanimate political and legal constructs that neither live nor die. Even if cities could perish, however, evidence from Flint and other communities near and far suggests that urban boosters will not let them. When renewers cannot lure new residents to struggling urban centers, enterprising officials and locals reinvent them—often as heritage sites, artistic communities, or "shrinking cities." If those efforts fail, then urban explorers, hipsters, and investors arrive to survey the ruins. Tellingly, even supposedly "lost cities" of antiquity—places such as Pompeii, in present-day Italy, and Mexico's Chichen Itza—have become major tourist attractions many centuries after their supposed deaths. Flint no doubt will prove to be equally resilient, because, in the parlance of urban renewers, *decline* is but a synonym for *opportunity*. The crucial questions, however, are the same as they have always been: Will the renewal campaigns of the future bring more opportunity, equity, and hope to the people of the Vehicle City and the Flint region? Or will the revitalization efforts of the new century mimic so many of those from the last by hardening and deepening social divisions? These are the central questions facing residents of the Flint region and other metropolitan centers in the twenty-first century.[62]

Back in 1945, when Americans celebrated the conclusion of World War II and looked forward to a future of peace and prosperity, Buick historian Carl Crow claimed that the United States consisted of a thousand Flints.[63] Even though many decades have since past, Crow's words still ring true. From coast to coast, the America of the twenty-first century is, in fact, a thousand Flints, but not at all in the whiggish capacity that Crow envisioned. There are Flints in the economically depressed neighborhoods of Decatur, Illinois; Camden, New Jersey; Erie, Pennsylvania, and other struggling cities once renowned for their industrial might. Flints also exist in hypersegregated ghettos on Chicago's south and west sides, in Miami's Overtown district, and in struggling suburbs such as Yonkers, New York; East Palo Alto, California; and Ferguson, Missouri, where the legacies of white supremacy and legal, popular, and administrative Jim Crow continue to abridge civil rights and economic opportunity. However, there are also a thousand Flints in the booming, affluent bastions of suburban capitalism surrounding high-tech metropolises such as San Francisco, Boston, Raleigh, Seattle, and Austin—places like Cupertino, California; Redmond, Washington; and Round Rock, Texas, all of them defined more by fragmentation and exclusion than by cooperation and inclusion. There are Flints on the Atlantic and Pacific Coasts as well as in the so-called Rust Belt and Sunbelt, for the conditions of racial, spatial, and economic inequality that took shape in the Vehicle City during the twentieth century know no regional boundaries. Indeed, Flints can be found anywhere in the world where the eternal quest for metropolitan growth and revitalization has buttressed social inequalities. Because it took the full weight of government at all levels along with the efforts of untold numbers of ordinary Americans to construct and fortify the walls that still surround the nation's Flints, it will require an equally concerted movement of millions to demolish them all and build anew.

Acknowledgments

It is my special privilege to thank all of the many teachers, colleagues, bene-factors, friends, and family members who helped make my dream of writing a book a reality. I could not have finished this project without them.

The journey leading to this book began when I was a teenager growing up in Cincinnati, Ohio. There a civil rights pioneer named Fred Shuttlesworth got me thinking about race, politics, and social justice for the first time, and his stories of struggle forever changed my life. Thank you for setting me on this path, Reverend Shuttlesworth. May you rest in peace.

Over the past twenty years, I have been lucky to be affiliated with six remarkable educational communities. As an undergraduate student at the College of William and Mary, I had the great fortune to work with Melvin Ely, who showed me how to become a better listener and a more diligent researcher. Upon graduating from college, I moved to Chicago to work as a teacher at DuSable High School. Members of the DuSable family deserve a special mention for their grit and determination in the face of extreme hardship and for reminding me that public schools are central to the process of community building. During my two stints as a graduate student, I had the privilege of working with a long list of esteemed academics. As a master's student at DePaul University, I studied with Larry Bennett and Howard Lindsey, who nurtured my interests in race, class, and urban education. At the University of Michigan, I was fortunate to work with Francis Blouin, Matthew Countryman, Robert Fishman, Kevin Gaines, Joseph Grengs, Scott Kurashige, Earl Lewis, Michele Mitchell, Maria Montoya, Regina Morantz-Sanchez, Julius Scott, J. Mills Thornton, Penny Von Eschen, and many other brilliant scholars and teachers. As an assistant professor in the Department of Public Administration and a faculty affiliate in the Department of History and the Program in Urban and Regional Planning at the University of Texas at San Antonio, it has been my honor to work with an extremely talented and supportive group of scholars, teachers, administrators, and staff members that includes Jennifer Alexander, Tansu Demir, Kirsten Gardner, Rhonda Gonzales, Gabriela González, Anne Hardgrove, Patricia Jaramillo, David Johnson, Barbara McCabe, Karen Metz, Gregg Michel, Catherine Nolan-Ferrell, Branco Ponomariov, Christopher Reddick, Jo Reyes-Boitel, John Reynolds, Francine Sanders Romero, Rogelio Saenz, Heywood Sanders, Richard Tangum, Ivy Taylor, and Maggie Valentine. It has also been my good fortune to work with several smart and energetic graduate research assistants. Lucy Barbosa-Ramirez, Jessica Foreman, and Karlerik Naslund deserve special thanks for doing more to support my research than they will ever know. Finally, to all of my new colleagues at the University of California, Irvine, thank you for hiring me. I can't wait to join you!

I also owe a debt of gratitude to all of the individuals who helped to transform my unfinished manuscript into a book. I am especially grateful for the support I have received from the editors and staff members at the University of Chicago Press. To former editor Robert Devens, current editor Timothy Mennel, and series editors Timothy Gilfoyle, James Grossman, and Becky Nicolaides, thank you very much for believing in this project and seeing it through to the finish. Thanks also to Russell Damian, Nora Devlin, Ruth Goring, Ashley Pierce, and other editorial associates and staff members at the Press, who provided assistance with permissions, copyediting, and many other undertakings. The two anonymous peer reviewers selected for this project deserve additional praise for their many valuable suggestions on how to improve the manuscript. For making two maps and improving fifteen others, I owe a sincere thanks to Gordon Thompson. Likewise, I want to thank my friend Clayton Howard for reading the entire manuscript very late in the process and offering excellent feedback. And for making the index, copyediting the manuscript, and, most importantly, being a good friend for over twenty years, many thanks to David Courtenay-Quirk.

Portions of this work appeared earlier, in somewhat different form, in "Demolition Means Progress: Urban Renewal, Local Politics, and State-Sanctioned Ghetto Formation in Flint, Michigan," *Journal of Urban History* 35, no. 3 (March 2009): 348–68; "Decline and Renewal in North American Cities," *Journal of Urban History* 37, no. 4 (July 2011): 619–26; "Prelude to the Subprime Crash: Beecher, Michigan, and the Origins of the Suburban Crisis," *Journal of Policy History* 24, no. 4 (October 2012): 572–611; and "Beyond Corporate Abandonment: General Motors and the Politics of Metropolitan

Capitalism in Flint, Michigan," *Journal of Urban History* 40, no. 1 (January 2014): 31–47, and are reprinted here with permission. I would like to thank the editors of the *Journal of Urban History* and the *Journal of Policy History* for granting permission to reprint this material.

This project is rooted in many years of research. None of it would have been possible without help from the librarians, archivists, and staff members at the Bentley Historical Library and the Harlan Hatcher Graduate Library at the University of Michigan; the *Detroit News* Archive; the Downtown Campus Library and the John Peace Library at the University of Texas at San Antonio; the Flint City Clerk's Office; the *Flint Journal* Archive; the Flint Public Library; the Frances Willson Thompson Library and the Genesee Historical Collections Center at the University of Michigan–Flint; the Library of Michigan; the National Archives and Records Administration facilities in Chicago, Washington, DC, and College Park, Maryland; the Perry Archives of the Alfred P. Sloan Museum; the Richard P. Scharchburg Archives at Kettering University; and the Walter P. Reuther Library of Labor and Urban Affairs at Wayne State University. I am especially appreciative of the efforts of Jeremy Dimick of the Sloan Museum; Paul Gifford, the archivist at the University of Michigan–Flint; Dave Larzelere, formerly of the *Flint Journal* Archive; Jeff Taylor, former curator of collections at the Sloan Museum; and David White from the Scharchburg Archives.

While conducting research for this book, I spent years reading about Olive Beasley, Aileen Butler, Paul Cabell, Woody Etherly Jr., Ben Hamper, John Hightower, Edgar Holt, Alex Kotlowitz, Michael Moore, and other activists, leaders, and visionaries whose struggles, past and present, have made Flint a more equitable city. They deserve a special acknowledgment for providing me with information and inspiration. Thanks also to Ananthakrishnan Aiyer, Erin Caudell, Shawn Chittle, Jerome Chou, Connor Coyne, Dawn Demps, Thomas Henthorn, Christina Kelly, Jason Kosnoski, Adam Lutzker, Laura Gillespie MacIntyre, Adrian Montague, Joel Rash, Nayyirah Shariff, Jan Worth-Nelson, Gordon Young, and Stephen Zacks for your commitment to improving Flint. Your efforts inspire me.

This project received generous financial support from a number of different sources. Grants and fellowships from the American Council of Learned Societies, the Andrew W. Mellon Foundation, the Department of Public Administration at the University of Texas at San Antonio, the Eisenberg Institute for Historical Studies and the Institute for the Humanities at the University of Michigan, the National Academy of Education, and the Spencer Foundation allowed me to spend extra time and money on this project. These funds also enabled me to attend numerous meetings and conferences, where I met and collaborated with a talented pool of scholars from across the disciplinary spectrum. Those individuals include Eric Avila, Christopher Bonastia, Kevin Boyle, Tomiko Brown-Nagin, Andy Clarno, Lizabeth Cohen, Claire Decoteau, Margie Dewar, Angela Dillard, Richardson Dilworth, Ruben Donato, Jack Dougherty, Phil Ethington, David Freund, Brett Gadsden, Scott Gelber, Kim Greenwell, Robbie Gross, Edin Hajdarpasic, Richard Harris, Khalil Anthony Johnson Jr., Carl Kaestle, Ed Kelly, Nora Krinitsky, Kevin Kruse, Elaine Lewinnek, Guian McKee, Hilary Moss, Donna Murch, Gary Orfield, Russell Rickford, Mark Rose, Robert Self, Jon Shelton, Thomas Sugrue, Heather Thompson, Maris Vinovskis, and Victoria Wolcott. My interactions with these colleagues and friends have vastly improved the quality of my work.

Out of all the wonderful scholars I have encountered while working on this book, I would like to single out one special person: Matthew Lassiter. For almost fifteen years now, Matt has mentored and guided me in ways that I can only begin to acknowledge here. He has read this manuscript on more occasions than I can count, each time offering new ideas about how to make it better. When I have veered off course, he has been the first to deliver critical feedback, but when I have needed a sympathetic ear, he has always been there for me. Matt, I would not be where I am without you. Neither this book nor my career in academia would exist without your support and guidance. Please accept my deepest and sincerest thanks for everything you have done for me.

I am extremely fortunate to have an amazing group of smart and sympathetic friends. For hanging out with me, listening to my stories about Flint, and showing me what really matters in life, sincere thanks to Lynda Barnes, Tamar Carroll, Aaron Cavin, Sherri Charleston, Dan Cieslik, Joe Pat Clayton, Nathan Connolly, Michan Connor, Allen Dieterich-Ward, Ansley Erickson, Lily Geismer, Todd Getz, Jerry González, Matt Ides, Hillary Jenks, Andrew Kahrl, Bella Muntz Kirchner, Gabe Kirchner, Rosina Lozano, Natalia Mehlman-Petrzela, Drew Meyers, Stephanie Minich, Tim Minich, Josh Mound, Andrew Needham, Alyssa Picard, Tim Retzloff, Anthony Ross, Pete Soppelsa, Mandana Varahrami, Kathleen Williams, Luke Williams, Dale Winling, Stephen Wisniewski, Laura Gerstler Wright, Marvin Wright, and Mike Zilliox.

Members of my family have been a constant source of love and strength throughout my life. Without them, my world would be incomplete. Suvinder Chadha, William Fearer, Jagmohan Singh, and my grandparents—Gerald Fearer, Rogene Fearer, Ferris Highsmith, and Pauline Highsmith—all passed away before I finished this book, but my memories of them continue to bring me joy and inspiration. May you all rest in peace. For your love, laughs, and accepting me with open arms, thanks to Arjun Chada, Esha Chada, Neha Chada, Meena Chadha, Elisa Nehls, Ethan Nehls, Neelu Nehls, Eric Romine, Comilla Sasson, and all of my extended family in India. This acknowledgment would not be complete without a special thank you to Satwant Sasson, my incomparable mother-in-law, for her love and encouragement.

Although I moved away from Cincinnati two decades ago, a piece of my heart will always remain there with my family. Many thanks to Allegra Highsmith, David Highsmith, Evan Highsmith, and Maya Highsmith for their love and inspiration. My parents, Martha and Robert Highsmith, deserve more gratitude than I can possibly communicate here. For over four decades now, they have offered me their unconditional love and support. Without those priceless gifts I would not be where I am today. Mom and Dad, thank you for putting food on the table, setting a positive example by working hard at your jobs, coaching my sports teams, always living in integrated neighborhoods and sending me to integrated public schools, encouraging me to dream big, reading to me for hours on end, and more, so much more. I love you both with all of my heart.

In closing, I would like to acknowledge my wife, Bobby Sasson, and my children, Asha May Highsmith, Mira Pauline Highsmith, and Aneel Gerald Highsmith. Bobby, we have been on this journey now for sixteen years. During that time we have gone from being total strangers to sharing the deepest love imaginable. We have celebrated more triumphs and milestones than I can even remember, but we've also suffered through the most difficult of times and the most painful losses. But here we are now—still standing, still together. Ten years ago, while we were sitting on our porch in Mott Park, we first talked about my plan to write a book about Flint. Since then you have been the most important supporter of and contributor to this project. Beyond reading drafts of chapters, helping me conduct research, and printing out reams of research notes, you have been my biggest cheerleader and the first person I turn to when I fail. You have also given me the three most amazing children in the world and shown me by your example how to be a better spouse, parent, and person. And you have done all of this while working outrageously long hours as a primary care physician. I don't know how you manage to do it all, Bobby, but I love you and respect you for it. You are the love of my life and the center of our family. Thank you for everything.

I would also like to thank my three stupendous children. Asha, you came along not long after I started working on this book. Today you are the prettiest, smartest, most incredible little eight-year-old in the whole world. You have the sweetest, gentlest heart of anyone I know, and your compassion for others warms my soul. I am also amazed and inspired by your work ethic and your willingness to face difficult challenges. A year and a half ago you agreed to enroll in a Spanish-language immersion program at your elementary school. Since then you have done virtually all of your schoolwork in another language. Not everyone could take on a challenge like that, Asha, and few could excel in the classroom as you have over the past eighteen months. I couldn't be more proud of you for that and a million other things. You may not realize this, but your love, support, and inspiration have helped me to accomplish my goal of writing a book. Just seeing your smiling face at the end of a hard day of writing is enough to brighten my mood. And, of course, you are my favorite companion in the world for San Antonio Spurs games. Asha, thank you for being such a breathtakingly lovely young lady. Te amo con todo mi corazón. Tú eres lo mejor.

Mira Pauline Highsmith, you are a force of nature. You are without question the naughtiest, most mischievous member of the Highsmith family, and you have given me lots of gray hair. But you are also the toughest, smartest, and most beautiful little four-year-old girl in all of Texas. In your short time on this earth you have suffered through three major surgeries, countless pokes and prods, and more pain than anyone should ever have to endure, but you have triumphed over all of it. Your bravery in the face of adversity is truly inspiring. I am also amazed by the power of your mind. Like your mother, you are so innately intelligent that most things just seem to come easily to you. Even when they don't, you have the tenacity and creativity to solve even the toughest of life's problems. You are also by far the funniest member of our family. I could spend hours watching you prance around the house in your costumes, and your karaoke routines are more entertaining than just about anything. In short, you are an utterly delightful person, Mira, and I cannot wait to see what life has in store for you. Today and always, I love you more than you will ever know.

Finally, I want to thank my son Aneel Gerald Highsmith, the third and *last* of our children. Aneel, you made your surprise appearance just as I was finishing this book. Before you were born, your mother and I were terrified at the prospect of raising three children. We didn't know how we were going to balance the many challenges of caring for a new baby with the hard work of raising two energetic young girls. But then you came along and put all of our fears to rest. Over the past year and a half, you have amazed us all with your sweetness, playfulness, curiosity, sensitivity, and, most of all, patience. In the morning while everyone is getting ready for work and school, you sit calmly and happily in your high chair, watching all the madness unfold. On weekends when we are out running errands, you lounge patiently in your car seat while the rest of us complain about traffic, the heat, and other minor nuisances. After dinner while your mom and I scramble to wash the dishes, you play quietly with your toys until one of us can get free to hold you. All of this is to say that you are the sweetest, most considerate boy in the world, Aneel. Your beautiful smile warms my heart, and your giggles are some of the greatest gifts imaginable. Thank you for completing our family. I love you, your sisters, and your beautiful mama with all of the love my heart can hold. That is why I dedicate this book to the four of you.

Gruene, Texas, February 2015

Abbreviations in the Notes

Newspapers and Magazines

ACS	*AC Sparkler*
BR	*Bronze Reporter*
CD	*Chicago Defender*
CM	*Clio Messenger*
DFP	*Detroit Free Press*
DI	*Davison Index*
FE	*Flint Enquirer*
FI	*Fenton Independent*
FJ	*Flint Journal*
FM	*Flint Mirror*
FNA	*Flint News-Advertiser*
FO	*Flushing Observer*
FS	*Flint Spokesman*
FV	*Flint Voice*
FWR	*Flint Weekly Review*
GBN	*Grand Blanc News*
GCH	*Genesee County Herald*
MV	*Michigan Voice*
NYT	*New York Times*
SCN	*Swartz Creek News*
USNWR	*U.S. News and World Report*
WAW	*Ward's Auto World*
WP	*Washington Post*
WSJ	*Wall Street Journal*

Archival Records

AAP	Albert Applegate Papers, Bentley Historical Library, University of Michigan, Ann Arbor
ACFP	Alexander C. Findlay Papers, Genesee Historical Collections Center, University of Michigan–Flint

AEP	Arthur Elder Papers, Walter P. Reuther Library of Labor and Urban Affairs, Wayne State University, Detroit
AWCF	AutoWorld Clipping Files, Perry Archives, Buick Gallery and Research Center, Alfred P. Sloan Museum, Flint, MI
BHF	Black History Files, Perry Archives, Buick Gallery and Research Center, Alfred P. Sloan Museum, Flint, MI
BPOHPF	Bronze Pillars Oral History Project Files, Perry Archives, Buick Gallery and Research Center, Alfred P. Sloan Museum, Flint, MI
CDN	Community Development Newsletters, Genesee Historical Collections Center, University of Michigan–Flint
CMCF	Chevrolet Motor Company Files, Perry Archives, Buick Gallery and Research Center, Alfred P. Sloan Museum, Flint, MI
CMF	City Manager Files, Office of the Flint City Clerk, Flint, MI
CSMP	Charles Stewart Mott Papers, Genesee Historical Collections Center, University of Michigan–Flint
DWRP	Donald W. Riegle Jr. Papers, Genesee Historical Collections Center, University of Michigan–Flint
EBHP	Edgar B. Holt Papers, Genesee Historical Collections Center, University of Michigan–Flint
FACIF	Flint Area Conference, Inc. Files, Perry Archives, Buick Gallery and Research Center, Alfred P. Sloan Museum, Flint, MI
FDF	Flint Development Files, Perry Archives, Buick Gallery and Research Center, Alfred P. Sloan Museum, Flint, MI
FJMP	Frank J. Manley Papers, Richard P. Scharchburg Archives, Kettering University, Flint, MI
GFHP	Gerald F. Healy Papers, Genesee Historical Collections Center, University of Michigan–Flint
GLP	Garland Lane Papers, Walter P. Reuther Library of Labor and Urban Affairs, Wayne State University, Detroit
GRP	George Romney Papers, Gubernatorial Collection, Bentley Historical Library, University of Michigan, Ann Arbor
JAANP	Joseph A. Anderson Papers, Perry Archives, Buick Gallery and Research Center, Alfred P. Sloan Museum, Flint, MI
MAFP	Manufacturers Association of Flint Papers, Richard P. Scharchburg Archives, Kettering University, Flint, MI
MWF	Mike Westfall Files, Genesee Historical Collections Center, University of Michigan–Flint
MWP	Mike Westfall Papers, Walter P. Reuther Library of Labor and Urban Affairs, Wayne State University, Detroit
OBP	Olive Beasley Papers, Genesee Historical Collections Center, University of Michigan–Flint
OH	Oral Histories, 1987–1989, Genesee Historical Collections Center, University of Michigan–Flint
OMFBE	*Official Minutes of the Flint Board of Education*
PCC	*Proceedings of the Flint City Council*
PFCC	*Proceedings of the Flint City Commission*

RG 31	Record Group 31, Records of the Federal Housing Administration, National Archives II, College Park, MD
RG 195	Record Group 195, Records of the Federal Home Loan Bank Board, National Archives II, College Park, MD
RG 207	Record Group 207, Records of the Department of Housing and Urban Development, National Archives II, College Park, MD
RMHP	Robert M. Hamady Papers, Bentley Historical Library, University of Michigan, Ann Arbor
ROWP	Ronald O. Warner Papers, Richard P. Scharchburg Archives, Kettering University, Flint, MI
UAW581	United Automobile Workers, Local 581 Collection, Walter P. Reuther Library of Labor and Urban Affairs, Wayne State University, Detroit
UAWCAP	United Automobile Workers, Community Action Program Department Collection, Walter P. Reuther Library of Labor and Urban Affairs, Wayne State University, Detroit
UAWFF	United Automobile Workers, Foundry and Forge Department Collection, Walter P. Reuther Library of Labor and Urban Affairs, Wayne State University, Detroit
UAWFP	United Automobile Workers, Fair Practices Department Collection, Walter P. Reuther Library of Labor and Urban Affairs, Wayne State University, Detroit
UAWGM	United Automobile Workers, General Motors Department Collection, Walter P. Reuther Library of Labor and Urban Affairs, Wayne State University, Detroit
UAWPA	United Automobile Workers, Political Action Department Collection, Walter P. Reuther Library of Labor and Urban Affairs, Wayne State University, Detroit
UAWR	United Automobile Workers, Research Department Collection, Walter P. Reuther Library of Labor and Urban Affairs, Wayne State University, Detroit
ULFP	Urban League of Flint Papers, Genesee Historical Collections Center, University of Michigan–Flint
UM-F/HDSP	University of Michigan–Flint, History Department, Student Papers, Genesee Historical Collections Center, University of Michigan–Flint

Libraries and Archives

BHL	Bentley Historical Library, University of Michigan, Ann Arbor
F-GC/FPL	Flint-Genesee County Reference Collection, Flint Public Library
GHCC	Genesee Historical Collections Center, University of Michigan–Flint
LM	Library of Michigan, Lansing
PA	Perry Archives, Buick Gallery and Research Center, Alfred P. Sloan Museum, Flint, MI
SA	Richard P. Scharchburg Archives, Kettering University, Flint, MI
WPRL	Walter P. Reuther Library of Labor and Urban Affairs, Wayne State University, Detroit

Government Agencies, Private Organizations, and Miscellaneous Entities

CERP	United Automobile Workers, Cost of Living Adjustment for Retirees, Early Retirement, Paid Personal Holidays Committee

CIO	Congress of Industrial Organizations
CPA	City Planning Associates
CRCF	Civic Research Council of Flint
CRCM	Citizens Research Council of Michigan
CSF	Records Relating to the City Survey File
CSMF	Charles Stewart Mott Foundation
DCD	Department of Community Development
DRS	Division of Research and Statistics
FACI	Flint Area Conference, Inc.
FBE	Flint Board of Education
FCC	Flint City Commission
FCPC	Flint City Planning Commission
FCSG	Flint Citizens' Study Group
FEPC	Fair Employment Practices Commission
FFEPC	Flint Fair Employment Practices Commission
FHA	Federal Housing Administration
FHC	Flint Housing Commission
FRHD	Flint Renewal and Housing Department
GCDC	Genesee Community Development Conference
GCMPC	Genesee County Metropolitan Planning Commission
GM	General Motors Corporation
GNRP	General Neighborhood Renewal Plan
HEW	Department of Health, Education, and Welfare
HOA	Home Ownership Assistance
HOLC	Home Owners' Loan Corporation
HOME	Housing Opportunities Made Equal
HSG	Housing Surveys, General
HUD	Department of Housing and Urban Development
IHA	Institute for Human Adjustment, University of Michigan
MAF	Manufacturers Association of Flint
MCRC	Michigan Civil Rights Commission
MESC	Michigan Employment Security Commission
MSHD	Michigan State Highway Department
NAACP	National Association for the Advancement of Colored People
NAREB	National Association of Real Estate Boards
PURA	Project for Urban and Regional Affairs
RHMA	Reports of Housing Market Analysis
RVPDC	Region V Planning and Development Commission
SEMCOG	Southeast Michigan Council of Governments
SOS	Save Our Schools
SPA	Social Planning Associates
UAW	United Automobile Workers
ULF	Urban League of Flint
ULI	Urban Land Institute
USBC	United States Department of Commerce, Bureau of the Census
USGPO	United States Government Printing Office

Notes

Introduction

1. "50 Millionth GM Auto Touches Off Big Celebration," *FJ*, November 23, 1954; "Curtice Praises Employees of GM," *FJ*, November 23, 1954.

2. On the postwar economy, see Robert M. Collins, *More: The Politics of Economic Growth in Postwar America* (New York: Oxford University Press, 2000).

3. Clarence H. Young and William A. Quinn, *Foundation for Living: The Story of Charles Stewart Mott and Flint* (New York: McGraw-Hill, 1963); Alfred P. Sloan Jr., *My Years with General Motors* (Garden City, NY: Doubleday, 1964); Timothy Jacobs, *A History of General Motors* (Greenwich, CT: Brompton Books, 1992); Axel Madsen, *The Deal Maker: How William C. Durant Made General Motors* (New York: John Wiley and Sons, 1999); David Farber, *Sloan Rules: Alfred P. Sloan and the Triumph of General Motors* (Chicago: University of Chicago Press, 2002).

4. "Symbolic Auto Touches Off Big Celebration Here," *FJ*, November 23, 1954.

5. "Spectacle Is Climaxed by 50 Millionth Car," *FJ*, November 23, 1954; "Brilliant Parade Tells General Motors Story," *FJ*, November 23, 1954; "Windows, Trees and Roofs Crowded along Carnival's Line of March," *FJ*, November 23, 1954.

6. "Curtice Announces $3,000,000 Flint Cultural Center Grant," *FJ*, November 23, 1954.

7. "Consensus in Flint: Golden Carnival Great," *FJ*, November 25, 1954.

8. Kenneth T. Jackson, *Crabgrass Frontier: The Suburbanization of the United States* (New York: Oxford University Press, 1985).

9. On regionalism, see Jon C. Teaford, *City and Suburb: The Political Fragmentation of Metropolitan America, 1850–1970* (Baltimore: Johns Hopkins University Press, 1979); David Rusk, *Cities without Suburbs* (Washington, DC: Woodrow Wilson Center Press, 1993); Manuel Pastor Jr. et al., *Regions That Work: How Cities and Suburbs Can Grow Together* (Minneapolis: University of Minnesota Press, 2000); Myron Orfield, *American Metropolitics: The New Suburban Reality* (Washington, DC: Brookings Institution Press, 2002); Peter Dreier, John Mollenkopf, and Todd Swanstrom, *Place Matters: Metropolitics for the Twenty-First Century*, 2nd ed. (Lawrence: University Press of Kansas, 2004); Jon C. Teaford, *The Metropolitan Revolution: The Rise of Post-Urban America* (New York: Columbia University Press, 2006).

10. *FJ*, October 8, 1974; Bryan D. Jones and Lynn W. Bachelor, *The Sustaining Hand: Community Leadership and Corporate Power* (Lawrence: University Press of Kansas, 1986); Ronald Edsforth, *Class Conflict and Cultural Consensus: The Making of a Mass Consumer Society in Flint,*

Michigan (New Brunswick, NJ: Rutgers University Press, 1987), 191–228; James D. Ananich, Neil O. Leighton, and Charles T. Webber, *Economic Impact of Plant Closings in Flint, Michigan* (Flint: Regents of the University of Michigan, 1989); Steven P. Dandaneau, *A Town Abandoned: Flint, Michigan, Confronts Deindustrialization* (Albany: SUNY Press, 1996); Theodore J. Gilman, *No Miracles Here: Fighting Urban Decline in Japan and the United States* (Albany: SUNY Press, 2001).

11. On economic retrenchment in the 1970s and 1980s, see William Serrin, *The Company and the Union: The "Civilized Relationship" of the General Motors Corporation and the United Automobile Workers* (New York: Random House, 1973); Mike Davis, *Prisoners of the American Dream: Politics and Economy in the History of the US Working Class* (London: Verso, 1986); Kim Moody, *An Injury to All: The Decline of American Unionism* (London: Verso, 1988), 95–126; Barbara Ehrenreich, *Fear of Falling: The Inner Life of the Middle Class* (New York: Pantheon Books, 1989); Robert Zieger, *American Workers, American Unions, 1920–1985* (Baltimore: Johns Hopkins University Press, 1995), 201–11; Nelson Lichtenstein, *The Most Dangerous Man in Detroit: Walter Reuther and the Fate of American Labor* (New York: Basic Books, 1995), 439–45; Heather Ann Thompson, *Whose Detroit? Politics, Labor, and Race in a Modern American City* (Ithaca, NY: Cornell University Press, 2001), 192–223; Bruce J. Schulman, *The Seventies: The Great Shift in American Culture, Society, and Politics* (New York: Free Press, 2001), 121–43; Jefferson Cowie, "'Vigorously Left, Right, and Center': The Crosscurrents of Working-Class America in the 1970s," in *America in the Seventies*, ed. Beth Bailey and David Farber (Lawrence: University Press of Kansas, 2004), 75–106; Judith Stein, *Pivotal Decade: How the United States Traded Factories for Finance in the Seventies* (New Haven, CT: Yale University Press, 2010); Jefferson Cowie, *Stayin' Alive: The 1970s and the Last Days of the Working Class* (New York: New Press, 2010).

12. Greg Gardner, "Buick City's Demise," *WAW*, June 1997, 23–25; David Leonhardt, "Even under a Cloud, G.M. Is Predicting Sunshine," *NYT*, June 2, 2009; Melissa Burden, "Flint's Unemployment Rate Hits 27.3 Percent in May," *FJ*, July 16, 2009, http://www.mlive.com/business/mid-michigan/index.ssf/2009/07/flints_unemployment_rate_hits.html. On deindustrialization in the late twentieth century, see Barry Bluestone and Irving Bluestone, *Negotiating the Future: A Labor Perspective on American Business* (New York: Basic Books, 1992); Kim Moody, *Workers in a Lean World: Unions in the International Economy* (London: Verso, 1997); Jefferson Cowie, *Capital Moves: RCA's Seventy-Year Quest for Cheap Labor* (New York: New Press, 1999), 127–209; Naomi Klein, *No Logo* (New York: Picador, 2000). On the 1990s, see Thomas Frank, *One Market under God: Extreme Capitalism, Market Populism, and the End of Economic Democracy* (New York: Anchor Books, 2000).

13. The most influential articulation of this thesis is in Thomas J. Sugrue, *The Origins of the Urban Crisis: Race and Inequality in Postwar Detroit* (Princeton, NJ: Princeton University Press, 1996).

14. Margaret Weir, Anna Shola Orloff, and Theda Skocpol, eds., *The Politics of Social Policy in the United States* (Princeton, NJ: Princeton University Press, 1983); Richard Feldman and Michael Betzold, eds., *End of the Line: Autoworkers and the American Dream* (New York: Weidenfeld and Nicolson, 1988); Steve Fraser and Gary Gerstle, eds., *The Rise and Fall of the New Deal Order, 1930–1980* (Princeton, NJ: Princeton University Press, 1989); Kevin Boyle, *The UAW and the Heyday of American Liberalism* (Ithaca, NY: Cornell University Press, 1995); Lichtenstein, *Most Dangerous Man in Detroit*; Alan Brinkley, *The End of Reform: New Deal Liberalism in Recession and War* (New York: Vintage Books, 1995); Michael Moore, *Downsize This! Random Threats from an Unarmed American* (New York: Crown, 1996). On industrial cities as "arsenals of democracy," see Sugrue, *Origins of the Urban Crisis*, 17–31. On postwar consumer culture

and politics, see Lizabeth Cohen, *A Consumers' Republic: The Politics of Mass Consumption in Postwar America* (New York: Alfred A. Knopf, 2003); Meg Jacobs, *Pocketbook Politics: Economic Citizenship in Twentieth-Century America* (Princeton, NJ: Princeton University Press, 2004).

15. For more on the commitment to revitalization among urban Americans, see Heather Ann Thompson, "Rethinking the Politics of White Flight in the Postwar City: Detroit, 1945–1980," *Journal of Urban History* 25, no. 2 (January 1999): 163–98; Thompson, *Whose Detroit?*; Howard Gillette Jr., *Camden after the Fall: Decline and Renewal in a Post-Industrial City* (Philadelphia: University of Pennsylvania Press, 2006); Suleiman Osman, *The Invention of Brownstone Brooklyn: Gentrification and the Search for Authenticity in Postwar New York* (New York: Oxford University Press, 2011).

16. Robert O. Self, *American Babylon: Race and the Struggle for Postwar Oakland* (Princeton, NJ: Princeton University Press, 2003); Eric Avila, *Popular Culture in the Age of White Flight: Fear and Fantasy in Suburban Los Angeles* (Berkeley: University of California Press, 2004); Alison Isenberg, *Downtown America: A History of the Place and the People Who Made It* (Chicago: University of Chicago Press, 2005); Gillette, *Camden after the Fall*; Guian A. McKee, *The Struggle for Jobs: Liberalism and Local Activism in Postwar Philadelphia* (Chicago: University of Chicago Press, 2008); Samuel Zipp, *Manhattan Projects: The Rise and Fall of Urban Renewal in Cold War New York* (New York: Oxford University Press, 2010); Christopher Klemek, *The Transatlantic Collapse of Urban Renewal: Postwar Urbanism from New York to Berlin* (Chicago: University of Chicago Press, 2011).

17. Arnold R. Hirsch, *Making the Second Ghetto: Race and Housing in Chicago, 1940–1960* (Cambridge: Cambridge University Press, 1983); Sugrue, *Origins of the Urban Crisis*; James N. Gregory, *The Southern Diaspora: How the Great Migrations of Black and White Southerners Transformed America* (Chapel Hill: University of North Carolina Press, 2005); Matthew J. Countryman, *Up South: Civil Rights and Black Power in Philadelphia* (Philadelphia: University of Pennsylvania Press, 2006).

18. Thomas J. Sugrue, "Crabgrass-Roots Politics: Race, Rights, and the Reaction against Liberalism in the Urban North, 1940–1964," *Journal of American History* 82, no. 2 (September 1995): 551–78.

19. On racial and economic inequalities in metropolitan America, see Gilbert Osofsky, *Harlem: The Making of a Ghetto; Negro New York, 1890–1930* (New York: Harper and Row, 1966); Allan H. Spear, *Black Chicago: The Making of a Negro Ghetto, 1890–1920* (Chicago: University of Chicago Press, 1967); Hirsch, *Making the Second Ghetto*; Jackson, *Crabgrass Frontier*, 190–230; Douglas S. Massey and Nancy A. Denton, *American Apartheid: Segregation and the Making of the Underclass* (Cambridge, MA: Harvard University Press, 1993); Sugrue, *Origins of the Urban Crisis*; Becky M. Nicolaides, *My Blue Heaven: Life and Politics in the Working-Class Suburbs of Los Angeles, 1920–1965* (Chicago: University of Chicago Press, 2002); Wendell Pritchett, *Brownsville, Brooklyn: Blacks, Jews, and the Changing Face of the Ghetto* (Chicago: University of Chicago Press, 2002); Self, *American Babylon*; Josh Sides, *L.A. City Limits: African American Los Angeles from the Great Depression to the Present* (Berkeley: University of California Press, 2003); Avila, *Popular Culture in the Age of White Flight*; Bryant Simon, *Boardwalk of Dreams: Atlantic City and the Fate of Urban America* (New York: Oxford University Press, 2004); Kevin M. Kruse, *White Flight: Atlanta and the Making of Modern Conservatism* (Princeton, NJ: Princeton University Press, 2005); Amanda I. Seligman, *Block by Block: Neighborhoods and Public Policy on Chicago's West Side* (Chicago: University of Chicago Press, 2005); Robert M. Fogelson, *Bourgeois Nightmares: Suburbia, 1870–1930* (New Haven, CT: Yale University Press, 2005); Kevin M. Kruse and Thomas J. Sugrue, eds., *The New Suburban History* (Chicago: University of Chi-

cago Press, 2006); David M. P. Freund, *Colored Property: State Policy and White Racial Politics in Suburban America* (Chicago: University of Chicago Press, 2007); Colin Gordon, *Mapping Decline: St. Louis and the Fate of the American City* (Philadelphia: University of Pennsylvania Press, 2008); Thomas J. Sugrue, *Sweet Land of Liberty: The Forgotten Struggle for Civil Rights in the North* (New York: Random House, 2008); Scott Kurashige, *The Shifting Grounds of Race: Black and Japanese Americans in the Making of Multiethnic Los Angeles* (Princeton, NJ: Princeton University Press, 2008); Charlotte Brooks, *Alien Neighbors, Foreign Friends: Asian Americans, Housing, and the Transformation of Urban California* (Chicago: University of Chicago Press, 2009); LeeAnn Lands, *The Culture of Property: Race, Class, and Housing Landscapes in Atlanta, 1880−1950* (Athens: University of Georgia Press, 2009); Monica Perales, *Smeltertown: Making and Remembering a Southwest Border Community* (Chapel Hill: University of North Carolina Press, 2010); Daniel Martinez HoSang, *Racial Propositions: Ballot Initiatives and the Making of Postwar California* (Berkeley: University of California Press, 2010); Mark Brilliant, *The Color of America Has Changed: How Racial Diversity Shaped Civil Rights Reform in California, 1941−1978* (New York: Oxford University Press, 2010); Carl H. Nightingale, *Segregation: A Global History of Divided Cities* (Chicago: University of Chicago Press, 2012); N. D. B. Connolly, *A World More Concrete: Real Estate and the Remaking of Jim Crow South Florida* (Chicago: University of Chicago Press, 2014); Andrew Needham, *Power Lines: Phoenix and the Making of the Modern Southwest* (Princeton, NJ: Princeton University Press, 2014).

20. Andrew R. Highsmith and Ansley T. Erickson, "The Strange Career of De Facto Segregation: Race and Region in the Scholarly Imagination," unpublished manuscript, October 1, 2014.

21. *Brown v. Board of Education*, 347 U.S. 483 (1954).

22. Gary Orfield, *Must We Bus? Segregated Schools and National Policy* (Washington, DC: Brookings Institution, 1978); Arnold R. Hirsch, "Containment on the Home Front: Race and Federal Housing Policy from the New Deal to the Cold War," *Journal of Urban History* 26, no. 2 (January 2000): 159; Martha Biondi, *To Stand and Fight: The Struggle for Civil Rights in Postwar New York City* (Cambridge, MA: Harvard University Press, 2003); Arnold R. Hirsch, "Less Than *Plessy*: The Inner City, Suburbs, and State-Sanctioned Residential Segregation in the Age of *Brown*," in *New Suburban History*, ed. Kruse and Sugrue, 33−56; Sugrue, *Sweet Land of Liberty*. For the most definitive critique of the mythology of de facto segregation, see Matthew D. Lassiter, *The Silent Majority: Suburban Politics in the Sunbelt South* (Princeton, NJ: Princeton University Press, 2005); Matthew D. Lassiter, "De Jure / De Facto Segregation: The Long Shadow of a National Myth," in *The Myth of Southern Exceptionalism*, ed. M. D. Lassiter and Joseph Crespino (New York: Oxford University Press, 2010), 25−48. On northern racial exceptionalism, see Jeanne Theoharis and Komozi Woodard, eds., *Freedom North: Black Freedom Struggles outside the South* (New York: Palgrave Macmillan, 2003); Countryman, *Up South*.

23. For examples of studies that emphasize housing over education, see Hirsch, *Making the Second Ghetto*; Jackson, *Crabgrass Frontier*; Massey and Denton, *American Apartheid*; Sugrue, *Origins of the Urban Crisis*; Self, *American Babylon*; Cohen, *Consumers' Republic*; Freund, *Colored Property*; Nightingale, *Segregation*. For a more in-depth articulation of this point, see Andrew R. Highsmith and Ansley T. Erickson, "Segregation as Splitting, Segregation as Joining: Schools, Housing, and the Many Modes of Jim Crow," *American Journal of Education* (forthcoming).

24. Orfield, *Must We Bus?*; Jennifer L. Hochschild, *The New American Dilemma: Liberal Democracy and School Desegregation* (New Haven, CT: Yale University Press, 1984); Jeffrey Mirel, *The Rise and Fall of an Urban School System: Detroit, 1907−1981* (Ann Arbor: University of Michigan Press, 1993); Jack Dougherty, *More than One Struggle: The Evolution of Black School Reform*

in Milwaukee (Chapel Hill: University of North Carolina Press, 2004); Davison M. Douglas, *Jim Crow Moves North: The Battle over Northern School Desegregation, 1865–1954* (New York: Cambridge University Press, 2005); Kruse, *White Flight*; Lassiter, *Silent Majority*; Jack Dougherty, "Bridging the Gap between Urban, Suburban, and Educational History," in *Rethinking the History of American Education*, ed. William J. Reese and John L. Rury (New York: Palgrave Macmillan, 2007); Lassiter, "De Jure / De Facto Segregation," 25–48; Matthew D. Lassiter, "Schools and Housing in Metropolitan History: An Introduction," *Journal of Urban History* 38, no. 2 (March 2012): 195–204; Ansley T. Erickson, "Building Inequality: The Spatial Organization of Schooling in Nashville, Tennessee, after *Brown*," *Journal of Urban History* 38, no. 2 (March 2012): 247–70; Karen Benjamin, "Suburbanizing Jim Crow: The Impact of School Policy on Residential Segregation in Raleigh," *Journal of Urban History* 38, no. 2 (March 2012): 225–46; Jack Dougherty, "Shopping for Schools: How Public Education and Private Housing Shaped Suburban Connecticut," *Journal of Urban History* 38, no. 2 (March 2012): 205–24.

25. See, for example, Jackson, *Crabgrass Frontier*, 190–230; Sugrue, *Origins of the Urban Crisis*, 33–88; Freund, *Colored Property*; Nightingale, *Segregation*.

26. On the relationship between state power and popular segregation, see Cheryl I. Harris, "Whiteness as Property," *Harvard Law Review* 106, no. 8 (June 1993): 1707–91; Risa Goluboff, *The Lost Promise of Civil Rights* (Cambridge, MA: Harvard University Press, 2010).

27. Jackson, *Crabgrass Frontier*; Freund, *Colored Property*.

28. On the array of factors that HOLC officers considered, see James Greer, "The Home Owners' Loan Corporation and the Development of the Residential Security Maps," *Journal of Urban History* 39, no. 2 (January 2013): 275–96. On working-class suburbs, see Richard Harris, *Unplanned Suburbs: Toronto's American Tragedy, 1900–1950* (Baltimore: Johns Hopkins University Press, 1996); Nicolaides, *My Blue Heaven*; Andrew Wiese, *Places of Their Own: African American Suburbanization in the Twentieth Century* (Chicago: University of Chicago Press, 2004).

29. On suburban utilities and services, see Adam Rome, *The Bulldozer in the Countryside: Suburban Sprawl and the Rise of American Environmentalism* (New York: Cambridge University Press, 2001), 87–118. On the urbanization of suburbia, see Jon C. Teaford, *Post-Suburbia: Government and Politics in the Edge Cities* (Baltimore: Johns Hopkins University Press, 1996). On redlining in other suburban areas, see CSF, 1935–1940, boxes 18, 39, 84, RG 195. As the HOLC's records confirm, lenders in many other parts of the country were already disinvesting from working-class suburbs and other "risky" areas prior to the creation of the security maps. In fact, HOLC surveyors often justified their low ratings for working-class suburbs by pointing to the existing practices of local lenders. If lenders had already disinvested from an area, HOLC officers typically assigned it a grade of C or D.

30. Disinvestment from working-class suburbs was not unique to the United States. See Richard Harris and Doris Forrester, "The Suburban Origins of Redlining: A Canadian Case Study, 1935–54," *Urban Studies* 40, no. 13 (December 2003): 2661–86.

31. Scholars have addressed the subjects of sprawl and metropolitan fragmentation from a variety of points of view. Some have argued in favor of sprawl and fragmentation by suggesting that they breed greater governmental efficiency and higher living standards. See, for example, Charles M. Tiebout, "A Pure Theory of Local Expenditures," *Journal of Political Economy* 64, no. 5 (October 1956): 416–24; Vincent Ostrom, Charles M. Tiebout, and Robert Warren, "The Organization of Government in Metropolitan Areas: A Theoretical Inquiry," *American Political Science Review* 55, no. 4 (1961): 831–42; Robert Bruegmann, *Sprawl: A Compact History* (Chicago: University of Chicago Press, 2006). By contrast, other scholars have deplored sprawl and fragmentation for being environmentally destructive, economically inefficient, and

socially divisive. For examples of this scholarship, see Teaford, *City and Suburb*; Rusk, *Cities without Suburbs*; Teaford, *Post-Suburbia*; Gerald E. Frug, *City Making: Building Communities without Building Walls* (Princeton, NJ: Princeton University Press, 1999); Ann Durkin Keating, *Building Chicago: Suburban Developers and the Creation of a Divided Metropolis* (Urbana: University of Illinois Press, 2002); Self, *American Babylon*; Richardson Dilworth, *The Urban Origins of Suburban Autonomy* (Cambridge, MA: Harvard University Press, 2005); Dreier, Mollenkopf, and Swanstrom, *Place Matters*; Michan Andrew Connor, "Public Benefits from Public Choice: Producing Decentralization in Metropolitan Los Angeles, 1954–1973," *Journal of Urban History* 39, no. 1 (January 2013): 79–100. On the racial politics of suburban secession, see Kruse, *White Flight*.

32. Sam Bass Warner, *Streetcar Suburbs: The Process of Growth in Boston, 1870–1900* (Cambridge, MA: Harvard University Press, 1978); Teaford, *City and Suburb*; Jackson, *Crabgrass Frontier*, 138–56.

33. The literature on suburban separatism and exclusion is vast. See, for instance, Seymour I. Toll, *Zoned American* (New York: Grossman Publishers, 1969); Michael N. Danielson, *The Politics of Exclusion* (New York: Columbia University Press, 1976); Teaford, *City and Suburb*; Jackson, *Crabgrass Frontier*; David L. Kirp, John P. Dwyer, and Larry A. Rosenthal, *Our Town: Race, Housing, and the Soul of Suburbia* (New Brunswick, NJ: Rutgers University Press, 1997); Nicolaides, *My Blue Heaven*; Self, *American Babylon*; Gillette, *Camden after the Fall*; Sides, *L.A. City Limits*; Kruse, *White Flight*; Lassiter, *Silent Majority*; Kruse and Sugrue, eds., *New Suburban History*; Freund, *Colored Property*; Gordon, *Mapping Decline*.

34. Barry Bluestone and Bennett Harrison, *The Deindustrialization of America: Plant Closings, Community Abandonment, and the Dismantling of Basic Industry* (New York: Basic Books, 1982), chap. 5.

35. *Thirty-Ninth Annual Report of the General Motors Corporation* (Detroit: GM, 1948); Amos H. Hawley and Basil Zimmer, "Resistance to Unification in a Metropolitan Community," in *Community Political Systems*, ed. Morris Janowitz (Glencoe, IL: Free Press, 1961), 153; *Chevrolet Motor Division Scrapbook*, vol. 1 (LaCrosse, WI: Brookhaven, n.d.), F-GC/FPL; Donald Mosher, ed., *We Make Our Own History: The History of UAW Local 659* (Flint, MI: UAW Local 659, 1993), 59–61.

36. For examples of works that emphasize corporate abandonment and capital flight, see Bluestone and Harrison, *Deindustrialization of America*; Sugrue, *Origins of the Urban Crisis*; William Julius Wilson, *When Work Disappears: The World of the New Urban Poor* (New York: Vintage Books, 1997); Cowie, *Capital Moves*; Joseph Heathcott and Maire A. Murphy, "Corridors of Flight, Zones of Renewal: Industry, Planning, and Policy in the Making of Metropolitan St. Louis, 1940–1980," *Journal of Urban History* 31, no. 2 (January 2005): 151–89; Steven High and David W. Lewis, *Corporate Wasteland: The Landscape and Memory of Deindustrialization* (Ithaca, NY: Cornell University Press, 2007); Gordon, *Mapping Decline*. On spatial mismatch, see John F. Kain, "Housing Segregation, Negro Employment, and Metropolitan Decentralization," *Quarterly Journal of Economics* 82, no. 2 (May 1968): 175–97; Kevin Boyle, "There Are No Union Sorrows That the Union Can't Heal: The Struggle for Racial Equality in the United Automobile Workers, 1940–1960," *Labor History* 35 (1995): 5–23; Sugrue, *Origins of the Urban Crisis*, 125–52; Sides, *L.A. City Limits*, 57–130. In many cases, the shifting of capital from cities to suburbs helped to fuel suburban separatism. Eager to shield their rising tax bases from neighboring areas, suburbanites often looked to incorporation and other strategies to resist annexation to central cities. See, for example, Nicolaides, *My Blue Heaven*; Self, *American Babylon*; Robert Lewis, ed., *Manufacturing Suburbs: Building Work and Home on the Metropolitan Fringe*

(Philadelphia: Temple University Press, 2004); Kruse and Sugrue, eds., *New Suburban History*; Freund, *Colored Property*. On the racial politics of suburban secession, see Kruse, *White Flight*.

37. Heather B. Barrow, "'The American Disease of Growth': Henry Ford and the Metropolitanization of Detroit," in *Manufacturing Suburbs*, ed. Robert Lewis, 200–220; Mark Binelli, *Detroit City Is the Place to Be: The Afterlife of an American Metropolis* (New York: Picador, 2013), 181–82.

38. See, for instance, Sugrue, *Origins of the Urban Crisis*; Self, *American Babylon*.

39. Sean Safford, *Why the Garden Club Couldn't Save Youngstown: The Transformation of the Rust Belt* (Cambridge, MA: Harvard University Press, 2009); S. Paul O'Hara, *Gary, the Most American of All Cities* (Bloomington: Indiana University Press, 2010). On the significance of the 1970s, see Thompson, *Whose Detroit?*, 192–223; Schulman, *Seventies*, 121–43; Cowie, "Vigorously Left, Right, and Center," 75–106; Stein, *Pivotal Decade*; Cowie, *Stayin' Alive*; Daniel T. Rodgers, *Age of Fracture* (Cambridge, MA: Belknap Press of Harvard University Press, 2011), 41–76.

40. The larger point here is that globalization and neoliberalism were not the only causes of deindustrialization in Flint and elsewhere. While there can be little doubt that these macroeconomic forces loom large in Flint's story, ordinary citizens and local policy makers have also played key roles in driving the spatial reorganization of capital, work, and poverty. On neoliberalism and globalization, see David Harvey, *A Brief History of Neoliberalism* (New York: Oxford University Press, 2007); Jason Hackworth, *The Neoliberal City: Governance, Ideology, and Development in American Urbanism* (Ithaca, NY: Cornell University Press, 2007).

41. See, for example, Theoharis and Woodard, eds., *Freedom North*; Countryman, *Up South*; Lassiter, *Silent Majority*; Sugrue, *Sweet Land of Liberty*; Lassiter and Crespino, eds., *Myth of Southern Exceptionalism*.

42. On the Rust Belt, see Bluestone and Harrison, *Deindustrialization of America*; Steven High, *Industrial Sunset: The Making of North America's Rust Belt, 1969–1984* (Toronto: University of Toronto Press, 2003); Robert A. Beauregard, *Voices of Decline: The Postwar Fate of US Cities* (Oxford: Blackwell, 1993); Sugrue, *Origins of the Urban Crisis*; Cowie, *Capital Moves*; Jefferson Cowie and Joseph Heathcott, eds., *Beyond the Ruins: The Meanings of Deindustrialization* (Ithaca, NY: Cornell University Press, 2003). On the Sunbelt, see Carl Abbott, *The New Urban America: Growth and Politics in Sunbelt Cities* (Chapel Hill: University of North Carolina Press, 1981); Bruce J. Schulman, *From Cotton Belt to Sunbelt: Federal Policy, Economic Development, and the Transformation of the South* (New York: Oxford University Press, 1991); Ann R. Markusen et al., *The Rise of the Gunbelt: The Military Remapping of Industrial America* (New York: Oxford University Press, 1991); Margaret Pugh O'Mara, *Cities of Knowledge: Cold War Science and the Search for the Next Silicon Valley* (Princeton, NJ: Princeton University Press, 2005); Elizabeth Tandy Shermer, *Sunbelt Capitalism: Phoenix and the Transformation of American Politics* (Philadelphia: University of Pennsylvania Press, 2013).

43. On the mythology of the Sunbelt, see Lassiter, *Silent Majority*.

44. Lynn Reuster, *Good Old Days at the Buick: Memories of the Men and Women Who Worked There* (Linden, MI: McVey Marketing and Advertising, 1990); Dandaneau, *Town Abandoned*, esp. xxi; Frederick F. Siegel, *The Future Once Happened Here: New York, D.C., L.A., and the Fate of America's Big Cities* (New York: Free Press, 1997); Gillette, *Camden after the Fall*; Edward McClelland, *Nothin' but Blue Skies: The Heyday, Hard Times, and Hopes of America's Industrial Heartland* (New York: Bloomsbury, 2013). My work here builds upon the accounts of other scholars who have challenged the postwar myth. See, for instance, Hirsch, *Making the Second Ghetto*; Sugrue, *Origins of the Urban Crisis*; Simon, *Boardwalk of Dreams*.

45. Cowie, *Capital Moves*, 73–99; Sides, *L.A. City Limits*; Self, *American Babylon*, 23–60; O'Mara, *Cities of Knowledge*, 182–222.

46. Self, *American Babylon*; Lassiter, *Silent Majority*; Kruse and Sugrue, eds., *New Suburban History*. See also Teaford, *City and Suburb*; Rusk, *Cities without Suburbs*; Pastor et al., *Regions That Work*; Orfield, *American Metropolitics*; Dreier, Mollenkopf, and Swanstrom, *Place Matters*; Teaford, *Metropolitan Revolution*.

47. On white flight, see Jonathan Rieder, *Canarsie: The Jews and Italians of Brooklyn against Liberalism* (Cambridge, MA: Harvard University Press, 1985); Gerald H. Gamm, *Urban Exodus: Why the Jews Left Boston and the Catholics Stayed* (Cambridge, MA: Harvard University Press, 1999); Avila, *Popular Culture in the Age of White Flight*; Kruse, *White Flight*. For critiques of the white flight framework, see Self, *American Babylon*, 1–2; Seligman, *Block by Block*, 209–21; Osman, *Invention of Brownstone Brooklyn*.

48. USBC, *United States Census of Population, 1950: Selected Population and Housing Characteristics; Flint, Michigan* (Washington, DC: USGPO, 1952); USBC, *Profile of General Population and Housing Characteristics: 2010 Demographic Profile Data; Flint, Michigan*, http://factfinder2 .census.gov/faces/tableservices/jsf/pages/productview.xhtml?pid=DEC_10_DP_DPDP1.

49. Arnold R. Hirsch and Joseph Logsdon, eds., *Creole New Orleans: Race and Americanization* (Baton Rouge: Louisiana State University Press, 1992), 199; Seligman, *Block by Block*; Thomas C. Henthorn, "A Catholic Dilemma: White Flight in Northwest Flint," *Michigan Historical Review* 31, no 2 (Fall 2005): 1–42.

50. On the importance of local politics and actors, see Ronald H. Bayor, *Race and the Shaping of Twentieth-Century Atlanta* (Chapel Hill: University of North Carolina Press, 1996); Sugrue, *Origins of the Urban Crisis*, 6, 125–52; Cowie, *Capital Moves*, 1–40; J. Mills Thornton III, *Dividing Lines: Municipal Politics and the Struggle for Civil Rights in Montgomery, Birmingham, and Selma* (Tuscaloosa: University of Alabama Press, 2002); Self, *American Babylon*, 23–60; Patrick D. Jones, *The Selma of the North: Civil Rights Insurgency in Milwaukee* (Cambridge, MA: Harvard University Press, 2009).

51. On suburban "tax revolts," see Peter Schrag, *Paradise Lost: California's Experience, America's Future* (Berkeley: University of California Press, 1999); Lisa McGirr, *Suburban Warriors: The Origins of the New American Right* (Princeton, NJ: Princeton University Press, 2001), 238–39; Self, *American Babylon*.

52. Jane Jacobs, *The Death and Life of Great American Cities* (New York: Random House, 1961); Robert A. Caro, *The Power Broker: Robert Moses and the Fall of New York* (New York: Alfred A. Knopf, 1974); Hirsch, *Making the Second Ghetto*, 100–170; Jon C. Teaford, *The Rough Road to Renaissance: Urban Revitalization in America, 1940–1985* (Baltimore: Johns Hopkins University Press, 1990); Mark Rose, *Interstate: Express Highway Politics, 1939–1989* (Knoxville: University of Tennessee Press, 1990); Robert M. Fogelson, *Downtown: Its Rise and Fall, 1880–1950* (New Haven, CT: Yale University Press, 2003); Self, *American Babylon*; Isenberg, *Downtown America*; McKee, *Struggle for Jobs*; Zipp, *Manhattan Projects*; Klemek, *Transatlantic Collapse of Urban Renewal*.

53. On the persistence of segregation, see Massey and Denton, *American Apartheid*; Stephen Grant Meyer, *As Long as They Don't Move Next Door: Segregation and Racial Conflict in American Neighborhoods* (Lanham, MD: Rowman and Littlefield, 2000); Charles M. Lamb, *Housing Segregation in Suburban America since 1960: Presidential and Judicial Politics* (New York: Cambridge University Press, 2005); Nightingale, *Segregation*; Jeannine Bell, *Hate Thy Neighbor: Move-In Violence and the Persistence of Racial Segregation in American Housing* (New York: New York University Press, 2013).

54. Brian D. Boyer, *Cities Destroyed for Cash: The FHA Scandal at HUD* (Chicago: Follett, 1973); R. Allen Hays, *The Federal Government and Urban Housing: Ideology and Change in Public Policy* (Albany: SUNY Press, 1985), 107–36; Jill Quadagno, *The Color of Welfare: How Racism Undermined the War on Poverty* (New York: Oxford University Press, 1994), 100–115; Kevin Fox Gotham, "Separate and Unequal: The Housing Act of 1968 and the Section 235 Program," *Sociological Forum* 15, no. 1 (2000): 13–37, and *Race, Real Estate, and Uneven Development: The Kansas City Experience, 1900–2000* (Albany: SUNY Press, 2002), 127–42.

55. Louis Hyman, *Debtor Nation: The History of America in Red Ink* (Princeton, NJ: Princeton University Press, 2011); Andrew R. Highsmith, "Prelude to the Subprime Crash: Beecher, Michigan, and the Origins of the Suburban Crisis," *Journal of Policy History* 24, no. 4 (October 2012): 572–611.

56. This book's findings on the Section 235 program and the history of acquisitive mortgage lending are part of a wave of recent scholarship rethinking the nature of credit, debt, and subsidization in American society. For examples of that work, see Dolores Hayden, *Building Suburbia: Green Fields and Urban Growth, 1820–2000* (New York: Vintage Books, 2004), chap. 7; Freund, *Colored Property*, 118–28; Hyman, *Debtor Nation*, 45–72. For decades many scholars of American housing policy have concluded that credit—especially when it comes with a federal seal of approval—is a form of subsidy for the borrower. See, for instance, Jackson, *Crabgrass Frontier*, 190–218; Sugrue, *Origins of the Urban Crisis*; Paul S. Grogan and Tony Proscio, *Comeback Cities: A Blueprint for Urban Neighborhood Revival* (Boulder, CO: Westview, 2000), 107–26; Self, *American Babylon*, 3, 97, 130. However, the intertwined histories of the Section 235 program and the more recent subprime lending disaster illustrate that debt often serves as an impediment to racial and economic justice. Furthermore, an examination of these two integrally related calamities points to the fact that the primary beneficiaries of so-called federal housing subsidies have often been lenders, developers, and investors as much as, or even more than, ordinary debtors. For those reasons, this account emphasizes both the benefits and the burdens of debt.

Chapter 1

1. Carl Crow, *The City of Flint Grows Up: The Success Story of an American Community* (New York: Harper and Brothers, 1945), vii.

2. Ibid., 205.

3. John Ihlder, "Flint: When Men Build Automobiles Who Builds Their City?," *Survey*, September 2, 1916, 549–50; Donald G. Richards, "The Greater City and the Wishing Well," *Surveying and Land Information Systems* 59, no. 4 (December 1999): 221–29.

4. Richard White, *The Middle Ground: Indians, Empires, and Republics in the Great Lakes Region, 1650–1815* (New York: Cambridge University Press, 1991); Charles E. Cleland, *Rites of Conquest: The History and Culture of Michigan's Native Americans* (Ann Arbor: University of Michigan Press, 1992).

5. "Money-Brains Combination Returned Buick Here," *FJ*, August 14, 1958; Axel Madsen, *The Deal Maker: How William C. Durant Made General Motors* (New York: John Wiley and Sons, 1999), chaps. 6–8.

6. Clarence H. Young and William A. Quinn, *Foundation for Living: The Story of Charles Stewart Mott and Flint* (New York: McGraw-Hill, 1963), 1; Alfred P. Sloan Jr., *My Years with General Motors* (Garden City, NY: Doubleday, 1964); Timothy Jacobs, *A History of General Motors* (Greenwich, CT: Brompton Books, 1992); Madsen, *Deal Maker*; David Farber, *Sloan Rules: Alfred P. Sloan and the Triumph of General Motors* (Chicago: University of Chicago Press, 2002).

7. Ihlder, "Flint," 550; Ernest M. Fisher, *Real Estate Subdividing Activity and Population Growth in Nine Urban Areas* (Ann Arbor: Bureau of Business Research, University of Michigan, 1928); J. D. Carroll Jr., *Urban Land Vacancy: A Study of Factors Affecting Residential Building on Improved Vacant Lots in Flint, Michigan* (Ann Arbor: IHA, 1952), 11–16; Sidney Fine, *Sit-Down: The General Motors Strike of 1936–37* (Ann Arbor: University of Michigan Press, 1969), 102; Richards, "Greater City and the Wishing Well," 221–29.

8. Leo Donovan, "Flint Rides High on GM Overtime," *DFP*, April 1, 1955; *General Motors Division Scrapbook*, vols. 1–3 (LaCrosse, WI: Brookhaven, n.d.), F-GC/FPL.

9. Russell B. Porter, "Speed, Speed, and Still More Speed!—That Is Flint," *New York Times Magazine*, January 31, 1937.

10. USBC, *Fifteenth Census of the United States: 1930, Population* 3, no. 1 (Washington, DC: USGPO, 1932), 1147, 1158; Research and Statistics Division, State and City Data Re: Economic Conditions, 1934–1942, Comparative City Data, box 3, RG 31; Fine, *Sit-Down*, 102–3.

11. Ihlder, "Flint," 549–50; Walter Firey, *Social Aspects to Land Use Planning in the Country-City Fringe: The Case of Flint, Michigan* (East Lansing: Michigan State College Agricultural Experiment Station, 1946), 9.

12. Alexander C. Findlay, *The Housing Situation in Flint, Michigan* (Flint: Flint Institute of Research and Planning, 1938); Elroy S. Guckert, "The Housing Status in Flint, Michigan," n.d., 12–13, box 1, ACFP; Peter J. Weidner, "The Evils of a Housing Shortage," n.d., 2, box 1, ACFP; Peter J. Weidner, "Some Real and Imaginary Stumbling Blocks in the Way of House Building," n.d., 1, box 1, ACFP.

13. Becky M. Nicolaides, *My Blue Heaven: Life and Politics in the Working-Class Suburbs of Los Angeles, 1920–1965* (Chicago: University of Chicago Press, 2002), 9–119.

14. Tom Dinell, *The Influences of Federal, State, and Local Legislation on Residential Building in the Flint Metropolitan Area* (Ann Arbor: IHA, 1951), 22–24.

15. Findlay, *Housing Situation in Flint*; Peter J. Weidner, "The Housing Problem in Flint," n.d., 2, box 1, ACFP; Weidner, "Evils of a Housing Shortage," 2; Weidner, "Some Real and Imaginary Stumbling Blocks"; Richards, "Greater City and the Wishing Well," 226.

16. The literature in this area is vast. See, for instance, Jon C. Teaford, *City and Suburb: The Political Fragmentation of Metropolitan America, 1850–1970* (Baltimore: Johns Hopkins University Press, 1979); Kenneth T. Jackson, *Crabgrass Frontier: The Suburbanization of the United States* (New York: Oxford University Press, 1985); Richard Harris, *Unplanned Suburbs: Toronto's American Tragedy, 1900–1950* (Baltimore: Johns Hopkins University Press, 1996); Nicolaides, *My Blue Heaven*; Dolores Hayden, *Building Suburbia: Green Fields and Urban Growth, 1820–2000* (New York: Vintage Books, 2004); Margaret Garb, *City of American Dreams: A History of Home Ownership and Housing Reform in Chicago, 1871–1919* (Chicago: University of Chicago Press, 2005); Cecelia Bucki, *Bridgeport's Socialist New Deal, 1915–36* (Urbana: University of Illinois Press, 2006); Elaine Lewinnek, *The Working Man's Reward: Chicago's Early Suburbs and the Roots of American Sprawl* (Chicago: University of Chicago Press, 2014). On Flint, see Findlay, *Housing Situation in Flint*; Weidner, "Housing Problem in Flint."

17. Ihlder, "Flint," 549; Richard W. Judd, *Socialist Cities: Municipal Politics and the Grass Roots of American Socialism* (Albany: SUNY Press, 1989), 96.

18. Ihlder, "Flint," 549.

19. Guckert, "Housing Status," 15–16.

20. Ibid., 14–15.

21. Ibid., 16–17.

22. *FJ*, September 2, 1970.

23. Erdmann D. Beynon, *Characteristics of the Relief Case Load in Genesee County, Michigan* (Flint, MI: Genesee County Welfare Relief Commission, 1940), 47; Peirce Lewis, "Geography in the Politics of Flint" (PhD diss., University of Michigan, 1958); Janice K. Wilberg, "An Analysis of an Urban Neighborhood: St. John Street," n.d., UM-F/HDSP.

24. Guckert, "Housing Status," 13–14.

25. Peirce Lewis, "Impact of Negro Migration on the Electoral Geography of Flint, Michigan, 1932–1962: A Cartographic Analysis," *Annals of the Association of American Geographers* 55, no. 1 (March 1965): 6–7; Rhonda Sanders, *Bronze Pillars: An Oral History of African-Americans in Flint* (Flint: *Flint Journal* and Sloan Museum, 1995), 10–22; Ananthakrishnan Aiyer, ed., *Telling Our Stories: Legacy of the Civil Rights Movement in Flint* (Flint, MI: Flint ColorLine Project, 2007), 26.

26. Dick Shappell, "Looks at Flint and Its People," *FJ*, n.d., ca. 1938.

27. Raab Realty Company, *General Motors Homesites* (Flint, MI: Raab Realty Company, n.d.); Scott Peters, "History of Mott Park Community," unpublished manuscript, n.d.

28. *Corrigan v. Buckley*, 271 U.S. 323 (1926).

29. David M. P. Freund, *Colored Property: State Policy and White Racial Politics in Suburban America* (Chicago: University of Chicago Press, 2007), 45–98; Robert M. Fogelson, *Bourgeois Nightmares: Suburbia, 1870–1930* (New Haven, CT: Yale University Press, 2005).

30. For the quotation, see BHF.

31. J. A. Welch Company, *The Weight of Evidence* (Flint, MI: J. A. Welch Company, n.d., ca. 1921).

32. *Flint Real Estate and Home Builder*, April 1930, BHF.

33. Donald O. Cowgill and Mary S. Cowgill, "An Index of Segregation Based on Block Statistics," *American Sociological Review* 16, no. 6 (December 1951): 825–31. The Cowgills used block-level census data to determine the residential segregation indices of 187 American cities.

34. Allan H. Spear, *Black Chicago: The Making of a Negro Ghetto, 1890–1920* (Chicago: University of Chicago Press, 1967); Arnold R. Hirsch, *Making the Second Ghetto: Race and Housing in Chicago, 1940–1960* (Cambridge: Cambridge University Press, 1983); Thomas J. Sugrue, *The Origins of the Urban Crisis: Race and Inequality in Postwar Detroit* (Princeton, NJ: Princeton University Press, 1996); Bryant Simon, *Boardwalk of Dreams: Atlantic City and the Fate of Urban America* (New York: Oxford University Press, 2004).

35. ULF, *Race Relations: A Problem of American Democracy* (Flint, MI: ULF, 1944); *Urban League Recorder*, box 1, folder 5, 1943–1950, ULFP; "Southernism in Flint," *FS*, June 29, 1946; *PFCC*, July 27, 1953, F-GC/FPL; William H. Oliver to Robert A. Carter, December 17, 1953, box 14, folder 29, Region 1C, Flint, Correspondence, Newspapers, Agendas, 1951–1954, UAWFP; "Charges Discrimination in Bowling Alleys," *FWR*, September 14, 1961; Rhonda Sanders, "Good, Bad Times in Flint: A History of Two Families," *FJ*, September 30, 1984; Albert Harris, interview by Robert Schafer, September 9, 1987, OH; Alvin Loving and Mary Helen Loving, interview by Robert Schafer, September 12, 1988, OH; John Hightower, interview by Rhonda Sanders and Wanda Howard, March 22, 1994, BPOHPF.

36. Reverend Eugene Simpson, interview by Wanda Howard, May 27, 1993, BPOHPF.

37. Charlotte Williams, interview by Wanda Howard, n.d., ca. 1993, BPOHPF; Dolores Ennis, interview by Wanda Howard, September 22, 1993, BPOHPF.

38. "Cemetery Ordered to Accept Negroes," *FJ*, June 16, 1964; Betty Brenner, "Burial in Area Formerly a Black or White Issue," *FJ*, February 25, 1993.

39. Loving, interview.

40. Simpson, interview.

41. ULF, "1944 in Review," box 1, folder 5, 1943–1950, ULFP; *Urban League Recorder*, box 1, folder 5, 1943–1950, ULFP; Lois Holt and Ruth Van Zandt, interview by Robert Schafer, October 27, 1988, OH.

42. Max Brandon, interview by Wanda Howard, April 13, 1993, BPOHPF; Annalea Bannister, interview by Wanda Howard, November 2, 1993, BPOHPF; Jeff Wiltse, *Contested Waters: A Social History of Swimming Pools in America* (Chapel Hill: University of North Carolina Press, 2007), 121–53.

43. George Oscar Bowen, *Book of Songs* (Flint, MI: Flint Board of Commerce, 1940).

44. *FJ*, January 6, April 9, 14, 25, 1933, January 4, 21, 28–29, February 1, 7, 1934, January 3, February 6, 14, 1935. The quotation is from "Kiwanis Paraders Ask Minstrel Show Support," *FJ*, April 23, 1933.

45. Simpson, interview.

46. Loving, interview. See also W. T. Lhamon Jr., *Raising Cain: Blackface Performance from Jim Crow to Hip Hop* (Cambridge, MA: Harvard University Press, 2000).

47. Ronald Edsforth, *Class Conflict and Cultural Consensus: The Making of a Mass Consumer Society in Flint, Michigan* (New Brunswick, NJ: Rutgers University Press, 1987), 127–55.

48. William Chafe, "Flint and the Great Depression," *Michigan History* 53 (Fall 1969): 234.

49. Jackson, *Crabgrass Frontier*, 187.

50. Weidner, "Housing Problem in Flint"; FHA, "An Analysis of the Flint, Michigan SMA (Genesee County) as of January 1953," n.d., RHMA, 1937–1963, box 10, RG 31.

51. Findlay, *Housing Situation in Flint*; *FJ*, April 4, 1937; Edmund N. Bacon, "A Diagnosis and Suggested Treatment of an Urban Community's Land Problems," *Journal of Land and Public Utility Economics* 16, no. 1 (February 1940): 72–88; Fine, *Sit-Down*, 102–6.

52. Jackson, *Crabgrass Frontier*, 187–89; Harris, *Unplanned Suburbs*; Nicolaides, *My Blue Heaven*, 39–64.

53. RVPDC, *1985–2010: Population Allocation Study* (Flint, MI: RVPDC, 1984).

54. Freund, *Colored Property*, 111–12.

55. Jackson, *Crabgrass Frontier*, 195–203; Amy E. Hillier, "Redlining and the Home Owners' Loan Corporation," *Journal of Urban History* 29, no. 4 (May 2003): 394–420; Freund, *Colored Property*, 111–18; Greer, "Home Owners' Loan Corporation," 275–96.

56. Clark Waters, "Confidential Report of a Survey in Flint, Michigan, and Its Suburban Area for the Home Owners' Loan Corporation," n.d., 15, CSF, 1935–1940, box 24, RG 195.

57. Historian James Greer has cast some doubt on the influence of the HOLC's residential security maps. See Greer, "Home Owners' Loan Corporation," 275–96.

58. CSF, 1935–1940, box 23, RG 195.

59. Ibid.

60. Ibid., box 24, RG 195; Kenneth B. West Papers, box 1, folder 24, GHCC.

61. CSF, 1935–1940, box 24, RG 195.

62. Ibid., box 23, RG 195. In other cities, especially those with larger immigrant populations, HOLC officers typically assigned D ratings to neighborhoods containing Mexican or Asian residents. See CSF, 1935–1940, RG 195.

63. CSF, 1935–1940, box 23, RG 195.

64. Ibid.

65. Ibid.

66. Williams, interview.

67. Freund, *Colored Property*, 117–18. As Freund notes, the decision to refinance homes in

C and D neighborhoods stemmed in large part from the fact that "low-income borrowers had a better record of repayment."

68. USBC, *Census of Housing, 1940, Housing—Nonfarm Mortgages* (Washington, DC: USGPO, 1942), tables f-1, f-2, f-3, g-4.

69. CSF, 1935–1940, box 24, RG 195.

70. On debates over federal redlining, see Robert A. Beauregard, "Federal Policy and Postwar Urban Decline: A Case of Government Complicity," *Housing Policy Debate* 17, no. 1 (2001): 129–51; Hillier, "Redlining and the Home Owners' Loan Corporation," 394–420; Robert Bruegmann, *Sprawl: A Compact History* (Chicago: University of Chicago Press, 2006), 102–3; Freund, *Colored Property*, 111–21.

71. See area descriptions for Mt. Morris and Grand Blanc, Michigan, in CSF, 1935–1940, box 23, RG 195.

72. See area descriptions for neighborhoods D-20 and D-22 in ibid.

73. CSF, 1935–1940, box 18, RG 195.

74. Ibid., box 39, RG 195.

75. Ibid., box 84, RG 195.

76. Ibid., boxes 35, 86, 87, RG 195. For the New York data, see Research and Statistics Division, Records Relating to Housing Market Analysis, 1935–42, Survey Reports, box 19, RG 31.

77. Jackson, *Crabgrass Frontier*, 205.

78. *FJ*, May 7, 1948; J. D. Carroll Jr., *Housing Characteristics of Flint in 1950* (Ann Arbor: IHA, 1952). This section builds upon an impressive body of literature on the FHA. See, for instance, Robert C. Weaver, *The Negro Ghetto* (New York: Harcourt Brace, 1948); Charles Abrams, *Forbidden Neighbors: A Study of Prejudice in Housing* (New York: Harper, 1955); Henry J. Aaron Jr., *Shelter and Subsidies: Who Benefits from Federal Housing Policies?* (Washington, DC: Brookings Institution, 1972); Jackson, *Crabgrass Frontier*; Douglas S. Massey and Nancy A. Denton, *American Apartheid: Segregation and the Making of the Underclass* (Cambridge, MA: Harvard University Press, 1993); Freund, *Colored Property*.

79. Freund, *Colored Property*, 45–139.

80. Homer Hoyt, *One Hundred Years of Land Values in Chicago: The Relationship of the Growth of Chicago to the Rise in Its Land Values, 1830–1933* (Chicago: University of Chicago Press, 1933), and "Instructions for Dividing the City into Neighborhoods," n.d., HSG, 1964–1968, box 3, RG 31; Freund, *Colored Property*, 120, 129, 210–11.

81. Hoyt, "Instructions for Dividing the City," and "Tests of a Reject Neighborhood," n.d., HSG, 1964–1968, box 3, RG 31.

82. FHA, *Underwriting Manual: Underwriting and Valuation Procedure under Title II of the National Housing Act* (Washington, DC: USGPO, 1939); Commissioners Correspondence and Subject File, 1938–1958, box 4, Minority Group Housing, 1938–1947, RG 31.

83. Memorandum, Homer Hoyt to Dr. Fisher, "Instructions as to Procedure in Rating Neighborhoods," n.d., HSG, 1964–1968, box 3, RG 31.

84. Part 1, box 15, folder 12, Housing, 1943–, UAWR.

85. FHA, "Analysis of the Flint, Michigan SMA (Genesee County) as of January 1953."

86. Bacon, "Diagnosis and Suggested Treatment," 79–81; Carroll, *Housing Characteristics of Flint in 1950*; FHA, "Analysis of the Flint, Michigan SMA (Genesee County) as of January 1953."

87. On the earlier efforts, see Jackson, *Crabgrass Frontier*, 138–56; Marc A. Weiss, *The Rise of the Community Builders: The American Real Estate Industry and Urban Land Planning*

(New York: Columbia University Press, 1987); Greg Hise, *Magnetic Los Angeles: Planning the Twentieth-Century Metropolis* (Baltimore: Johns Hopkins University Press, 1997), 14–55.

88. Dinell, *Influences of Federal, State, and Local Legislation.*

89. *FJ*, October 24, 1949.

90. FHA, "Current Housing Situation: Flint, Michigan," March 31, 1941, Housing Monographs, box 6, RG 207.

91. *FJ*, May 7, 1948; Carroll, *Housing Characteristics of Flint in 1950*; FHA, "Analysis of the Flint, Michigan SMA (Genesee County) as of January 1953."

92. Carroll, *Housing Characteristics of Flint in 1950.*

93. W. P. Atkinson et al., eds., *Housing U.S.A: As Industry Leaders See it* (New York: Simmons-Boardman, 1954); *FJ*, June 18, 1958; Gerald F. Healy, interview by Robert Schafer, August 7, 1987, OH.

94. Robert W. Schnuck, "Doubled Negro Population Creates Housing 'Squeeze,'" *FJ*, January 3, 1947; J. D. Carroll Jr., *Housing Conditions of Flint Metropolitan Area, April 1948* (Ann Arbor: IHA, 1948). On the nation as a whole, see Jackson, *Crabgrass Frontier*, 190–218; Andrew Wiese, *Places of Their Own: African American Suburbanization in the Twentieth Century* (Chicago: University of Chicago Press, 2004), chaps. 1–4; Freund, *Colored Property.*

95. "Democracy Is Out in Flint," *Brownsville Weekly News*, August 9, 1941; Schnuck, "Doubled Negro Population"; FCPC, *Recreation and Park Plan for Flint* (Flint, MI: FCPC, 1949), 53; James L. Rose, "Civil Rights in Housing in Michigan," n.d., 5, box 9, folder 14, OBP.

96. USBC, *Census of Housing, 1940, Housing—Nonfarm Mortgages* (Washington, DC: USGPO, 1941), table f-1.

97. Box 1, folder 5, ULFP; *FJ*, April 8, 1944.

98. "Negro Homes Start Soon," *FJ*, December 18, 1949.

99. Frank J. Corbett and Arthur J. Edmunds, *The Negro Housing Market—An Untapped Resource?* (Flint, MI: ULF, 1954), 22.

100. "Ratio of Home Ownership Climbs 14 Percent in 5 Years," *FJ*, May 7, 1948.

101. Part 1, box 15, folder 11, Housing, UAWR.

102. "City Officials to Investigate Shortage of Houses for Negroes," *FJ*, October 22, 1943.

103. *FJ*, November 12, 1943.

104. *FJ*, January 3, 1947; part 1, box 15, folder 11, Housing, UAWR.

105. Hirsch, *Making the Second Ghetto*, 1–39; Stephen Grant Meyer, *As Long as They Don't Move Next Door: Segregation and Racial Conflict in American Neighborhoods* (Lanham, MD: Rowman and Littlefield, 2000), 64–78; Colin Gordon, *Mapping Decline: St. Louis and the Fate of the American City* (Philadelphia: University of Pennsylvania Press, 2008), 69–111.

106. Hugh Downing and Peter Borwick, "Post-War Housing Trends in the Flint Metropolitan Area and Future Outlook for the Sales Market," April 27, 1949, RHMA, 1937–1963, box 10, RG 31; Carroll, *Housing Characteristics of Flint in 1950*; *Housing Conditions in Flint, Michigan: Proceedings of a City-Wide Conference on Housing, November, 1955* (Flint, MI: ULF, 1955), 8; *FJ*, June 10, 1959.

107. ULF, "Monthly Activities Report, November 1943," box 1, 77-7.9-8, CSMP.

108. *Shelley v. Kraemer*, 334 U.S. 1 (1948); Freund, *Colored Property*, 207–11.

109. "Address by Franklin D. Richards, Commissioner, FHA, before the Detroit Mortgage Bankers Association," December 8, 1949, Speeches, box 31, RG 31.

110. "Address by Walter L. Greene, Commissioner, FHA, before National Urban League, Cleveland, Ohio," September 2, 1952, 2, Commissioner's Correspondence and Subject File, 1938–1958, box 4, RG 31.

111. Corbett and Edmunds, *Negro Housing Market*, vi; Frank J. Corbett and Arthur J. Edmunds, *Some Characteristics of Real Property Maintenance by Negro Home Occupants* (Flint, MI: ULF, 1954); Allan R. Wilhelm, "Urban League Survey Boosts Negro Housing," *FJ*, January 18, 1955.

112. Other scholars have made a similar point. See Arnold R. Hirsch, "Containment on the Home Front: Race and Federal Housing Policy from the New Deal to the Cold War," *Journal of Urban History* 26, no. 2 (January 2000): 163; Freund, *Colored Property*, 99–240.

113. "Address by Walter L. Greene," 6.

114. Corbett and Edmunds, *Negro Housing Market*, 23; "Bad Housing Threatens City's Future," *Amplifier*, Winter 1956, box 1, folder 7, ULFP.

115. Richard Margolis and Diane Margolis, *How the Federal Government Builds Ghettos* (New York: National Committee against Discrimination in Housing, 1967), 14–30.

Chapter 2

1. On the connection between segregation and community building, see Andrew R. Highsmith and Ansley T. Erickson, "Segregation as Splitting, Segregation as Joining: Schools, Housing, and the Many Modes of Jim Crow," *American Journal of Education* (forthcoming). On segregation as "city-splitting," see Carl H. Nightingale, *Segregation: A Global History of Divided Cities* (Chicago: University of Chicago Press, 2012).

2. Clarence H. Young and William A. Quinn, *Foundation for Living: The Story of Charles Stewart Mott and Flint* (New York: McGraw-Hill, 1963), 8–40.

3. "Who Holds GM?" *Time*, August 31, 1931; "Many Happy Returns," *Time*, November 13, 1964.

4. *FJ*, April 2, 1912.

5. Young and Quinn, *Foundation for Living*, vii. See also box 7, 77-7.10-1.1, CSMP.

6. "Memorandum Regarding Charles Stewart Mott Foundation," n.d., box 3, 78-8.1-59a, FJMP; Young and Quinn, *Foundation for Living*, 97; Dawn Olmsted, "Mott: The Man Revealed," UM-F/HDSP. See also "Charles Stewart Mott: Flint's Benefactor?," *Searchlight*, June 18, 1959.

7. Webb Waldron, "The City That Found Itself," *Reader's Digest*, July 1937; "Flint: Where It's Fun to Be People," *Rotarian*, June 1957, 28–31; Peter L. Clancy, "The Contributions of the Charles Stewart Mott Foundation in the Development of the Community School Program in Flint, Michigan" (PhD diss., Michigan State University, 1963), 43–57; Frank J. Manley, Bernard W. Reed, and Robert K. Burns, *The Community School in Action: The Flint Program* (Chicago: Education-Industry Service, 1967), 20–59; Paul W. Kearney, "Dancing in the Streets: Any City Can Use This Life-Saving, Blues-Curing Plan," *Family Circle*, n.d., box 1, 78-8.1-12a, FJMP; Patrick Charles Manley, "Frank J. Manley: The Man and the Idea" (PhD diss., University of Michigan, 1978).

8. Highsmith and Erickson, "Segregation as Splitting, Segregation as Joining."

9. "Tide of Socialism Stemmed by C. S. Mott," *FJ*, April 1, 1952. See also Clarence A. Perry, *The School as a Factor in Neighborhood Development* (New York: Russell Sage Foundation, 1914), and *School Center Gazette, 1919–1920* (New York: Russell Sage Foundation, 1920).

10. Wilbur Pardon Bowen and Elmer Dayton Mitchell, *The Theory of Organized Play; Its Nature and Significance* (New York: A. S. Barnes, 1923), 87.

11. Young and Quinn, *Foundation for Living*, 101; Manley, Reed, and Burns, *Community School in Action*, 22; Sidney Fine, *Sit-Down: The General Motors Strike of 1936–37* (Ann Arbor: University of Michigan Press, 1969), 102; William Chafe, "Flint and the Great Depression,"

Michigan History 53 (Fall 1969): 230–37; Kearney, "Dancing in the Streets." On public education during the Depression, see David Bruce Tyack, Robert Lowe, and Elisabeth Hansot, *Public Schools in Hard Times: The Great Depression and Recent Years* (Cambridge, MA: Harvard University Press, 1984).

12. "Tide of Socialism"; Young and Quinn, *Foundation for Living*, 118. On community education as a national movement, see Mary Jean Seubert, "The Origin, Development, and Issues of the Community Education Movement in the United States, 1935–1995" (PhD diss., Florida Atlantic University, 1995); Kevin Mattson, *Creating a Democratic Public: The Struggle for Urban Participatory Democracy during the Progressive Era* (University Park: Pennsylvania State University Press, 1998), 48–67.

13. *NYT*, October 25, 1936.

14. Henry Kraus, *The Many and the Few: A Chronicle of the Dynamic Auto Workers* (Los Angeles: Plantin, 1947); Fine, *Sit-Down*; Ronald Edsforth, *Class Conflict and Cultural Consensus: The Making of a Mass Consumer Society in Flint, Michigan* (New Brunswick, NJ: Rutgers University Press, 1987); Steven P. Dandaneau, *A Town Abandoned: Flint, Michigan, Confronts Deindustrialization* (Albany: SUNY Press, 1996). For the national context, see Robert Zieger, *The CIO, 1935–1955* (Chapel Hill: University of North Carolina Press, 1997).

15. Studs Terkel, *Hard Times: An Oral History of the Great Depression* (New York: Pantheon Books, 1970), 134–35.

16. See, for example, Kimberly Phillips-Fein, *Invisible Hands: The Making of the Conservative Movement from the New Deal to Reagan* (New York: W. W. Norton, 2009), 3–67.

17. "Flint: A True Report," n.d., 2, box 1, folder 4, UAWCAP.

18. Young and Quinn, *Foundation for Living*, 137.

19. "Mr. Flint at Work," *Time*, September 22, 1952. According to one anonymous trade unionist, Mott and GM saw community education as a means to control "every phase of the worker's life." See "Flint: A True Report," 3. Mott and Manley also hoped to weaken the local chapter of the American Federation of Teachers. See box 13, folder 1, Flint, Mich.—History of Cases, AEP.

20. Frank J. Manley to Ray Cromley, July 27, 1959, box 17, 78-8.2-319, FJMP.

21. Box 4, folder 15, Flint Public Schools, 1952–1960, UAW581.

22. Samuel Simmons and Robert Greene, "Flint Community Survey," June 20, 1956, box 10, folder 31, OBP; "Model Use of Money," *Time*, April 12, 1968.

23. Frank J. Manley, "Long Range Program: Immediate Results," August 1946, box 4, 78-8.1-59b, FJMP; *FJ*, April 15, 1951; Young and Quinn, *Foundation for Living*, 165; Minutes, CSMF Program Coordinating Committee, 1947–1963, quoted in Clancy, "Contributions of the Charles Stewart Mott Foundation," 168.

24. CSMF, *Annual Report for 1947–48* (Flint, MI: CSMF, 1948); Young and Quinn, *Foundation for Living*, 172.

25. Minutes, CSMF Advisory Council Meeting, September 27, 1963, box 10, 78-8.2-97d-1, FJMP.

26. W. Fred Totten and Frank J. Manley, *The Community School: Basic Concepts, Function, and Organization* (Galien, MI: Allied Education Council, 1969).

27. Manley, Reed, and Burns, *Community School in Action*, 58.

28. Minutes, CSMF Coordinating Committee, 1947–1963.

29. "Practical Solutions for School Problem," n.d., box 13, folder 1, Flint, Mich.—History of Cases, AEP. Also, see box 13, folder 14, Flint 1945–1946, AEP; James E. Larson, *The City of Flint*

Tax Limitation Experiment (Ann Arbor: IHA, 1948); "Flint's Capital Improvement Program," March 9, 1953, box 1, 87-12, MAFP; John L. Riegle, *Day before Yesterday: An Autobiography and History of Michigan Schools* (Minneapolis: T. S. Denison, 1971), 300; *OMFBE*, September 18, 1974, 235, F-GC/FPL.

30. Franklin K. Killian, "Flint's Fiscal Capacity to Support Secondary and Advanced Education" (PhD diss., University of Michigan, 1949); John Guy Fowlkes, "Excerpts from Report of the Expenditures, Business Management, and Salary Policies of the Public Schools of Flint," n.d., box 1, 78-8.1-5b, FJMP.

31. Manley, Reed, and Burns, *Community School in Action*, 71–75.

32. Ibid., 69–76; CSMF, *Designing, Constructing, and Financing Facilities for a Community School* (Flint, MI: CSMF, n.d.), box 19, 78-8.2-436b, FJMP.

33. *FJ*, October 3, 1956, September 19, 1957, March 8, 1959, October 8, 1970. See also box 2, 78-8.1-16, box 3, 78-8.1-44 and 78-8.1-59a, box 17, 78-8.2-319, and box 24, 78-8.3-52, FJMP; Karl Detzer, "Flint's Gone Crazy over Culture," *Reader's Digest*, March 1959, 184–88; Kearney, "Dancing in the Streets"; box 326, Program Information—Community Schools, GRP.

34. "A Description of the Activities of the Mott Foundation Program of the Flint Board of Education," *Journal of Educational Sociology* 33, no. 4 (December 1959): 153; Manley, Reed, and Burns, *Community School in Action*, 31; "Divisions of the Mott Program of the Flint Board of Education," n.d., FJMP.

35. Totten and Manley, *Community School*, 12.

36. *DFP*, July 5, 1949; FBE, *The Work We Live By*, vols. 1–20 (Flint, MI: FBE, 1958–65), F-GC/FPL; Joseph A. Anderson, "The Economic Value of the Community School Concept to Local Business," *Community School and Its Administration* 7, no. 9 (May 1969).

37. "Mr. Flint," *Time*, June 28, 1963; Manley, Reed, and Burns, *Community School in Action*, 61; Frank J. Manley, "Some Basic Assumptions," n.d., box 16, 78-8.2-366b, FJMP; Totten and Manley, *Community School*, xxvii. On big business and the battle against liberalism, see Phillips-Fein, *Invisible Hands*; Bethany Moreton, *To Serve God and Wal-Mart: The Making of Christian Free Enterprise* (Cambridge, MA: Harvard University Press, 2009).

38. "Mr. Flint."

39. "Philosophy of Flint Community Schools," n.d., box 14, 78-8.2-270c-2, FJMP; W. Fred Totten, *The Power of Community Education* (Midland, MI: Pendell, 1970). For an insightful discussion of the significance of schools to local politics and community building, see Michelle M. Nickerson, *Mothers of Conservatism: Women and the Postwar Right* (Princeton, NJ: Princeton University Press, 2012), 32–102.

40. Davison M. Douglas, *Jim Crow Moves North: The Battle over Northern School Desegregation, 1865–1954* (New York: Cambridge University Press, 2005); Joan Singler et al., *Seattle in Black and White: The Congress of Racial Equality and the Fight for Equal Opportunity* (Seattle: University of Washington Press, 2011), chap. 7. On the rise of administrative school segregation in the South, see Ansley T. Erickson, *Making the Unequal Metropolis: School Desegregation and Its Limits* (Chicago: University of Chicago Press, 2016).

41. The FBE adopted high school attendance boundaries in 1960.

42. *OMFBE*, July 14, 1949, F-GC/FPL.

43. Kenneth A. Mines to Peter L. Clancy, August 29, 1975, box 16, folder 9, OBP; Rhonda Sanders, *Bronze Pillars: An Oral History of African-Americans in Flint* (Flint, MI: *Flint Journal* and Sloan Museum, 1995), 1–22.

44. Mines to Clancy. See also box 28, folder 12, box 40, folders 22–23, OBP.

45. "Racial Distribution by School, K–12—1950–1968," n.d., box 23, 78-8.3-31f, FJMP.

46. Ibid.

47. Ansley T. Erickson, "Building Inequality: The Spatial Organization of Schooling in Nashville, Tennessee, after *Brown*," *Journal of Urban History* 38, no. 2 (March 2012): 247–70; Karen Benjamin, "Suburbanizing Jim Crow: The Impact of School Policy on Residential Segregation in Raleigh," *Journal of Urban History* 38, no. 2 (March 2012): 225–46; Jack Dougherty, "Shopping for Schools: How Public Education and Private Housing Shaped Suburban Connecticut," *Journal of Urban History* 38, no. 2 (March 2012): 205–24.

48. Raymond L. Gover, "38 Million: Construction Program Had 'Golden' 10 Years," *FJ*, December 27, 1959.

49. William J. Early, "Presentation to Civil Rights Commission of the State of Michigan," December 1, 1966, box 29, folder 21, OBP.

50. "Primary Units Partial Solution to Flint's Classroom Shortage," *Michigan Association of School Boards Journal*, May 1955, box 1, 78-8.1-12a, FJMP; "Primary Units—An Answer to the Cost Problem," n.d., box 2, 78-8.2-97b, FJMP; Lydia A. Giles, "Flint Portable Classrooms Aided Segregation," *FJ*, April 16, 1977.

51. J. A. Welch Company, *The Weight of Evidence* (Flint, MI: J. A. Welch Company, n.d., ca. 1921); Raab Realty Company, *General Motors Homesites* (Flint, MI: Raab Realty Company, n.d.); CSMF, *Designing, Constructing, and Financing Facilities for a Community School* (Flint, MI: CSMF, n.d.), box 19, 78-8.2-436b, FJMP.

52. On other cities, see Douglas, *Jim Crow Moves North*, 116, 146, 208, 244.

53. Van G. Sauter and John L. Dotson, "How Flint Deals with Racial Problems," *DFP*, June 14, 1965.

54. *OMFBE*, May 23, 1961, 330a–c, F-GC/FPL.

55. Sanders, *Bronze Pillars*, 117.

56. FBE, *Frequency Distribution of Pupils by Race* (Flint, MI: FBE, n.d.), F-GC/FPL; Mines to Clancy.

57. *OMFBE*, July 19, 1955, 17–28, F-GC/FPL.

58. Affidavit of Edgar B. Holt, *Holman et al. v. School District of the City of Flint et al.*, Civil Action No. 76-40023, United States District Court, Eastern District of Michigan, Southern Division (1976). The Stewart case was not an isolated example. See Ananthakrishnan Aiyer, ed., *Telling Our Stories: Legacy of the Civil Rights Movement in Flint* (Flint, MI: Flint ColorLine Project, 2007), 18.

59. Holt, affidavit, 2.

60. On "imagined communities," see Benedict R. Anderson, *Imagined Communities: Reflections on the Origin and Spread of Nationalism* (London: Verso, 1983).

61. *OMFBE*, February 4, 1954, F-GC/FPL; Mines to Clancy.

62. Robert A. Carter, Norman W. Bully, and Earl Crompton to FBE, September 9, 1955, box 25, folder 18, FEPC, Flint, Region 1C, Correspondence, Notes, 1955, UAWFP.

63. *OMFBE*, September 10, October 8, November 5, 1953, February 4, 1954, F-GC/FPL; Mines to Clancy; Douglas, *Jim Crow Moves North*.

64. Paul Street, *Segregated Schools: Educational Apartheid in Post–Civil Rights America* (New York: Routledge, 2005), 22–23; Thomas J. Sugrue, *Sweet Land of Liberty: The Forgotten Struggle for Civil Rights in the North* (New York: Random House, 2008), 452–53.

65. "Primary Units Partial Solution to Flint's Classroom Shortage"; "Primary Units—An Answer to the Cost Problem"; "Racial Distribution by School, K–12—1950–1968"; "Information on Flint Public School's Primary Units," August 1957, box 4, 78-8.1-93, FJMP; "Location of

Primary Units in 1957–58," n.d., box 4, 78-8.1-93, FJMP; Giles, "Flint Portable Classrooms Aided Segregation."

66. Arnold R. Hirsch, *Making the Second Ghetto: Race and Housing in Chicago, 1940–1960* (Cambridge: Cambridge University Press, 1983), 1–39; James N. Gregory, *The Southern Diaspora: How the Great Migrations of Black and White Southerners Transformed America* (Chapel Hill: University of North Carolina Press, 2005), 153–236.

67. William P. Walsh, *The Story of Urban Dynamics* (Flint, MI: W. P. Walsh, 1967); SPA, *Diagnostic Survey for St. John Street, Flint, Michigan* (Flint, MI: SPA, 1969); Sanders, *Bronze Pillars*, 22–26. On land contracts, see Beryl Satter, *Family Properties: How the Struggle over Race and Real Estate Transformed Chicago and Urban America* (New York: Metropolitan Books, 2009), 3–7, 64–97, 111–16.

68. "Racial Distribution by School, K–12—1950–1968"; *OMFBE*, March 9, 1954, F-GC/FPL; *BR*, March 29, 1958.

69. Harold Hayden to George Romney, June 2, 1965, box 92, Civil Rights, G–L, 1965, GRP.

70. "Racial Distribution by School, K–12—1950–1968"; *OMFBE*, September 23, 1964, F-GC/FPL.

71. *OMFBE*, January 8, 1953, 122, F-GC/FPL; "Flint NAACP Fact Sheet," n.d., box 4, folder 13, EBHP; "Community School Program Proposal" and H. C. McKinney Jr. to Lt. Governor William G. Milliken, February 12, 1968, box 326, Program Information—Community Schools, GRP; Early, "Presentation to Civil Rights Commission"; Lewis Morrissey, "Physical Improvements Abound at Flint Schools," *FJ*, August 31, 1969. On segregation and compensatory education, see Jack Dougherty, *More than One Struggle: The Evolution of Black School Reform in Milwaukee* (Chapel Hill: University of North Carolina Press, 2004), 51–103.

72. Jeffrey Mirel, *The Rise and Fall of an Urban School System: Detroit, 1907–1981* (Ann Arbor: University of Michigan Press, 1993), 151–398; Martha Biondi, *To Stand and Fight: The Struggle for Civil Rights in Postwar New York City* (Cambridge, MA: Harvard University Press, 2003), 223–49; Sugrue, *Sweet Land of Liberty*, 163–99.

73. According to legal scholar Davison Douglas, state-sanctioned school segregation "persisted in open defiance of state law in many northern communities until the late 1940s and early 1950s." See Douglas, *Jim Crow Moves North*, i. In Flint, however, segregationist policies continued throughout the civil rights era. See, for example, ULF, "Quality Education and Busing," December 21, 1971, box 40, folder 2, OBP.

74. Flint NAACP Newsletter, "Small Boy Kept in School Closet," n.d., box 4, folder 4, EBHP.

75. For more on segregation within integrated schools, see Dougherty, *More than One Struggle*, 71–103.

76. Memorandum, "Strawman Procedure on Grouping," December 11, 1958, box 9, 78-8.2-51k, FJMP; box 1, 78-8.1-5d, FJMP; Vivien Ingram, "Flint Evaluates Its Primary Cycle," *Elementary School Journal* 61, no. 2 (November 1960): 76–80; Rita A. Scott, "The Status of Equal Opportunity in Michigan's Public Schools," n.d., box 9, folder 57, EBHP; Education Committee, Flint Branch, NAACP, "Fact Sheet," n.d., ca. 1965, box 92, Civil Rights, G–L, 1965, GRP.

77. "Negroes Charge Plot against Race in Two Community Programs," *FJ*, January 14, 1965; John R. Davis, "NAACP Officials Renew Charges on Mott Program, Flint Schools," *FJ*, February 11, 1965; Eleanor L. Elliott, "'Better Tomorrow for Urban Child' Is Outlined at Kiwanis Meeting," *FJ*, March 26, 1965.

78. Sanders, *Bronze Pillars*, 112.

79. "Happiest Town in Michigan," *Coronet*, June 1956.

80. Detzer, "Flint's Gone Crazy over Culture"; Kearney, "Dancing in the Streets."

Chapter 3

1. Box 6, folder 14, EBHP.

2. Elizabeth Chapelski, Wilfred Marston, and Mari Molseed, *1990 Demographic Profile of the Flint Urban Area in Comparative Perspective* (Flint, MI: PURA, 1992), 4.

3. Nancy MacLean, *Freedom Is Not Enough: The Opening of the American Workplace* (New York: Russell Sage Foundation and Harvard University Press, 2006); Anthony S. Chen, *The Fifth Freedom: Jobs, Politics, and Civil Rights in the United States, 1941–1972* (Princeton, NJ: Princeton University Press, 2009).

4. Donald Mosher, ed., *We Make Our Own History: The History of UAW Local 659* (Flint, MI: UAW Local 659, 1993), 65; Sol Dollinger and Genora Johnson Dollinger, *Not Automatic: Women and the Left in the Forging of the Auto Workers' Union* (New York: Monthly Review Press, 2000), 143. On the sit-down strike, see Henry Kraus, *The Many and the Few: A Chronicle of the Dynamic Auto Workers* (Los Angeles: Plantin, 1947); Sidney Fine, *Sit-Down: The General Motors Strike of 1936–37* (Ann Arbor: University of Michigan Press, 1969); Ronald Edsforth, *Class Conflict and Cultural Consensus: The Making of a Mass Consumer Society in Flint, Michigan* (New Brunswick, NJ: Rutgers University Press, 1987).

5. Lloyd H. Bailer, "The Negro Automobile Worker," *Journal of Political Economy* 51, no. 5 (October 1943): 415–28. On black workers at Ford, see David M. Lewis-Colman, *Race against Liberalism: Black Workers and the UAW in Detroit* (Urbana: University of Illinois Press, 2008), 5–6; Beth Tompkins Bates, *The Making of Black Detroit in the Age of Henry Ford* (Chapel Hill: University of North Carolina Press, 2012).

6. Thomas J. Sugrue, *The Origins of the Urban Crisis: Race and Inequality in Postwar Detroit* (Princeton, NJ: Princeton University Press, 1996), 96.

7. Herbert R. Northrup, ed., *Negro Employment in Basic Industry: A Study of Racial Policies in Six Industries* (Philadelphia: Wharton School of Finance and Commerce, University of Pennsylvania, 1970), 52.

8. Ruth Milkman, "Redefining 'Women's Work': The Sexual Division of Labor within the Auto Industry during World War II," *Feminist Studies* 8, no. 2 (Summer 1982): 343–47.

9. Robert C. Weaver, "Detroit and Negro Skill," *Phylon* 4, no. 2 (1943): 131–43.

10. Ananthakrishnan Aiyer, ed., *Telling Our Stories: Legacy of the Civil Rights Movement in Flint* (Flint, MI: Flint ColorLine Project, 2007), 38–39.

11. Ibid., 39. Black workers in Detroit felt much the same as their counterparts in Flint. On Detroit, see Sugrue, *Origins of the Urban Crisis*, chap. 4.

12. Lynn Reuster, *Good Old Days at the Buick: Memories of the Men and Women Who Worked There* (Linden, MI: McVey Marketing and Advertising, 1990), 156. On Taylor, see Frederick Winslow Taylor, *The Principles of Scientific Management* (New York: Harper and Brothers, 1911); Robert Kanigel, *The One Best Way: Frederick Winslow Taylor and the Enigma of Efficiency* (Cambridge, MA: MIT Press, 2005).

13. Reuster, *Good Old Days at the Buick*, 147; FWR, April 24, 1953.

14. Roger Townsend, interview by Michael Marve, May 1979, box 5, University of Michigan–Flint Labor History Project, GHCC.

15. My claim here builds upon Risa Goluboff's work on employment discrimination. See Goluboff, *The Lost Promise of Civil Rights* (Cambridge, MA: Harvard University Press, 2010).

16. Sugrue, *Origins of the Urban Crisis*, 17–32.

17. *AC Division Scrapbook*, vol. 1, *1927–1948* (LaCrosse, WI: Brookhaven, n.d.), F-GC/FPL.

18. MacLean, *Freedom Is Not Enough*, 22; Chen, *Fifth Freedom*, 32–87.

19. Bailer, "The Negro Automobile Worker," 424–25; Sugrue, *Origins of the Urban Crisis*, 26–28. On the Double V campaign, see MacLean, *Freedom Is Not Enough*, 23.

20. Chen, *Fifth Freedom*, 32–87.

21. "Employment of Negroes in Some of the General Motors Plants in Michigan—June 1944," part 1, box 9, folder 24, Discrimination against Negroes in Employment, 1942–47, UAWR.

22. Rhonda Sanders, *Bronze Pillars: An Oral History of African-Americans in Flint* (Flint, MI: Flint Journal and Sloan Museum, 1995), 59.

23. Ibid., 61. On black workers and World War II, see Thomas J. Sugrue, *Sweet Land of Liberty: The Forgotten Struggle for Civil Rights in the North* (New York: Random House, 2008), 59–84; Neil A. Wynn, *The African American Experience during World War II* (Lanham, MD: Rowman and Littlefield, 2010).

24. On the "Second Great Migration," see James N. Gregory, *The Southern Diaspora: How the Great Migrations of Black and White Southerners Transformed America* (Chapel Hill: University of North Carolina Press, 2005). On the expanding opportunities for black workers, see Chen, *Fifth Freedom*, 32–169.

25. Sanders, *Bronze Pillars*, 57.

26. Jacquelyn Dowd Hall, "The Long Civil Rights Movement and the Political Uses of the Past," *Journal of American History* 91, no. 4 (March 2005): 1233–63.

27. Sanders, *Bronze Pillars*, 60.

28. On the UAW's early history, see Kraus, *Many and the Few*; Fine, *Sit-Down*; Nelson Lichtenstein, *The Most Dangerous Man in Detroit: Walter Reuther and the Fate of American Labor* (New York: Basic Books, 1995), 47–131. On the UAW and racial liberalism, see box 10, folder 24, box 11, folders 6–7, box 17, folder 10, box 18, folder 6, UAWCAP; August Meier and Elliott M. Rudwick, *Black Detroit and the Rise of the UAW* (New York: Oxford University Press, 1979); Kevin Boyle, *The UAW and the Heyday of American Liberalism* (Ithaca, NY: Cornell University Press, 1995); Lewis-Colman, *Race against Liberalism*.

29. "Flint: A True Report," n.d., box 1, folder 4, UAWCAP; "How the UAW Practices What It Preaches: Powerful UAW-CIO Sets Example for Nation," n.d., box 6, folder 1, UAWCAP.

30. "How the UAW Practices What It Preaches"; Herbert Hill, *Black Labor and the American Legal System: Race, Work, and the Law* (Madison: University of Wisconsin Press, 1977), 270; Lewis-Colman, *Race against Liberalism*, 52–71.

31. Sugrue, *Origins of the Urban Crisis*, 91–123; Lewis-Colman, *Race against Liberalism*.

32. Herbert Hill and James E. Jones, *Race in America: The Struggle for Equality* (Madison: University of Wisconsin Press, 1993), 288; Boyle, *UAW and the Heyday of American Liberalism*, 117; Sugrue, *Origins of the Urban Crisis*, 100–102; Andrew Edmund Kersten, *Race, Jobs, and the War: The FEPC in the Midwest, 1941–1946* (Urbana: University of Illinois Press, 2000), 96–97; Lewis-Colman, *Race against Liberalism*, 17, 54; Colleen Doody, *Detroit's Cold War: The Origins of Postwar Conservatism* (Urbana: University of Illinois Press, 2012), 69–70.

33. Meier and Rudwick, *Black Detroit and the Rise of the UAW*, 162–72; Sugrue, *Origins of the Urban Crisis*, 91–124; Boyle, *UAW and the Heyday of American Liberalism*, 113–31; Lewis-Colman, *Race against Liberalism*, 14–17.

34. Fair Practices and Anti-Discrimination Department, UAW-CIO, "Report of Findings on Appeal," December 10, 1946, box 14, folder 30, UAWFP.

35. Richard T. Leonard to L. T. Heller, December 30, 1946, Region 1C, Flint, Correspondence, 1946–1948, box 14, folder 31, UAWFP; box 5, folder 2, Bowling, Statements, Reports, Resolutions, 1940s, UAWFP; box 1, folder 47, UAWFP; L. T. Heller to Walter P. Reuther, January 13, 1948, Region 1C, Flint, Correspondence, 1947–1950, box 14, folder 30, UAWFP. For more

on segregated bowling, see Lewis-Colman, *Race against Liberalism*, 20–49; Victoria Wolcott, *Race, Riots, and Roller Coasters: The Struggle over Segregated Recreation in America* (Philadelphia: University of Pennsylvania Press, 2012).

36. Box 5, folder 2, UAWFP.

37. Leonard to Heller, December 30, 1946; box 5, folder 2, UAWFP; box 1, folder 47, UAWFP; Heller to Reuther, January 13, 1948.

38. *FWR*, November 15, 1946, June 11, 25, 1948, July 21, 28, November 10, 24, 1950, November 7, 1952, August 6, 1954; Carroll H. Clark, "Some Aspects of Voting Behavior in Flint, Michigan—A City with Nonpartisan Municipal Elections" (PhD diss., University of Michigan, 1952); box 1, 87-12, MAFP; box 33, folder 21, Region 1C—Robert Frost, July 1954–1955, UAWPA; Norman W. Bully to Roy Reuther, October 26, 1954, Roy Reuther, 1947–August 15, 1957, box 50, folder 14, Michigan-Flint Industrial Council, UAWPA; Robert Frost to Roy L. Reuther, February 15, 1955, box 33, folder 21, Region 1C—Robert Frost, July 1954–1955, UAWPA; Roy Reuther, 1947–August 15, 1957, box 55, folder 21—1952 Elections—Flint City Statistics, UAWPA.

39. For the most comprehensive source of information on GM's deep involvement in local politics, see Minutes, MAFP.

40. Clark, "Some Aspects of Voting Behavior," 15–55; box 5, 87-12.4-12, Annual Meeting Folders, 1968, MAFP.

41. Box 2, 87-12, 87-12.2-1d, MAFP.

42. Arthur A. Elder, "Some Problems in Michigan Taxation" (master's thesis, Wayne State University, 1946), 1–42; Minutes, MAFP; "Flint: A True Report," 4; Ed Purdy to Robert Frost, September 2, 1954, Roy Reuther, box 33, folder 21, UAWPA; Robert Frost to Roy Reuther, February 9, 1956, Roy Reuther, box 33, folder 22, UAWPA; Robert A. Carter to Roy Reuther, February 9, 1956, Roy Reuther, box 33, folder 22, UAWPA.

43. Winston L. Livingston to William H. Oliver, November 29, 1949, box 14, folder 30, Region 1C, Flint, Correspondence, Reports, 1947–1950, UAWFP; William H. Oliver to Earl Crompton, June 21, 1951, box 14, folder 29, Region 1C, Flint, Correspondence, Newspapers, Agendas, 1951–1954, UAWFP; Robert J. Chase to Roy Reuther, September 28, 1951, box 14, folder 29, Region 1C, Flint, Correspondence, Newspapers, Agendas, 1951–1954, UAWFP; "City Commission Kills FEPC Measure," *FJ*, October 31, 1951; Robert A. Carter, "City Commission Uses Phoney [*sic*] Excuses to Justify Their Votes against FEP," *Searchlight*, November 1, 1951.

44. "Tie Vote Kills FEP; General Motors Supervisors Pack Commission Chamber," *FWR*, April 14, 1955. Also, see *FWR*, March 31, 1955.

45. *PFCC*, April 5, 1955, F-GC/FPL.

46. Although city commissioners never passed the FFEPC, the state's fair employment law, enacted in 1955, still legally protected many of Flint's workers. However, because representatives of the Michigan FEPC could not pursue discrimination cases unless a formal claim had been filed, the law's authority was quite limited. See "Michigan FEPC, 1962 Annual Report," box 12, FEPC, 1963, GRP.

47. *BR*, December 26, 1953, January 9, 1954.

48. Walter R. Greene, "Problems in Equal Employment Opportunity," box 10, folder 8, UAWFF.

49. Box 16, folder 67, Local 599, Region 1C, Correspondence, 1953–1958, UAWFP; box 25, folder 18, FEPC, Region 1C, Flint, Correspondence, Notes, 1955, UAWFP.

50. Frank J. Manley, "A Summary of the Development of the FCSG as of 3-25-1959," March 25, 1959, box 10, 78-8.2-84b, FJMP.

51. Box 18, 78-8.2-376, FJMP.

52. Manley, "Summary of the Development."

53. On other cities, see Touré F. Reed, *Not Alms but Opportunity: The Urban League and the Politics of Racial Uplift, 1910–1950* (Chapel Hill: University of North Carolina Press, 2008), 59–80, 139–68.

54. Minutes, Meeting of the FCSG, June 17, 1959, box 10, 78-8.2-84b, FJMP.

55. Richard A. English, *Survey of the Number of Negroes Employed as Salesclerks in Flint during the 1959 Christmas Shopping Period* (Flint, MI: ULF, 1960), 6.

56. Ibid., 2.

57. Report, May 14, 1953, box 14, folder 29, Region 1C, Flint, Correspondence, Newspapers, Agendas, 1951–1954, UAWFP.

58. Earl Crompton to William H. Oliver, July 2, 1953, box 14, folder 29, Region 1C, Flint, Correspondence, Newspapers, Agendas, 1951–1954, UAWFP.

59. Box 6, folder 9, Apprentice Aptitude Tests, 1950–1954, UAW581; box 6, folder 16, Apprenticeship Program, 1950s and 1960s, UAW581. On gender discrimination in American workplaces, see MacLean, *Freedom Is Not Enough*, 117–54.

60. Box 14, folder 29, Region 1C, Flint, Correspondence, Newspapers, Agendas, 1951–1954, UAWFP.

61. Earl Crompton to William H. Oliver, June 9, 1953, box 14, folder 29, Region 1C, Flint, Correspondence, Newspapers, Agendas, 1951–1954, UAWFP.

62. Mike Westfall, "The Minority Situation," *UAW CERP Newsletter*, no. 22, June 1983, box 1, folder 7, UAW National CERP Committee, Public Awareness Letters, 1982–1984, MWP.

63. Box 16, folder 67, Local 599, Region 1C, Correspondence, 1953–58, UAWFP.

64. Box 5, folder 10, EBHP.

65. MacLean, *Freedom Is Not Enough*.

66. "Motor Vehicle Industry: Male White Collar Workers by Race and Occupation, 1960," n.d., part 1, box 72, folder 28, UAWR; Sugrue, *Origins of the Urban Crisis*, 91–124.

67. "Suggested Course of Study in Applied Science I for the Public Schools of Flint, Michigan," June 20, 1952, box 4, folder 15, UAW581; "Minutes of the Meeting," November 12, 1952, box 4, folder 15, UAW581; "Proposed Co-op Program Involving Flint Public Schools," July 22, 1959, box 4, folder 15, UAW581; "A Proposed Cooperative Training Program in the Technical Occupations Involving Flint Area General Motors Plants and the Flint Public Schools," n.d., box 4, folder 15, UAW581.

68. Sugrue, *Origins of the Urban Crisis*, 91–123; MacLean, *Freedom Is Not Enough*, 76–113.

69. Part 2, box 39, folder 10, UAWGM; box 74, folders 11, 19.

70. Robert W. Cherny, William Issel, and Kieran Walsh Taylor, eds., *American Labor and the Cold War: Grassroots Politics* (New Brunswick, NJ: Rutgers University Press, 2004); Doody, *Detroit's Cold War*.

71. US Congress, House Committee on Un-American Activities, *Investigation of Communist Activities in the State of Michigan* (Washington, DC: USGPO, 1954); Jay Feldman, *Manufacturing Hysteria: A History of Scapegoating, Surveillance, and Secrecy in Modern America* (New York: Pantheon Books, 2011), 241.

72. *DFP*, June 19, 1954; part 2, box 78, folder 9, UAWGM.

73. "Statement of Charles Shinn," part 2, box 74, folder 11, UAWGM.

74. "Bulletin: G.M. Worker Beaten Second Time by Goons in Flint," n.d., part 2, box 78, folder 9, UAWGM; *DN*, March 19, 1952, June 17, July 16, 1954.

75. Part 2, box 78, folder 9, UAWGM; part 2, box 74, folder 11, UAWGM.

76. Martin Halpern, *UAW Politics in the Cold War Era* (Albany: SUNY Press, 1988), 95–172;

Lichtenstein, *Most Dangerous Man in Detroit*, 299–326; Boyle, *UAW and the Heyday of American Liberalism*, 32–82.

77. Raymond Walker, "Discrimination and Jim-Crow Tactics in Chevrolet-Flint," *Searchlight*, April 16, 1953; "GM Worker Asks GM Corp.: Is Jimcrow 'Good for Nation?'" *Daily Worker*, April 30, 1953; Motor Division Minutes, 1950–1964, CMCF.

78. W. J. McConnell and J. William Fehnel, "Health Hazards in the Foundry Industry," *Journal of Industrial Hygiene* 16, no. 4 (June 1934): 227–51; International Metalworkers' Federation, "Survey of Silicosis among Foundry Workers," May 1957, box 1, folder 12, UAWFF; "Silicosis—A Continuing Problem in Foundries," *Michigan's Occupational Health*, Fall 1963, box 7, folder 1, Health and Safety, 1963–1972, UAWFF; "How to Keep a Foundry Clean: Part I," *Michigan's Occupational Health*, Winter 1963–1964, box 7, folder 1, Health and Safety, 1963–1972, UAWFF; "How to Keep a Foundry Clean: Part II," *Michigan's Occupational Health*, Spring 1964, box 7, folder 1, Health and Safety, 1963–1972, UAWFF; "Foundry Working Conditions and Health Hazards," n.d., ca. 1970, box 1, folders 12–13, UAWFF; "UAW to Launch Battle against Safety and Health Hazards in Grey Iron Foundries," and "Effects of Silica Dust on the Lungs," *UAW Occupational Safety and Health Newsletter*, August 1972, box 1, folder 13, UAWFF; "The Foundry Department," n.d., box 1, folder 12, UAWFF.

79. Albert Christner to Leonard Woodcock, July 29, 1956, part 2, box 107, folder 2, UAWGM; Albert Christner to C. S. Meneer, July 1, 1965, part 2, box 107, folder 4, UAWGM.

80. Sanders, *Bronze Pillars*, 64.

81. "Hightower Fired," *BR*, January 7, 1956; John Hightower, interview by Rhonda Sanders and Wanda Howard, March 22, 1994, BPOHPF. On racial violence in the workplace, see Heather Ann Thompson, *Whose Detroit? Politics, Labor, and Race in a Modern American City* (Ithaca, NY: Cornell University Press, 2001).

82. Harold S. McFarland, "Minority Group Employment at General Motors," in *The Negro and Employment Opportunities: Problems and Practices*, ed. Herbert R. Northrup and Richard L. Rowan (Ann Arbor: University of Michigan, Bureau of Industrial Relations, 1965), 131–36.

83. Allan R. Wilhelm, "NAACP Pickets Buick Offices Here," *FJ*, April 25, 1964; box 2, folder 40, EBHP.

84. "NAACP Plans Protest of GM 'Discrimination,'" *FJ*, April 12, 1964; "Job Bias Denied by GM, Target for Protest Rally," *DFP*, May 1, 1964; "Hundreds Picket GM to Dramatize Charges of Bias," *DFP*, May 5, 1964; box 15, folder 6, EBHP.

Chapter 4

1. RVPDC, *1985–2010: Population Allocation Study* (Flint, MI: RVPDC, 1984); USBC, "Demographic Trends in the Twentieth Century" (Washington, DC: USGPO, 2002), 33, http://www.census.gov/prod/2002pubs/censr-4.pdf.

2. William Schneider, "The Suburban Century Begins," *Atlantic Monthly*, July 1992, 33; Kevin M. Kruse and Thomas J. Sugrue, eds., *The New Suburban History* (Chicago: University of Chicago Press, 2006), 1.

3. Kenneth T. Jackson, *Crabgrass Frontier: The Suburbanization of the United States* (New York: Oxford University Press, 1985), 190–218; David M. P. Freund, *Colored Property: State Policy and White Racial Politics in Suburban America* (Chicago: University of Chicago Press, 2007), 99–240.

4. Israel Harding Hughes Jr., *Local Government in the Fringe Area of Flint, Michigan* (Ann Arbor: IHA, 1947), i–v; Kenneth VerBurg et al., *Report to the Michigan State Boundary Commis-*

sion on the Proposed Cities of Burton, Carman, and Genesee (East Lansing: Institute for Community Development and Continuing Education Services, Michigan State University, 1970).

5. Betty Tableman, *Intra-Community Migration in the Flint Metropolitan District* (Ann Arbor: IHA, 1948), i–viii; Betty Tableman, *Governmental Organization in Metropolitan Areas* (Ann Arbor: University of Michigan Press, 1951); *Flint City-Fringe Survey* (Ann Arbor: IHA, 1955), 1–14; Basil Zimmer and Amos H. Hawley, "Home Owners and Attitude toward Tax Increase," *Journal of the American Institute of Planners*, Spring 1956, 66; Basil Zimmer, *Flint Area Study Report*, vol. 1 (Ann Arbor: IHA, 1957), chap. 2. On the nation as a whole, see Jackson, *Crabgrass Frontier*.

6. Becky M. Nicolaides, *My Blue Heaven: Life and Politics in the Working-Class Suburbs of Los Angeles, 1920–1965* (Chicago: University of Chicago Press, 2002), 9–119; Andrew Wiese, *Places of Their Own: African American Suburbanization in the Twentieth Century* (Chicago: University of Chicago Press, 2004), 11–93.

7. Walter Firey, *Social Aspects to Land Use Planning in the Country-City Fringe: The Case of Flint, Michigan* (East Lansing: Michigan State College Agricultural Experiment Station, 1946), 24–25.

8. Hughes, *Local Government*, i.

9. Peirce Lewis, "Geography in the Politics of Flint" (PhD diss., University of Michigan, 1958), 20.

10. Nicolaides, *My Blue Heaven*, 9–119; Wiese, *Places of Their Own*, 11–93; Becky M. Nicolaides and Andrew Wiese, eds., *The Suburb Reader* (New York: Routledge, 2006).

11. Hughes, *Local Government*.

12. Firey, *Social Aspects to Land Use Planning*, 27–37.

13. Erdmann D. Beynon, *Characteristics of the Relief Case Load in Genesee County, Michigan* (Flint, MI: Genesee County Welfare Relief Commission, 1940); Firey, *Social Aspects to Land Use Planning*, 27–37.

14. Beynon, *Characteristics of the Relief Case Load in Genesee County*; Firey, *Social Aspects to Land Use Planning*, 27–37; James N. Gregory, *The Southern Diaspora: How the Great Migrations of Black and White Southerners Transformed America* (Chapel Hill: University of North Carolina Press, 2005). On working-class suburbanization, see Nicolaides, *My Blue Heaven*, 9–119; Wiese, *Places of Their Own*, 11–93.

15. Firey, *Social Aspects to Land Use Planning*, 24, 27–37.

16. "FAS Says Practice May Develop Slums," *FJ*, September 28, 1955; Zimmer and Hawley, "Home Owners," 68.

17. Hughes, *Local Government*; Tableman, *Intra-Community Migration*, i–ix.

18. *Proceedings of the Board of Supervisors, Genesee County, Michigan*, April 1948, F-GC/FPL.

19. Hughes, *Local Government*; Tableman, *Governmental Organization*, 83; Basil Zimmer and Amos H. Hawley, "Approaches to the Solution of Fringe Problems: Preferences of Residents in the Flint Metropolitan Area," *Public Administration Review* 16, no. 4 (Autumn 1956): 258–68; Amos H. Hawley and Basil Zimmer, *Resistance to Governmental Unification in a Metropolitan Community* (Ann Arbor: IHA, 1957), 65; Emerson J. Elliott, Charles B. Hetrick, and Stanley W. Johnson, *Urban-Fringe Problem in the Flint Metropolitan Area: A Study of Services and Attitudes* (Ann Arbor: IHA, 1957).

20. Hughes, *Local Government*, 47–50; *FNA*, January 22, 1954; Homer Dowdy, "Time Has Come to Attack Problems of City's Growth," and "Area Population Growth Outruns Fringe Facilities," in *Where's Flint Going? The Metropolitan Area: A Look at Problems Confronting Flint*

and Its Neighbors (Flint, MI: *Flint Journal*, 1954); Elliott, Hetrick, and Johnson, *Urban-Fringe Problem*; Zimmer, *Flint Area Study Report*, vol. 1, chap. 3.

21. Lawrence Lader, "Chaos in the Suburbs," *Better Homes and Gardens*, October 1958, 10–17.

22. Hughes, *Local Government*, 47–50; Dowdy, "Time Has Come to Attack Problems of City's Growth" and "Area Population Growth Outruns Fringe Facilities"; Zimmer, *Flint Area Study Report*, vol. 1, chap. 3.

23. Homer Dowdy, "Flint Area Study Leads the Way for Medium-Sized Communities," in *What's Going on Here?* (Flint, MI: *Flint Journal*, 1954); FNA, January 22, 1954; Elliott, Hetrick, and Johnson, *Urban-Fringe Problem*; Hawley and Zimmer, *Resistance to Governmental Unification*, 44; Adam Rome, *The Bulldozer in the Countryside: Suburban Sprawl and the Rise of American Environmentalism* (New York: Cambridge University Press, 2001).

24. Tableman, *Intra-Community Migration*, ii.

25. GCH, May 21, 1947; Hawley and Zimmer, *Resistance to Governmental Unification*, 43–45.

26. RVPDC, *1985–2010*.

27. FO, December 31, 1953, February 4, 11, August 19, 1954.

28. FWR, June 26, 1953; Dowdy, "Flint Area Study Leads the Way"; Homer Dowdy, "Most Flint-Area Communities See Water Shortage in Future," FJ, March 18, 1956; Hawley and Zimmer, *Resistance to Governmental Unification*, 43; GCH, October 22, 1958.

29. FO, December 2, 1954. On the national scope of the problem, see Rome, *Bulldozer in the Countryside*.

30. "The Intermediate Public Works Stabilization Acceleration Plan," Commissioners Correspondence and Subject File, 1938–1958, box 1, Acceleration of Public Works, RG 31; Subject Correspondence, 1966–1973, box 58, PRO 6, Programs of Other Government Agencies, November 26, December 31, RG 207.

31. FI, May 1, 1952.

32. On the Flint area, see boxes 1–2, 87-12, MAFP. For more on how the FHA supported and subsidized suburban development projects nationally, see Marc A. Weiss, *The Rise of the Community Builders: The American Real Estate Industry and Urban Land Planning* (New York: Columbia University Press, 1987), 141–58; Freund, *Colored Property*.

33. FO, January 6, 1955.

34. FO, September 26, 1946, March 16, 1950.

35. FO, March 13, 1952.

36. FO, January 13, 20, 1955.

37. FO, January 3, 1957; "Dividends from Sanitary Sewers," GCH, September 13, 1961.

38. FO, June 7, 1956. On the FHA's support for suburban incorporation, see, for instance, "Spring Hill, Louisiana," and "Laurel, Maryland," DRS, Reports of Housing Market Analysts, 1937–1963, box 9, RG 31.

39. CM, November 21, 1951.

40. CM, December 26, 1956.

41. CM, February 6, 1957.

42. DI, July 5, 1946.

43. DI, April 19, 1946.

44. DI, October 4, 1946.

45. DI, July 8, 1949, December 30, 1954.

46. DI, October 29, 1948.

47. DI, May 17, 1951.

48. *DI*, August 5, December 30, 1954. On school consolidation, see *FO*, January 19, 1950.

49. *DI*, May 13, 1954.

50. *DI*, May 13, June 3, September 9, 1954. On local mobilizations in defense of the color line in other areas of the United States, see Arnold R. Hirsch, *Making the Second Ghetto: Race and Housing in Chicago, 1940–1960* (Cambridge: Cambridge University Press, 1983); Thomas J. Sugrue, *The Origins of the Urban Crisis: Race and Inequality in Postwar Detroit* (Princeton, NJ: Princeton University Press, 1996); Freund, *Colored Property*; Jeannine Bell, *Hate Thy Neighbor: Move-In Violence and the Persistence of Racial Segregation in American Housing* (New York: New York University Press, 2013).

51. *DI*, December 30, 1954.

52. RVPDC, *1985–2010*; *DI*, January 14, 1960.

53. Jackson, *Crabgrass Frontier*, 231–45.

54. RVPDC, *1985–2010*.

55. Zimmer, *Flint Area Study Report*, vol. 1, chap. 1.

56. *ACS*, June 12, 1958; RVPDC, *1985–2010*.

57. Zimmer, *Flint Area Study Report*, vol. 1, chap. 4.

58. *FO*, December 17, 1953, January 14, 1954; Basil Zimmer, *Flint Area Study Report*, vol. 2 (Ann Arbor: IHA, 1957), chap. 10.

59. Zimmer, *Flint Area Study Report*, vol. 2, chap. 10.

60. *FJ*, September 28, 1955. On exclusionary zoning and metropolitan fragmentation as national phenomena, see Seymour I. Toll, *Zoned American* (New York: Grossman, 1969); Michael N. Danielson, *The Politics of Exclusion* (New York: Columbia University Press, 1976), chaps. 2–4; Freund, *Colored Property*, chap. 2.

61. Freund, *Colored Property*, 45–98, 213–40.

62. Hughes, *Local Government*, 51.

63. Zimmer, *Flint Area Study Report*, vol. 2, chap. 9.

64. James Lloyd Sundquist, *Politics and Policy: The Eisenhower, Kennedy, and Johnson Years* (Washington, DC: Brookings Institution, 1968), 155–220.

65. Zimmer, *Flint Area Study Report*, vol. 1, chap. 1.

66. *FO*, May 1, 22, July 10, August 7, 1958.

67. Zimmer, *Flint Area Study Report*, vol. 2, chap. 9; *FO*, August 2, 1959.

68. Basil Zimmer, *Demographic Handbook of Flint Metropolitan Area* (Ann Arbor: IHA, 1955); *Flint City-Fringe Survey*, 1–14; Zimmer and Hawley, "Approaches to the Solution of Fringe Problems," 258–68; Basil Zimmer and Amos H. Hawley, "Property Taxes and Solutions to Fringe Problems: Attitudes of Residents of the Flint Metropolitan Area," *Land Economics* 32, no. 4 (November 1956): 369–76; Zimmer, *Flint Area Study Report*, vol. 1, chap. 3; Elliott, Hetrick, and Johnson, *Urban-Fringe Problem*, 1–46; Hawley and Zimmer, *Resistance to Governmental Unification*, 41–45; Basil Zimmer and Amos H. Hawley, "Suburbanization and Some of Its Consequences," *Land Economics* 37, no. 1 (February 1961): 88–93.

69. Zimmer, *Flint Area Study Report*, vol. 1, chap. 7; Hawley and Zimmer, *Resistance to Governmental Unification*, 49–51.

70. Tableman, *Intra-Community Migration*, ii.

71. Zimmer and Hawley, "Property Taxes and Solutions to Fringe Problems," 369–76; Basil Zimmer, *Flint Area Study Report*, vol. 3 (Ann Arbor: IHA, 1957), chap. 18.

72. Dowdy, "Area Population Growth Outruns Fringe Facilities"; Hawley and Zimmer, *Resistance to Governmental Unification*, 50.

73. On suburban boosterism and identity formation, see Nicolaides, *My Blue Heaven*; Rob-

ert O. Self, *American Babylon: Race and the Struggle for Postwar Oakland* (Princeton, NJ: Princeton University Press, 2003), 23−132; Freund, *Colored Property*, esp. 243−83.

74. Sam Bass Warner, *Streetcar Suburbs: The Process of Growth in Boston, 1870−1900* (Cambridge, MA: Harvard University Press, 1978); Jon C. Teaford, *City and Suburb: The Political Fragmentation of Metropolitan America, 1850−1970* (Baltimore: Johns Hopkins University Press, 1979); Jackson, *Crabgrass Frontier*, 138−56.

75. *FI*, n.d., ca. December 1948.

76. *FI*, n.d., ca. December 1949.

77. Hawley and Zimmer, *Resistance to Governmental Unification*, 48; Zimmer, *Flint Area Study Report*, vol. 3, chaps. 17−18.

Chapter 5

1. Kenneth T. Jackson, *Crabgrass Frontier: The Suburbanization of the United States* (New York: Oxford University Press, 1985), 266−69; Jefferson Cowie, *Capital Moves: RCA's Seventy-Year Quest for Cheap Labor* (New York: New Press, 1999); Robert O. Self, *American Babylon: Race and the Struggle for Postwar Oakland* (Princeton, NJ: Princeton University Press, 2003), 23−60.

2. Heather B. Barrow, "'The American Disease of Growth': Henry Ford and the Metropolitanization of Detroit," in *Manufacturing Suburbs: Building Work and Home on the Metropolitan Fringe*, ed. Robert Lewis (Philadelphia: Temple University Press, 2004), 200−220.

3. The philosophy of metropolitan capitalism had a broad appeal among corporate leaders in twentieth-century America. See, for example, Carl Abbott, *The New Urban America: Growth and Politics in Sunbelt Cities* (Chapel Hill: University of North Carolina Press, 1981); C. James Owen and York Willbern, *Governing Metropolitan Indianapolis: The Politics of Unigov* (Berkeley: University of California Press, 1985); Bob Ortega, *In Sam We Trust: The Untold Story of Sam Walton and How Wal-Mart Is Devouring the World* (New York: Times Books, 1998), 166; Bernard H. Ross and Myron A. Levine, *Urban Politics: Power in Metropolitan America*, 7th ed. (Belmont, CA: Thomson Wadsworth, 2006), 429−32; Matthew D. Lassiter, *The Silent Majority: Suburban Politics in the Sunbelt South* (Princeton, NJ: Princeton University Press, 2005); Terry Christensen, *Local Politics: A Practical Guide to Governing at the Grassroots*, 2nd ed. (New York: M. E. Sharpe, 2006), 358−59, 368−69; Louise Nelson Dyble, *Paying the Toll: Local Power, Regional Politics, and the Golden Gate Bridge* (Philadelphia: University of Pennsylvania Press, 2009), 191−95; Todd E. Robinson, *A City within a City: The Black Freedom Struggle in Grand Rapids, Michigan* (Philadelphia: Temple University Press, 2012), esp. xiv, 81; Elizabeth Tandy Shermer, *Sunbelt Capitalism: Phoenix and the Transformation of American Politics* (Philadelphia: University of Pennsylvania Press, 2013), 147−224.

4. I borrow the term "suburban strategy" from Matthew Lassiter, who has used it in a very different context to describe conservative politics in the American South. See Lassiter, *Silent Majority*.

5. On regional government in the South and West, see David Rusk, *Cities without Suburbs* (Washington, DC: Woodrow Wilson Center Press, 1993); Manuel Pastor Jr. et al., *Regions That Work: How Cities and Suburbs Can Grow Together* (Minneapolis: University of Minnesota Press, 2000).

6. This was not unique to Flint. See Self, *American Babylon*, 23−60; Sarah Jo Peterson, *Planning the Home Front: Building Bombers and Communities at Willow Run* (Chicago: University of Chicago Press, 2013).

7. "For Plants of Tomorrow, Country Club Air," *DFP*, n.d., ca. 1955.

8. Self, *American Babylon*, 23–60.

9. Basil Zimmer, *Flint Area Study Report*, vol. 1 (Ann Arbor: IHA, 1957), chap. 2. On GM's pre–World War II building strategy, see *The City Plan of Flint, Michigan* (Flint, MI: Flint City Planning Board, 1920), 62; Noland Rall Heiden, "A Land Use Sequence Study of the Dort Highway Area, Flint, Michigan" (PhD diss., University of Michigan, 1949); Leo Donovan, "Flint Fattening up Fast," *DFP*, April 1, 1955; *Michigan: Annual Survey of Industrial Expansions, 1955* (Lansing: Michigan Economic Development Department, 1956), LM; Ronald Larson and Eric Schenker, *Land and Property Values in Relation to Dort Highway Improvements* (East Lansing: Michigan State University Traffic Safety Center, 1960); Benjamin Klein, "Fisher-General Motors and the Nature of the Firm," *Journal of Law and Economics* 43, no. 1 (April 2000): 105–41; *AC Division Scrapbook*, vol. 1, *1927–1948* (LaCrosse, WI: Brookhaven, n.d.), F-GC/FPL.

10. C. S. Mott to Floyd A. Allen, April 2, 1942, box 29, 77-7.6-1.1b, CSMP; *Annual Report of General Motors Corporation, 1944* (Detroit: GM, 1945), F-GC/FPL; *Michigan: Annual Survey of Industrial Expansions, 1955*; Alan Clive, *State of War: Michigan in World War II* (Ann Arbor: University of Michigan Press, 1979); Richard P. Scharchburg, ed., *GM Story* (Flint, MI: General Motors Institute, 1981), 14; Richard M. Langworth and Jan P. Norbye, *The Complete History of General Motors, 1908–1986* (New York: Beekman House, 1986), 169; Alan K. Binder and Deebe Ferris, *General Motors in the Twentieth Century* (Southfield, MI: Ward's Communications, 2000), 9–10.

11. Ysabel Rennie and Robert Rennie, "Durant, Sloan Cited in GM Success Story," *WP*, March 22, 1950; Lawrence J. White, *The Automobile Industry since 1945* (Cambridge, MA: Harvard University Press, 1971), 10–18. On GM's emphasis on suburban facilities, see *FJ*, August 31, November 29, 1951, July 1, 1953, January 19, 1954, June 23, 1955; *FNA*, January 21, 1954; "GM Will Spend Billion on Expansion Plan," *FJ*, January 19, 1954; Public Relations Staff of GM, *1966–1967 Information Handbook*, Automotive Files, 70.26.3, PA. On investment in the city, see *Buick Motor Division Scrapbook*, vols. 1–3 (LaCrosse, WI: Brookhaven, n.d.), F-GC/FPL.

12. "Finally It's Official—Ternstedt Is Coming," *FNA*, March 27, 1953; *Ternstedt Division Scrapbook* (LaCrosse, WI: Brookhaven, n.d.), F-GC/FPL.

13. *AC Division Scrapbook*, vols. 1–3 (LaCrosse, WI: Brookhaven, n.d.), F-GC/FPL.

14. *FJ*, March 4, 1954.

15. *Chevrolet Motor Division Scrapbook*, vols. 1–3 (LaCrosse, WI: Brookhaven, n.d.), F-GC/FPL; *Thirty-Ninth Annual Report of the General Motors Corporation* (Detroit: GM, 1948).

16. *FJ*, August 31, November 29, 1951, July 1, 1953.

17. *Chevrolet Motor Division Scrapbook*, vol. 1; *Thirty-Ninth Annual Report of the General Motors Corporation*; Amos H. Hawley and Basil Zimmer, "Resistance to Unification in a Metropolitan Community," in *Community Political Systems*, ed. Morris Janowitz (Glencoe, IL: Free Press, 1961), 153; Donald Mosher, ed., *We Make Our Own History: The History of UAW Local 659* (Flint, MI: UAW Local 659, 1993), 59–61.

18. Gregory M. Miller, "Place, Space, Pace, and Power: The Struggle for Control of the Automobile Factory Shop Floor, 1896–2006" (PhD diss., University of Toledo, 2008), 111–47.

19. See War/Miscellaneous, 1946–1954, 95.97.113–118, CMCF; "For Plants of Tomorrow"; Larson and Schenker, *Land and Property Values*.

20. "Statement Prepared for Presentation by UAW President Walter P. Reuther before the Subcommittee on Production and Stabilization of the Banking and Currency Committee of the Senate in Detroit, Michigan," March 2, 1959, 20, part 1, box 90, folder 8, UAWR.

21. *Survey of Recent Industrial Move-Ins and Expansions in Michigan* (Lansing: Michigan Department of Commerce, 1967), LM.

22. *FWR*, October 3, 1953, January 15, December 10, 1954; part 1, box 72, folder 34, Plant Locations, 1955–1957, UAWR; Eva Mueller, Arnold Wilken, and Margaret Wood, *Location Decisions and Industrial Mobility in Michigan, 1961* (Ann Arbor: Institute for Social Research, University of Michigan, 1962); Michael T. Oravecz, *Locational Preferences of Manufacturing Firms: Flint-Genesee County, Michigan* (Flint, MI: Economic Development Commission of Genesee County, 1974); Jackson, *Crabgrass Frontier*, 266–71; Keith Chapman and David Walker, *Industrial Location: Principles and Politics* (New York: Blackwell, 1987); Self, *American Babylon*, 23–60.

23. Jackson, *Crabgrass Frontier*, 157–71, 246–71; Mark Rose, *Interstate: Express Highway Politics, 1939–1989* (Knoxville: University of Tennessee Press, 1990); Owen D. Gutfreund, *Twentieth-Century Sprawl: Highways and the Reshaping of the American Landscape* (New York: Oxford University Press, 2004).

24. Minutes, February 7, 1962, box 2, 87-12, MAFP; George R. Elges to George Ursuy, December 16, 1974, box 8, folder 36, DWRP.

25. Melissa Misekow, "The Highway Was Born in Flint: How I-475 Came to Be," UM-F/HDSP.

26. GM Public Relations Staff, *Adventures of the Inquiring Mind: Some General Motors Scientific and Engineering Contributions of the Last Half Century* (Detroit: GM, 1957); MSHD, *Highway Needs in Michigan: An Engineering Analysis* (Lansing: MSHD and the Michigan Good Roads Federation, n.d.); Frank Suggitt, *How Expressways Help Michigan and You* (Lansing: Michigan Good Roads Federation, n.d.), 6.

27. David Novick, Melvin Anshen, and W. C. Truppner, *Wartime Production Controls* (New York: Columbia University Press, 1949), 412; Morris L. Sweet, *Industrial Location Policy for Economic Revitalization: National and International Perspectives* (New York: Praeger, 1981), 11.

28. Joint Committee on the Economic Report, *The Need for Industrial Dispersal* (Washington, DC: USGPO, 1951), 1.

29. T. Brooks Brademas, "An Evacuation Study of the Critical Target Area of Flint, Michigan," University of Michigan, College of Architecture and Urban Planning, Student Papers, box 1, BHL; National Security Resources Board, *Is Your Plant a Target?* (Washington, DC: USGPO, 1951); Margaret Pugh O'Mara, *Cities of Knowledge: Cold War Science and the Search for the Next Silicon Valley* (Princeton, NJ: Princeton University Press, 2005), 28–45.

30. O'Mara, *Cities of Knowledge*, 31.

31. Between 1951 and 1984, the number of prime defense contracts per capita awarded to Michigan firms declined by approximately 75 percent. See Sweet, *Industrial Location Policy*, 13–15; Ann R. Markusen and Virginia Carlson, "Deindustrialization in the American Midwest: Causes and Responses," in *Deindustrialization and Regional Economic Transformation: The Experience of the United States*, ed. Lloyd Rodwin and Hidehiko Sazanami (Boston: Unwin Hyman, 1989), 44–45.

32. *Thirty-Seventh Annual Report of General Motors Corporation for the Year Ended December 31, 1945* (Detroit: GM, 1946), 15, F-GC/FPL.

33. *Fortieth Annual Report of General Motors Corporation for the Year Ended December 31, 1948* (Detroit: GM, 1949), 12, F-GC/FPL; "Statement Prepared for Presentation by UAW President Walter P. Reuther before the Special Senate Committee on Unemployment Problems," November 12, 1959, box 11, folder 6, Civil Rights, 1956–1959, UAWCAP; Charles E. Edwards, *Dynamics of the United States Automobile Industry* (Columbia: University of South Carolina Press, 1965).

34. *FJ*, November 20, 1957; Neil P. Hurley, "The Automotive Industry: A Study in Industrial Location," *Land Economics* 35, no. 1 (February 1959): 5; C. Estall and R. Ogilvie Buchanan,

Industrial Activity and Economic Geography: A Study of the Forces behind the Geographical Location of Productive Activity in Manufacturing Industry (London: Hutchinson University Library, 1968), 181.

35. Daniel H. Kruger, "The Labor Factor in Plant Location in Michigan," in *The Michigan Economy: Its Potentials and Its Problems*, ed. William Haber, Eugene C. McKean, and Harold C. Taylor (Kalamazoo, MI: W. E. Upjohn Institute for Employment Research, 1959), 257. On GM's international operations, see Frederic G. Donner, *The World-Wide Industrial Enterprise: Its Challenge and Promise* (New York: McGraw-Hill, 1967).

36. Mueller, Wilken, and Wood, *Location Decisions*, 53.

37. Hurley, "Automotive Industry," 12.

38. Chapman and Walker, *Industrial Location*, 116; Don Stillman, "The Devastating Impact of Plant Closures," *Working Papers for a New Society*, July–August 1978, F-GC/FPL; Barry Bluestone and Bennett Harrison, *The Deindustrialization of America: Plant Closings, Community Abandonment, and the Dismantling of Basic Industry* (New York: Basic Books, 1982).

39. "GM-Dixie Moves 'Concern' UAW," *FJ*, February 20, 1978.

40. Miller, "Place, Space, Pace, and Power," 111–47.

41. *FJ*, October 11, 1949.

42. *Searchlight*, April 17, 1952; *ACS*, October 29, 1953; *PFCC*, July 8, 1957, F-GC/FPL.

43. Memorandum, Herman D. Young to Ted Moss, August 6, 1953, CMF; "Commission Hints Shift in Policy," *FNA*, March 26, 1954.

44. *PFCC*, July 8, 1957, F-GC/FPL.

45. "Flint Taxpayers 'Pay the Shot,'" *Searchlight*, April 17, 1952.

46. *PFCC*, July 12, 1960, 1829–51, F-GC/FPL.

47. *PFCC*, February 10, 1960, 1581, F-GC/FPL.

48. *FJ*, November 22, 1954; Clarence H. Young, *Big-Crossing-Place* (Flint, MI: CSMF, 1962). Historically, the business leaders of the South and West have been among the most enthusiastic proponents of the metropolitan capitalist view. See Pastor et al., *Regions That Work*, 125–54; Shermer, *Sunbelt Capitalism*, 147–224.

49. Charles W. Minshall, *Genesee County Economic Conditions Conclusions Report* (Columbus, OH: Battelle Memorial Institute, 1969), 3–40.

50. James R. Custer, ed., *Applied Automation* (Philadelphia: Chilton, 1956); *PFCC*, March 4, 1963, 88, F-GC/FPL.

51. E. J. Moran to Jim Ogden, May 23, 1960, part 2, box 74, folder 18, Local 599, Correspondence, 1960–1961, UAWGM; *FWR*, March 2, 1961.

52. *FWR*, January 12, 1961.

53. Zimmer, *Flint Area Study Report*, vol. 1, chap. 4.

54. Minshall, *Genesee County Economic Conditions*, 3–40.

55. *Flint-Genesee County, Michigan: Data Profile* (Flint, MI: Flint-Genesee Corporation for Economic Growth, n.d., ca. 1983), 35.

56. Becky M. Nicolaides, *My Blue Heaven: Life and Politics in the Working-Class Suburbs of Los Angeles, 1920–1965* (Chicago: University of Chicago Press, 2002), 120–271; Self, *American Babylon*, 23–60, 135–76; O'Mara, *Cities of Knowledge*.

57. Box 1, folder 2, CRCF Files, GHCC.

58. Deil S. Wright, *Central Business District of Flint, Michigan: Changes in the Assessed Valuations of Real Property, 1930–1951* (Ann Arbor: IHA, 1953); Hawley and Zimmer, "Resistance to Unification," 155; USBC, *1963 Census of Business, Major Retail Centers, Flint, Michigan* (Washington, DC: USGPO, 1964), table 1.

59. Alison Isenberg, *Downtown America: A History of the Place and the People Who Made It* (Chicago: University of Chicago Press, 2005), 166–202.

60. "Motor Vehicle Registration," n.d., DRS, Housing Market Data, 1938–1952, box 12, Flint, Michigan, RG 31; Zimmer, *Flint Area Study Report*, vol. 1, chap. 14; Richard Hebert, *Highways to Nowhere: The Politics of City Transportation* (Indianapolis: Bobbs-Merrill, 1972), 8.

61. Proponents of the spatial mismatch model include John F. Kain, "Housing Segregation, Negro Employment, and Metropolitan Decentralization," *Quarterly Journal of Economics* 82, no. 2 (May 1968): 175–97; Kevin Boyle, "There Are No Union Sorrows That the Union Can't Heal: The Struggle for Racial Equality in the United Automobile Workers, 1940–1960," *Labor History* 35 (1995): 5–23; Thomas J. Sugrue, *The Origins of the Urban Crisis: Race and Inequality in Postwar Detroit* (Princeton, NJ: Princeton University Press, 1996), 125–52; Josh Sides, *L.A. City Limits: African American Los Angeles from the Great Depression to the Present* (Berkeley: University of California Press, 2003), 57–130.

62. Young, *Big-Crossing-Place*, 2.

63. Sugrue, *Origins of the Urban Crisis*, chap. 5; Colin Gordon, *Mapping Decline: St. Louis and the Fate of the American City* (Philadelphia: University of Pennsylvania Press, 2008).

64. On corporate support for annexation and regional government in the United States, see Abbott, *The New Urban America*; Owen and Willbern, *Governing Metropolitan Indianapolis*; Ortega, *In Sam We Trust*, 166; Ross and Levine, *Urban Politics*, 429–32; Lassiter, *Silent Majority*; Christensen, *Local Politics*, 358–59, 368–69; Dyble, *Paying the Toll*, 191–95; Robinson, *A City within a City*, esp. xiv, 81; Shermer, *Sunbelt Capitalism*, 147–224.

65. Kimberly Phillips-Fein, *Invisible Hands: The Making of the Conservative Movement from the New Deal to Reagan* (New York: W. W. Norton, 2009); Bethany Moreton, *To Serve God and Wal-Mart: The Making of Christian Free Enterprise* (Cambridge, MA: Harvard University Press, 2009).

66. E. S. Patterson to Edward T. Ragsdale, March 27, 1959, part 2, box 74, folder 17, Local 599, Correspondence, 1958–1960, UAWGM; Lynn Reuster, *Good Old Days at the Buick: Memories of the Men and Women Who Worked There* (Linden, MI: McVey Marketing and Advertising, 1990), 1–10.

67. "Last Layoff of 2200 Means More Than 50% of Buick Workers Idle," *FWR*, March 6, 1958; A. H. Raskin, "People behind Statistics: A Study of Unemployed," *NYT*, March 16, 1959; part 1, box 63, folder 13, Urban Renewal, 1959–1964, UAWR.

68. Otis Bishop to Leonard Woodcock, February 17, 1961, part 2, box 74, folder 18, Local 599, Correspondence, 1960–1961, UAWGM; Otis Bishop to E. S. Patterson, February 12, 1961, part 2, box 74, folder 18, Local 599, Correspondence, 1960–1961, UAWGM.

69. "Flint Classified as Depressed Area as Unemployment Rises," *FWR*, May 30, 1957.

70. Harlow Curtice, "Remarks at Buick Announcement Dinner," September 10, 1957, box 5, 91-1.3-9, Harlow Curtice Papers, SA.

71. "Annual Report of the Social Science Research Project, 1949," box 5, 78-8.1-124, FJMP.

72. By pointing to the efficiency of metropolitan government, Hawley and Zimmer broke sharply with influential scholars such as Charles Tiebout, who maintained that small, fragmented governments were more responsive than regional bodies. On the Tiebout thesis, see Charles M. Tiebout, "A Pure Theory of Local Expenditures," *Journal of Political Economy* 64, no. 5 (October 1956): 416–24.

73. Homer Dowdy, "Area Population Growth Outruns Fringe Facilities," in *Where's Flint Going? The Metropolitan Area: A Look at Problems Confronting Flint and Its Neighbors* (Flint, MI: *Flint Journal*, 1954).

74. "Presentation for the New City of Flint," n.d., box 4, 78-8.1-67, FJMP.

75. Basil Zimmer, *Flint Area Study Report*, vol. 3 (Ann Arbor: IHA, 1957), chap. 21.

76. On regionalism in the United States, see Jon C. Teaford, *City and Suburb: The Political Fragmentation of Metropolitan America, 1850–1970* (Baltimore: Johns Hopkins University Press, 1979); Rusk, *Cities without Suburbs*; Pastor et al., *Regions That Work*; Myron Orfield, *American Metropolitics: The New Suburban Reality* (Washington, DC: Brookings Institution Press, 2002); Peter Dreier, John Mollenkopf, and Todd Swanstrom, *Place Matters: Metropolitics for the Twenty-First Century*, 2nd ed. (Lawrence: University Press of Kansas, 2004); Jon C. Teaford, *The Metropolitan Revolution: The Rise of Post-Urban America* (New York: Columbia University Press, 2006).

77. Israel Harding Hughes Jr., *Local Government in the Fringe Area of Flint, Michigan* (Ann Arbor: IHA, 1947), 7.

78. Homer Dowdy, "Reasoning behind 'New Flint' Plan Given," *FJ*, October 27, 1957; "Committee for 'New Flint' Outlines Plans, Proposals," *ACS*, June 12, 1958.

79. "Presentation for the New City of Flint"; "Why a New City?" box 4, 78-8.1-67, FJMP.

80. "Three Tell New Flint Advantages," *FJ*, July 8, 1958; "Presentation for the New City of Flint"; "Why a New City?"

81. Dowdy, "Flint Area Study Leads the Way."

82. The quotation is from an undated speech in JAANP; box 1, 1956–1960, RMHP.

83. *PFCC*, March 10, May 19, 1958, F-GC/FPL; *FJ*, September 16, 1958.

84. "Personnel of New Flint Committee Represents Area Cross Section," *FJ*, n.d., ca. 1958.

85. This had already happened in other cities. See David Goldfield, *Still Fighting the Civil War: The American South and Southern History* (Baton Rouge: Louisiana State University Press, 2002), chap. 9; Andrew Wiese, *Places of Their Own: African American Suburbanization in the Twentieth Century* (Chicago: University of Chicago Press, 2004), 173–74.

86. Mark Baldassare, "The Image Problem of Regional Government: Factors Contributing to Suburban Opposition," in *Suburban Communities: Change and Policy Responses*, ed. M. Baldassare (Greenwich, CT: JAI, 1994), 195–208; Mark Baldassare, *When Government Fails: The Orange County Bankruptcy* (Berkeley: University of California Press, 1998), 19.

87. Dowdy, "Proposed Westhaven Much Like Livonia," in *Where's Flint Going? The Metropolitan Area: A Look at Problems Confronting Flint and Its Neighbors* (Flint, MI: Flint Journal, 1954); Homer Dowdy, "300 Flint Township Residents Hold Meeting in Utley School," *FJ*, August 6, 1956; Amos H. Hawley and Basil Zimmer, *Resistance to Governmental Unification in a Metropolitan Community* (Ann Arbor: IHA, 1957), 53–59.

88. "Area Opposition Develops to Proposal for Greater Flint," *FWR*, November 14, 1957.

89. *FJ*, March 19, 1958; "Carman Educator Vows to Fight 'New Flint,'" *FJ*, n.d., ca. April 1958.

90. Gordon Gapper, "Newness of 'New Flint' Stressed," *FJ*, April 11, 1958.

91. Lou Giampetroni, "Petitions Seek to Ban New Flint Election," *FJ*, June 10, 1958; "New Flint Foes Outline Opposition to Platform," *FJ*, October 8, 1958.

92. "Area Opposition Develops."

93. *FJ*, March 13, 1958.

94. *Genesee County Free Press*, December 5, 1957, January 9, March 27, April 3, 17, 24, May 1, 8, 22, 29, June 12, 19, 26, July 3, 10, 24, August 7, 1958.

95. "Bell Will Lead Battle against 'New Flint' Vote," *FJ*, August 6, 1958; "Supervisors Act against Legal Counsel," *FJ*, August 12, 1958.

96. *FJ*, August 13, 1958.

97. *FJ*, October 2, 1958.

98. *Taliaferro v. Genesee County Supervisors*, 354 Mich. 49, 92 N.W. 2d 319 (1958); "Plea to Force Nov. 4 Vote Is Rejected," *FJ*, October 8, 1958.

99. William O. Winter, "Annexation as a Solution to the Fringe Problem: An Analysis of Past and Potential Annexations of Suburban Areas to the City of Flint, Michigan" (PhD diss., University of Michigan, 1949); Jesse I. Bourquin and Albert G. Ballert, "Expansion of Michigan Cities and Villages in the 1940 Decade," *Papers of the Michigan Academy of Sciences, Arts and Letters* 36 (1950); Alvin D. Sokolow, *Annexation and Incorporation in Michigan: An Evaluation of the Boundary Commission Plan* (East Lansing: Institute for Community Development and Services, Michigan State University, 1965); Louis C. Andrews, ed., *Annexation Procedures in Michigan for Home Rule Cities and Villages* (Ann Arbor: Michigan Municipal League, 1968).

100. For more on the national context, see Rusk, *Cities without Suburbs*.

101. *FJ*, September 13, 1961.

102. Homer Dowdy, "Says GM for Annexation as Being Best for All," *FJ*, August 20, 1961.

103. Rudolph H. Pallotta, "Sections of Four Townships Are Involved in Move," *FJ*, November 14, 1961.

104. "Text of GM's Statement Endorsing Annexation," *FJ*, August 20, 1961; *FWR*, November 16, December 28, 1961; "Harding Mott Urges Approval on City Annexation Issues," *FWR*, February 15, 1962.

105. "AC Chief Says Annexation Would Keep Area Attractive to Industry," *FJ*, n.d., ca. September 14, 1961; "Let's Capitalize on the Growth Potential of the Flint Area," January 28, 1964, JAANP.

106. "Annexation Gains Seen for Township and City," *FJ*, n.d., ca. February 20, 1962; "Statement Made Regarding Annexation," February 19, 1962, CMF.

107. "Annexation Vote Is Called City's Last Chance," *FJ*, February 21, 1962.

108. "Flint Township Spokesman Calls Flint Annexation Move Thievery," *FJ*, December 21, 1961.

109. Jesse C. Hatcher, "Debate Flint Twp. Annexation," *FJ*, n.d., ca. March 1961.

110. *FJ*, November 8, 1963, January 8, 1964.

111. *FJ*, December 21, 1961.

112. *FO*, March 28, April 4, 1963; *FI*, April 4, 1963; Gail R. Light, "Areas outside Flint Can Expect More Growth in '70s," *FJ*, December 31, 1969.

113. Like metropolitan capitalism, suburban capitalism and secession were national phenomena. For more on the national context, see Robert C. Wood, *Suburbia: Its People and Their Politics* (Boston: Houghton Mifflin, 1959); Nicolaides, *My Blue Heaven*; Self, *American Babylon*; Kevin M. Kruse, *White Flight: Atlanta and the Making of Modern Conservatism* (Princeton, NJ: Princeton University Press, 2005); David M. P. Freund, *Colored Property: State Policy and White Racial Politics in Suburban America* (Chicago: University of Chicago Press, 2007), 46–48.

Chapter 6

1. James Grossman, *Land of Hope: Chicago, Black Southerners, and the Great Migration* (Chicago: University of Chicago Press, 1991); James N. Gregory, *The Southern Diaspora: How the Great Migrations of Black and White Southerners Transformed America* (Chapel Hill: University of North Carolina Press, 2005).

2. Elizabeth Chapelski, Wilfred Marston, and Mari Molseed, *1990 Demographic Profile of the Flint Urban Area in Comparative Perspective* (Flint, MI: PURA, 1992), 4. The city's Latino population also grew during this period. However, by 1960 there were just 687 Latinos living in

the city of Flint. See Jane B. Haney, "Chicanos of Michigan," in *Blacks and Chicanos in Urban Michigan*, ed. Homer C. Hawkins and Richard W. Thomas (Lansing: Michigan Department of State, 1979), 17–19.

3. Karl E. Taeuber and Alma F. Taeuber, *Negroes in Cities: Residential Segregation and Neighborhood Change* (Chicago: Aldine, 1965), 32–34. The Taeubers' segregation index represented the percentage of nonwhites citywide who would have had to relocate in order to achieve racial integration.

4. Memorandum, Olive Beasley to Burton I. Gordin, August 3, 1966, box 11, folder 32, OBP.

5. Nicholas Dagen Bloom, *Public Housing That Worked: New York in the Twentieth Century* (Philadelphia: University of Pennsylvania Press, 2009); Bradford D. Hunt, *Blueprint for Disaster: The Unraveling of Chicago Public Housing* (Chicago: University of Chicago Press, 2009).

6. See Roy Reuther, box 33, folder 14, UAWPAC; part 1, box 15, folder 11, Housing, UAWR.

7. *FJ*, January 22, 1939, January 1, 1940, November 23, 1949, September 13, 15, 1950, March 3, 1959. In the hopes of gaining control of the Flint City Commission, local UAW officials did not publicly endorse the public housing proposals. See Roy Reuther, box 33, folder 14, Region 1C—Robert Chase, August–December 1950, UAWPAC.

8. Part 1, box 15, folder 11, Housing, UAWR.

9. On socioeconomic diversity in black neighborhoods, see Earl Lewis, *In Their Own Interests: Race, Class, and Power in Twentieth-Century Norfolk, Virginia* (Berkeley: University of California Press, 1991); Michael C. Dawson, *Behind the Mule: Race and Class in African-American Politics* (Princeton, NJ: Princeton University Press, 1994); Mary Pattillo-McCoy, *Black Picket Fences: Privilege and Peril among the Black Middle Class* (Chicago: University of Chicago Press, 1999); Mary Pattillo, *Black on the Block: The Politics of Race and Class in the City* (Chicago: University of Chicago Press, 2007).

10. *PFCC*, February 17, May 5, 1958, F-GC/FPL; Ananthakrishnan Aiyer, ed., *Telling Our Stories: Legacy of the Civil Rights Movement in Flint* (Flint, MI: Flint ColorLine Project, 2007), 42.

11. On class segregation among African Americans, see Thomas J. Sugrue, *The Origins of the Urban Crisis: Race and Inequality in Postwar Detroit* (Princeton, NJ: Princeton University Press, 1996), 181–208; Pattillo-McCoy, *Black Picket Fences*.

12. *PFCC*, December 16, 1957, August 7, 1958, January 4, 1960, F-GC/FPL.

13. Part 1, box 15, folder 14, Housing-General, 1945–1946, UAWR; USBC, *United States Census of Population and Housing, 1960, Summary Population and Housing Characteristics: Flint, Michigan* (Washington, DC: USGPO, 1961), tables p-1, p-2, p-3, p-4, h-1, h-2; Annelise Orleck, *Storming Caesars Palace: How Black Mothers Fought Their Own War on Poverty* (Boston: Beacon, 2005); Felicia Kornbluh, *The Battle for Welfare Rights: Politics and Poverty in Modern America* (Philadelphia: University of Pennsylvania Press, 2007).

14. USBC, *Census of Housing, 1940, Supplement to the First Series Housing Bulletin for Flint, Michigan* (Washington, DC: USGPO, 1942), tables 1, 3; USBC, *United States Census of Population, 1950 Census Tract Statistics, Flint, Michigan, and Adjacent Area* (Washington, DC: USGPO, 1952), tables 1, 2, 3; USBC, *Census of Population and Housing, 1960*, tables p-1, p-2, p-3, p-4, h-1, h-2; Rhonda Sanders, *Bronze Pillars: An Oral History of African-Americans in Flint* (Flint, MI: *Flint Journal* and Sloan Museum, 1995); Aiyer, ed., *Telling Our Stories*.

15. Amanda I. Seligman, *Block by Block: Neighborhoods and Public Policy on Chicago's West Side* (Chicago: University of Chicago Press, 2005); Kevin M. Kruse, *White Flight: Atlanta and the Making of Modern Conservatism* (Princeton, NJ: Princeton University Press, 2005).

16. Mrs. E. Dycus to George Romney, September 19, 1963, box 12, Civil Rights—1963, A–F,

GRP; Michael W. Evanoff, *St. John St.: An Ethnic Feature of the St. John St. Community, Flint, Michigan, 1874–1974* (Flint, MI: Edelweiss, 1986), 324–25.

17. Sanders, *Bronze Pillars*, 14.

18. James Blakely, interview by Wanda Howard, August 11, 1993, BPOHPF.

19. USBC, *Census of Population and Housing, 1960*, tables h-1, h-2. According to a neighborhood survey conducted in the late 1960s, owners occupied 56 percent of the dwellings in St. John. See G. Dale Bishop and Janice K. Wilberg, "Occupied Housing Units by Tenure and Race," in *A Demographic Analysis of Neighborhood Development Project Areas: Flint, Michigan* (Flint: University of Michigan–Flint, 1972).

20. *FJ*, June 26, 1960; Harriet E. Tillock, "St. John Street Community Center," n.d., box 2, OBP.

21. William G. Grigsby, "Housing Markets and Public Policy," in *Urban Renewal: The Record and the Controversy*, ed. James Q. Wilson (Cambridge, MA: MIT Press, 1966), 24–49.

22. *FJ*, September 2, 1970.

23. Ailene Butler, "Testimony regarding Public Hearing on Equal Housing Opportunities in Flint, Michigan," November 1966, box 8, folder 21, OBP; *FJ*, May 29, 1971.

24. Olive Beasley, "Weekly Report from Flint," January 16, 1968, box 8, folder 18, OBP; Jon C. Teaford, *The Rough Road to Renaissance: Urban Revitalization in America, 1940–1985* (Baltimore: Johns Hopkins University Press, 1990), 31–32, 83–85; David Stradling, *Smokestacks and Progressives: Environmentalists, Engineers, and Air Quality in America, 1881–1951* (Baltimore: Johns Hopkins University Press, 1999); Joel A. Tarr, ed., *Devastation and Renewal: An Environmental History of Pittsburgh and Its Region* (Pittsburgh: University of Pittsburgh Press, 2003).

25. Memorandum, Thomas Kay to William Polk, n.d., CMF; *FJ*, February 24, 1966, March 4, 30, September 4, 1969.

26. NAREB, *Flint Faces the Future: An Advisory Report to the City of Flint, Michigan* (Flint, MI: NAREB, 1963), 16.

27. Thomas J. Sugrue, *Sweet Land of Liberty: The Forgotten Struggle for Civil Rights in the North* (New York: Random House, 2008), 356–448.

28. Homer Dowdy, "Big Car Called Negro Buyer's Only Outlet," *FJ*, October 3, 1952.

29. "Negroes Ask House Survey," *FJ*, November 17, 1953; "City Investigation of Flint's Bad Housing Conditions Requested," *BR*, November 21, 1953.

30. Frank J. Corbett and Arthur J. Edmunds, *The Negro Housing Market—An Untapped Resource?* (Flint, MI: ULF, 1954), v, 1–23.

31. *Housing Conditions in Flint, Michigan: Proceedings of a City-Wide Conference on Housing, November, 1955* (Flint, MI: ULF, 1955), 11–12; "Claim Negroes Not Interested in Homes," *FNA*, April 16, 1954.

32. *Housing Conditions in Flint*, 17–20.

33. Lee A. Kaake and Donn D. Parker to FCC, April 19, 1954, CMF.

34. *FJ*, December 3, 1954.

35. "Find 19 Persons Living in Basement; 14 Ousted," *FJ*, October 21, 1955.

36. *BR*, February 4, 1956.

37. *FWR*, August 23, 1956.

38. Part 1, box 76, folder 7, Urban Renewal, Cost of Slum Clearance, 1965, UAWR.

39. For more on other cities, see Irene Holliman, "From Crackertown to Model City? Urban Renewal and Community Building in Atlanta, 1963–1966," *Journal of Urban History* 35, no. 3 (March 2009): 369–86; N. D. B. Connolly, *A World More Concrete: Real Estate and the Remaking of Jim Crow South Florida* (Chicago: University of Chicago Press, 2014).

40. Dolores Ennis, interview by Wanda Howard, September 22, 1993, BPOHPF; box 6, folder 1, Civil Rights, 1953, 1955, 1957, UAWCAP.

41. *FM*, April 1964.

42. *Amplifier*, February 1962, box 1, folder 7, ULFP.

43. *FNA*, June 19, 1956; "Urban League Will Ask for Early Action on Slum Clearance," *FWR*, December 6, 1956; *Amplifier*, Winter 1957, box 1, folder 7, ULFP.

44. *FWR*, October 3, 1957.

45. SPA, *Diagnostic Survey for St. John Street, Flint, Michigan* (Flint, MI: SPA, 1969); box 8, folder 21, OBP.

46. *PFCC*, March 23, 1959, F-GC/FPL.

47. Ibid., August 8, 1955, F-GC/FPL.

48. "Flint Is Marching for Civil Rights," n.d., box 4, folder 5, EBHP; "Our Reasons for Picketing City Hall," n.d., box 4, folder 17, EBHP; *FJ*, July 7, 1967. For evidence from other cities, see Patrick D. Jones, *The Selma of the North: Civil Rights Insurgency in Milwaukee* (Cambridge, MA: Harvard University Press, 2009), chap. 2; Leonard N. Moore, *Black Rage in New Orleans: Police Brutality and African American Activism from World War II to Hurricane Katrina* (Baton Rouge: Louisiana State University Press, 2010), 1–69.

49. "Supplement to the October 15th Minutes of the Flint Study Group," n.d., box 10, 78-8.2-84b, FJMP.

50. Box 1, FEPC, 1952–1955, UAW Local 599 Collection, WPRL; Roy Reuther, box 33, folder 22, UAW Region 1C, Robert Frost, 1956, UAWPA.

51. On urban renewal, growth liberalism, and civil rights, see Harvey Molotch, "The City as a Growth Machine: Toward a Political Economy of Place," *American Sociological Review* 82, no. 2 (September 1976): 309–32; John Mollenkopf, *The Contested City* (Princeton, NJ: Princeton University Press, 1983); Teaford, *Rough Road to Renaissance*; George F. Lord and Albert Price, "Growth Ideology in a Period of Decline: Deindustrialization and Restructuring, Flint Style," *Social Problems* 39, no. 2 (May 1992): 155–69; Barbara Ferman, *Challenging the Growth Machine: Neighborhood Politics in Chicago and Pittsburgh* (Lawrence: University Press of Kansas, 1996); David Schuyler, *A City Transformed: Redevelopment, Race, and Suburbanization in Lancaster, Pennsylvania* (University Park: Pennsylvania State University Press, 2002); Robert M. Fogelson, *Downtown: Its Rise and Fall, 1880–1950* (New Haven, CT: Yale University Press, 2003); Robert O. Self, *American Babylon: Race and the Struggle for Postwar Oakland* (Princeton, NJ: Princeton University Press, 2003); Bryant Simon, *Boardwalk of Dreams: Atlantic City and the Fate of Urban America* (New York: Oxford University Press, 2004); Howard Gillette Jr., *Camden after the Fall: Decline and Renewal in a Post-Industrial City* (Philadelphia: University of Pennsylvania Press, 2006); Alison Isenberg, *Downtown America: A History of the Place and the People Who Made It* (Chicago: University of Chicago Press, 2005). On race, pollution, and urban politics, see Andrew Hurley, *Environmental Inequalities: Class, Race, and Industrial Pollution in Gary, Indiana, 1945–1980* (Chapel Hill: University of North Carolina Press, 1995). On the relationship between private enterprise and urban renewal, see Milford A. Vieser, "Urban Renewal: A Program for Private Enterprise," September 18, 1964, box 14, 78-8.2-270 c-2, FJMP; John D. Harper, "Private Enterprise's Public Responsibility," December 8, 1966, box 5, 87-12.4-14, MAFP.

52. "Segoe Planner Gives Report," *FJ*, June 10, 1959; Homer Dowdy, "St. John St. Area Has 'Top' Priority," *FJ*, June 26, 1960.

53. ULI, *A Greater Flint* (Flint, MI: ULI, n.d., ca. 1958); Larry Smith, *Flint, Michigan, Land Use and Market Absorption Capacity Study: Municipal Center GNRP* (Flint, MI: Larry Smith, 1964); Isenberg, *Downtown America*, 166–254. By the end of the 1950s, Flint's downtown ac-

counted for approximately one-third of the total retail business conducted in the metropolitan area. By contrast, downtown districts in comparably sized cities typically gained approximately 40 percent of the metropolitan retail market.

54. Ladislas Segoe and Associates, *Comprehensive Master Plan of Flint, Michigan, and Environs* (Cincinnati: Ladislas Segoe and Associates, 1960).

55. Part 4, box 21, folder 22, GLP. In 1966 city officials estimated that 89 percent of the structures in the neighborhood were "substandard." See Allan R. Wilhelm, "St. John Area Again Being Eyed for Renewal," *FJ*, December 11, 1966.

56. Olive Beasley to Brendan Sexton, August 3, 1965, box 1, OBP.

57. Community Civic League to George Romney, August 24, 1967, box 171, Department of Civil Rights, General, G–L, 1967, GRP.

58. "Flint NAACP Protests Racial Housing Ads," *BR*, April 6, 1963; "*Flint Journal* Picket Averted," *BR*, April 20, 1963; Minutes, Board of Directors, American Civil Liberties Union, Greater Flint Chapter, May 1, 1963, box 9, folder 27, EBHP.

59. "NAACP to Continue Drive for 'Occupancy' Law Here," *FJ*, October 4, 1963.

60. *PFCC*, March 14, 1955, 1059, F-GC/FPL. For more on blockbusting, see Seligman, *Block by Block*, chap. 6.

61. "Attempt Blockbusting in Evergreen," *FM*, October 1964; Memorandum, Olive Beasley to Arthur L. Johnson, January 25, 1965, box 8, folder 15, OBP.

62. Memorandum, Olive Beasley to Arthur L. Johnson, January 5, 1965; GCMPC, *Housing—Genesee County* (Flint, MI: GCMPC, 1971), table i-7.

63. On "community integrity," see Kruse, *White Flight*, esp. 79–97.

64. Memorandum, Olive Beasley to Don Holtrop, May 26, 1965, box 10, folder 36, OBP.

65. Burton I. Gordin, "Weekly Report," November 20, 1964, box 92, Civil Rights, Weekly Reports, 1964–1965, GRP.

66. Robert L. Feinberg, "Evergreen Valley Integration Smoothing," *FJ*, n.d.; DCD, *Flint: 1983 Profile* (Flint, MI: City of Flint, 1983), 8–26.

67. "NAACP Urges Negro Boycott of White Real Estate Dealers," *FJ*, October 14, 1965; Lewis Morrissey, "Enforcement of Housing Code, Realtors Hit at Rights Hearing," *FJ*, December 1, 1966.

68. *FJ*, August 26, 1966.

69. James L. Rose, "Civil Rights in Housing in Michigan," n.d., 4, box 9, folder 14, OBP.

70. Burton Levy to Burton I. Gordin, August 31, 1965, box 6, OBP; box 50, folder 29, OBP; Olive Beasley, "Weekly Report," May 4, 1966, box 8, folder 16, OBP; box 14, 95-1.1-322, ROWP. On the national scope of the problem, see David M. P. Freund, *Colored Property: State Policy and White Racial Politics in Suburban America* (Chicago: University of Chicago Press, 2007).

71. "HOME, Inc. to Intensify Operations," *Circle*, November 1965, box 1, folder 8, ULFP.

72. "30 Persons Picket Belle Meade Streets," *DI*, March 30, 1966.

73. "Davison Township, Genesee County, Michigan, Ordinance No. 27," box 10, folder 25, OBP; "Ordinance to Regulate Picketing Adopted by Davison Township; Subdivision Involved," *FJ*, May 10, 1966; "New Law to Halt Pickets," *DI*, May 11, 1966.

74. *FJ*, February 5, 1967.

75. Olive Beasley, "Weekly Report, Flint Activities Office," n.d., ca. November 17, 1964, box 8, folder 15, OBP.

76. Andrew Wiese, *Places of Their Own: African American Suburbanization in the Twentieth Century* (Chicago: University of Chicago Press, 2004).

77. George F. Plum to Burton I. Gordin, June 7, 1966, box 50, folder 29, OBP.

78. Chris Craig, "The Fight for Open Housing in Flint, Michigan," n.d., UM-F/HDSP; T. J. Bauswell, "We Won! Didn't We?" n.d., UM-F/HDSP; John Troy Williams, "Black Leadership in an Urban Setting" (PhD diss., Michigan State University, 1973), 209–45; Sanders, *Bronze Pillars*, 276–309; Aiyer, ed., *Telling Our Stories*.

79. Taeuber and Taeuber, *Negroes in Cities*; ULF, *Housing Committee Report* (Flint, MI: ULF, 1966), 1–8; box 8, folder 21, OBP; "Urban League Chief Raps Realtor Stand on 'Rights,'" *FJ*, July 28, 1966; "African Americans in Genesee County: General Historical Perspective," n.d., 6–9, box 1, OBP.

80. Olive Beasley to Arthur L. Johnson, May 7, 1965, box 8, folder 21, OBP; Memorandum, Olive Beasley to Arthur L. Johnson, August 26, 1965, box 8, folder 15, OBP; Olive Beasley, "Weekly Report," July 22, 1966, box 8, folder 16, OBP; Bruce A. Kelley to Burton I. Gordin, November 18, 1966, box 8, folder 21, OBP; "Testimony Alleging Discrimination in Housing against Gerholz Realty," November 29, 1966, box 8, folder 21, OBP; Fred Tucker, "Testimony on Drew Realty," November 29, 1966, box 8, folder 21, OBP; "Public Hearing on Equal Housing Opportunities, Flint, Michigan," n.d., box 8, folder 21, OBP; "Testimony Alleging Discrimination in Housing against General Motors Corporation," n.d., box 8, folder 21, OBP; "African Americans in Genesee County," 6–9; Charlotte Williams, interview by Wanda Howard, n.d., ca. 1993, BPOHPF; Craig, "The Fight for Open Housing in Flint, Michigan," 9.

81. Bertha R. Simms, "Testimony regarding Public Hearing on Equal Housing Opportunities, Flint, Michigan," December 1, 1966, box 10, folder 38, OBP; "Equal Housing Opportunities in Flint: Findings and Recommendations Based on a Public Hearing Conducted on November 29 and 30, and December 1 and 9, 1966," n.d., box 8, folder 21, OBP; Michael L. Kiefer, "Mack Tells Realtors Open Housing Still Isn't Reality," *FJ*, October 23, 1968; "African Americans in Genesee County," 6–9.

82. "Equal Housing Opportunities in Flint: Findings and Recommendations," 2–6; box 10, folder 38, OBP.

83. "Equal Housing Opportunities in Flint: Findings and Recommendations," 6–8.

84. Elliott Horton to Orville Pratt, February 19, 1965, box 50, folder 29, OBP; Simms, "Testimony regarding Public Hearing," 1–6; "Testimony Making Allegations against the Flint Service Office, Federal Housing Administration," December 1, 1966, box 8, folder 21, OBP.

85. Butler, "Testimony regarding Public Hearing"; "Equal Housing Opportunities in Flint: Findings and Recommendations," 8–10.

86. Butler, "Testimony regarding Public Hearing."

87. Cora Hollman, "Testimony Alleging Discrimination in Housing against Dr. J. Donald Wilson, DDS," November 1966, box 8, folder 21, OBP; "Equal Housing Opportunities in Flint: Findings and Recommendations," 8–10.

88. "Equal Housing Opportunities in Flint: Findings and Recommendations," 8–10; "Fire Department," n.d., box 8, folder 21, OBP.

89. Harold A. Draper Jr., "Statement on Equal Housing Opportunity in Genesee County," December 1, 1966, box 10, folder 38, OBP. On the suburban embrace of the de facto segregation narrative, see Matthew D. Lassiter, *The Silent Majority: Suburban Politics in the Sunbelt South* (Princeton, NJ: Princeton University Press, 2005).

90. Two exceptions were Charles Stewart Mott and Frank Manley, who testified in defense of their decision to avoid a stance on open housing. See Frank J. Manley to Mr. Damon Keith, December 1, 1966, box 10, folder 38, OBP; "Equal Housing Opportunities in Flint: Findings and Recommendations," 11–16; Lewis Morrissey, "Mott Foundation's Housing Stand Told," *FJ*, December 10, 1966.

91. Olive Beasley, "Weekly Report," November 2, 1966, box 8, folder 16, OBP; "Equal Housing Opportunities in Flint: Findings and Recommendations," 11–19; "Did Flint's Conscience Doze during Hearing on Housing?" *FJ*, December 8, 1966. On city leaders' reluctance to address open housing, see box 5, 87-12.4-13, Annual Meeting Folders, 1967, MAFP.

92. Kenneth B. Moore, "Open Occupancy Backers Make Pitch for Ordinance," *FJ*, February 14, 1967. On the composition of the city commission, see box 5, 87-12.4-13, Annual Meeting Folders, 1967, MAFP.

93. "Upset State Ruling Cited in Flint Action," *FJ*, June 19, 1965.

94. William E. Schmidt, "City Attorney Still Opposed to Housing Law," *FJ*, June 2, 1967.

95. Stephen Grant Meyer, *As Long as They Don't Move Next Door: Segregation and Racial Conflict in American Neighborhoods* (Lanham, MD: Rowman and Littlefield, 2000), 172–96; Sugrue, *Sweet Land of Liberty*, 356–448.

96. Allan R. Wilhelm, "NAACP Leader Says North Lags in Breaking Down Bias," *FJ*, April 24, 1967.

97. Marise Hadden to FCC, July 17, 1967, box 8, folder 12, OBP.

98. Sugrue, *Origins of the Urban Crisis*, 259–68; Heather Ann Thompson, *Whose Detroit? Politics, Labor, and Race in a Modern American City* (Ithaca, NY: Cornell University Press, 2001), 44–47; Kevin J. Mumford, *Newark: A History of Race, Rights, and Riots in America* (New York: New York University Press, 2007).

99. Box 319, Flint—Emergency, July 1967, GRP.

100. *FJ*, July 25–27, 1967; James W. Rutherford to Colonel Fredrick E. Davids, July 31, 1967, box 319, Miscellaneous File—General, Emergency, Riots, Etc., 1966–1967, GRP.

101. *PFCC*, July 27, 1967, 423–29, F-GC/FPL; Kenneth B. Moore, "Need for Open Housing Is Urgent, Speakers Tell City Commission," *FJ*, July 28, 1967.

102. *PFCC*, August 14, 1967, F-GC/FPL.

103. William E. Schmidt, "Open Occupancy Is Defeated; McCree Says He Will Resign," *FJ*, August 15, 1967; "Democracy for Everyone except Negro," *FJ*, August 15, 1967; Memorandum, Burton I. Gordin to MCRC, August 18, 1967, box 5, OBP; "McCree Resignation," n.d., box 21, folder 16, OBP.

104. "Flint Needs Floyd McCree and Open Occupancy Law," *FJ*, August 15, 1967.

105. William E. Schmidt, "Open Occupancy Advocates Huddle at Soggy Sleep-In," *FJ*, August 19, 1967; Woody Etherly Jr., interview by Wanda Howard, April 1, 1994, BPOHPF.

106. *FJ*, August 21, 1967.

107. "Mott Foundation Asks Open Housing," *FJ*, August 20, 1967.

108. Kenneth B. Moore, "McCree Will Remain as Mayor," *FJ*, August 28, 1967; Olive Beasley to Burton Levy, August 29, 1967, box 8, folder 17, OBP.

109. Kenneth B. Moore, "Occupancy Fight Forges New Power Bloc," *FJ*, September 6, 1967; *FJ*, September 7–October 28, 1967; *ACS*, February 9, 1968.

110. *PFCC*, October 23, 30, 1967, F-GC/FPL.

111. Kenneth B. Moore, "Referendum Sought on Open Occupancy," *FJ*, October 31, 1967.

112. *FJ*, October 17, 1967.

113. Memorandum, Olive Beasley to Burton Levy, November 13, 1967, box 8, folder 17, OBP; Kenneth B. Moore, "Petitions Filed Seeking Repeal of Open Housing," *FJ*, November 24, 1967; *FJ*, November 24–28, 1967; Olive Beasley, "Weekly Report," December 31, 1967, box 8, folder 17, OBP.

114. "Group Formed to Push Support for Fair Housing," *FJ*, January 5, 1968; Olive Beasley,

"Weekly Report," January 19, 1968, box 8, folder 18, OBP; Lou Giampetroni, "Housing Suit Pits Civil vs. Political Rights," *FJ*, January 21, 1968.

115. Beryl Satter, *Family Properties: How the Struggle over Race and Real Estate Transformed Chicago and Urban America* (New York: Metropolitan Books, 2009), 193.

116. *FJ*, February 21–28, 1968.

117. Stanley D. Brunn and Wayne L. Hoffman, "The Spatial Response of Negroes and Whites toward Open Housing: The Flint Referendum," *Annals of the Association of American Geographers* 60, no. 1 (March 1970): 18–36; MCRC, "A Summary of Municipal Fair Housing Ordinances," n.d., box 9, folder 56, EBHP.

118. "Housing Vote Is Praised by Reuther," *FJ*, February 23, 1968.

119. Allan R. Wilhelm, "Vote in Flint May Boost State Open Housing Effort," *FJ*, February 22, 1968.

120. "Flint League Chief Raps Cronin Claim," *FJ*, April 11, 1969.

121. CRCF, "Analysis of Open Housing Referendum," February 20, 1968, box 11, 95-1.1-277, ROWP; *FJ*, February 21, 1968; Brunn and Hoffman, "Spatial Response of Negroes and Whites," 18–36.

122. "Housing Referendum Proves Voting Is Vital to Negroes," *FJ*, February 27, 1968.

Chapter 7

1. *Report of the National Advisory Commission on Civil Disorders* (New York: Bantam Books, 1968); Memorandum, Olive Beasley to Richard Wilberg, June 7, 1973, box 21, folder 51, OBP.

2. Robert O. Self, *American Babylon: Race and the Struggle for Postwar Oakland* (Princeton, NJ: Princeton University Press, 2003), 135–76; Bryant Simon, *Boardwalk of Dreams: Atlantic City and the Fate of Urban America* (New York: Oxford University Press, 2004), 132–70.

3. Olive Beasley, "Weekly Activity Report," January 20, 1974, box 8, folder 20, OBP; Thomas C. Henthorn, "A Catholic Dilemma: White Flight in Northwest Flint," *Michigan Historical Review* 31, no. 2 (Fall 2005): 1–42.

4. Milford A. Vieser, "Urban Renewal: A Program for Private Enterprise," September 18, 1964, box 14, 78-8.2-270 c-2, FJMP; John D. Harper, "Private Enterprise's Public Responsibility," December 8, 1966, box 5, 87-12.4-14, MAFP; Mandi Isaacs Jackson, *Model City Blues: Urban Space and Organized Resistance in New Haven* (Philadelphia: Temple University Press, 2007).

5. On "blight," see Colin Gordon, "Blighting the Way: Urban Renewal, Economic Development, and the Elusive Definition of Blight," *Fordham Law Journal* 31, no. 2 (2004): 305–37.

6. On the purported economic benefits of freeways and interchanges, see MSHD, Office of Planning, "The New F Corners Interchange Areas," n.d., box 60, Highway Department, Mackie File, 1964, GRP.

7. NAREB, *Flint Faces the Future*, 5; part 4, box 21, folder 22, GLP.

8. E. S. Patterson to Leonard Woodcock, November 29, 1965, part 2, box 107, folder 4, UAWGM; Albert Christner to Leonard Woodcock, July 28, 1966, part 2, box 107, folder 2, UAWGM.

9. Box 1, 1956–60, RMHP; Minutes, February 7, 1962, box 2, 87-12, MAFP.

10. On the UAW's support for redevelopment, see "Statement by R. J. Thomas before the House Building and Grounds Committee on Disposition of War Housing," n.d., part 1, box 15, folder 14, Housing-General, 1945–1946, UAWR; "Annual Meeting of the Great Lakes States Industrial Development Council," January 9, 1964, part 1, box 63, folder 13, UAWR; "Statement of

Walter P. Reuther before the Subcommittee on Executive Reorganization of the Senate Committee on Government Operations, on the Problems of Cities," December 5, 1966, box 63, folder 14, Urban Renewal, 1959–1964, UAWR.

11. Many of the progrowth arguments that Flint residents offered in support of renewal were similar to those made in other cities. See Christopher Klemek, *The Transatlantic Collapse of Urban Renewal: Postwar Urbanism from New York to Berlin* (Chicago: University of Chicago Press, 2011), 48–78.

12. On the ambivalent feelings of African Americans toward St. John Street and redevelopment, see Dolores Ennis, interview by Wanda Howard, September 22, 1993, BPOHPF.

13. Box 11, folder 23, OBP. On GM's financial support for urban renewal, see box 5, 87-12.4-12, Annual Meeting Folder, 1968, MAFP.

14. Kenneth Peterson, "Mott Cooperation in Urban-Renewal Survey Is Sought," *FJ*, January 17, 1963; FRHD, *Annual Report for 1965* (Flint, MI: FRHD, 1965), 2; SPA, *Diagnostic Survey for St. John Street, Flint, Michigan* (Flint, MI: SPA, 1969); GCDC, "The Genesee Community Development Conference: A Broad Plan of Development," October 28, 1971, box 7, folder 33, EBHP.

15. MSHD, *Freeways for Flint: Proposals for Location of I-475, M-78/21 Freeways in Flint* (Lansing: MSHD, 1963), 8–24, F-GC/FPL; "A Report to the Flint City Commission on Proposed Expressway Construction," n.d., CMF.

16. MSHD, *Freeways for Flint*, 24.

17. In 1966 Leo Wilensky, Flint's urban renewal director, acknowledged that the city sought to bring whites back to the city through redevelopment. See John A. Ferris, "Staff Report and Recommendations," May 2, 1966, 17, box 10, folder 45, OBP.

18. MAF Minutes, February 7, 1962; "Manufacturers' Association Clarifies Stand on City X-Ways," *FJ*, November 28, 1962.

19. Kenneth B. Moore, "Flint Told It Must Act in 30 Days," *FJ*, January 17, 1963; "Funds Allotted to Flint X-Ways Eyed by Lansing," *FJ*, February 7, 1963.

20. FRHD, *Annual Report for 1965*, 3; Laura R. Wascha, "Public Housing Efforts in Flint Beset by Problems since 1949," *FJ*, September 27, 1970.

21. CPA, *General Neighborhood Renewal Plan for the Municipal Center of Flint, Michigan* (Flint, MI: City of Flint, 1964); *FJ*, June 20, 1964, November 19, 1972.

22. Homer Dowdy, "St. John St. Area Has 'Top' Priority," *FJ*, June 26, 1960.

23. "Need for Housing Authority Forecast," *FJ*, March 22, 1963; Kenneth B. Moore, "Public Housing Issue Is Discussed by Flint Officials," *FJ*, November 12, 1964; box 8, folder 21, OBP; SPA, *Diagnostic Survey*; Thomas Yeotis, "Projection 1969," n.d., box 15, 95-1.1-340, ROWP.

24. Robert M. Fogelson, *Downtown: Its Rise and Fall, 1880–1950* (New Haven, CT: Yale University Press, 2003), 370–80; Sylvan H. Cohen, "The Politics of Urban Renewal and the Politics of Ecology," in *The Environment: Critical Factors in Strategy Development*, ed. Richard A. Gabriel and Cohen (New York: MSS Information, 1973), 177–78.

25. Bernard H. Ross and Myron A. Levine, *Urban Politics: Power in Metropolitan America*, 7th ed. (Belmont, CA: Thomson Wadsworth, 2006), 99.

26. "Revenue Possibilities," n.d., ca. 1963, CMF.

27. Robert A. Carter to Roy Reuther, February 9, 1956, Roy Reuther, box 33, folder 22, UAWPA.

28. Rudolph H. Pallotta, "Service Demands Exceed Revenue, City Officials Say," *FJ*, August 26, 1964.

29. *FJ*, May 17, July 16, 1963, January 7, 31, February 18, April 7, August 26, 29, 31, September 3, 15, 1964, March 9, 1965.

30. FRHD, *Annual Report for 1965*, 8; *FJ*, December 11, 1966, November 19, 1972.

31. ULI, *Flint, Michigan: A Report to Greater Flint Committee* (Washington, DC: ULI, 1967). On GM's support for urban renewal, see NAREB, *Flint Faces the Future*; MSHD, *Freeways for Flint*; Peterson, "Mott Cooperation."

32. ULI, *Flint, Michigan*, 38.

33. As a strategic decision, the citizens' committee used only the term "Oak Park" in its official title, but the bond issue it endorsed was to fund renewal actions in both Oak Park and St. John. See *FJ*, February 19, 1969.

34. Citizens' Committee in Support of the Oak Park Bond Issue, "Plant Seeds of Hope in Flint," n.d., box 9, folder 31, EBHP.

35. *FJ*, February 18, March 11, 18, 25–26, 30, April 1–5, 1969.

36. See FCC, *Charter of the City of Flint, Michigan, and Rules of the City Commission* (Flint, MI: FCC, 1930), chap. 12 (55), sec. 9.

37. *PCC*, August 14, 1972, F-GC/FPL; box 9, folder 31, EBHP.

38. "The Case of the People vs. I-475—Express Ways or Walk Ways in 1990," *Sound Off*, May 1974, CDN.

39. *FJ*, November 21–22, 1967, March 10, 1968, August 27, 1972, November 11, 1973, February 26, March 19, June 30, 1974, February 23, 1975; Brian W. Rapp to William G. Milliken, March 14, 1974, part 4, box 21, folder 23, GLP; "I-475 Acceleration Task Force Report," October 25, 1974, 91.136.94, FACIF; box 8, folder 36, DWRP; box 16, 95-1.1-373, ROWP.

40. James Sharp, interview by Wanda Howard, n.d., 43, BPOHPF.

41. Donald R. Cronin to Edward N. Cole, December 18, 1968, box 9, 95-1.1-169, ROWP; "1968: Let's Face It: A Memorandum from Flint to General Motors," n.d., box 22, 78-8.3-22, FJMP; box 5, 87-12.4-11, Annual Meeting Folders, 1969–1970, MAFP.

42. Part 4, box 21, folder 22, GLP.

43. *FJ*, March 2, 1972, February 10, 1974.

44. NAREB, *Flint Faces the Future*, 13–14; Kenneth Peterson, "X-Ways Called Key to Urban Renewal," *FJ*, March 29, 1963.

45. *FJ*, March 4, 30, September 4, 1969, March 12, 1972; Michael J. Riha, "Report Reveals Flint Deficiencies: Black Neighborhoods Share Poorest Quality of Living," *FJ*, August 2, 1973.

46. ULF, "Position Paper on Martin-Jefferson Urban Renewal District," n.d., box 38, folder 11, OBP; Joseph Heathcott, "The City Quietly Remade: National Programs and Local Agendas in the Movement to Clear Slums, 1942–1952," *Journal of Urban History* 34, no. 2 (January 2008): 221–42.

47. *FJ*, December 21, 1966.

48. *PCC*, April 26, 1976, F-GC/FPL. Also, see FCPC, *Urban Renewal Development Plan: St. John Street Renewal Area, Neighborhood Development Program 6* (Flint, MI: City of Flint, n.d., ca. 1970), F-GC/FPL.

49. SPA, *Diagnostic Survey*.

50. Mrs. Everett R. Cunningham to George Romney, October 13, 1965, box 101, Highway Department, A–F, 1965, GRP.

51. William P. Walsh, *The Story of Urban Dynamics* (Flint, MI: W. P. Walsh, 1967), 25.

52. Olive Beasley, "Weekly Report," May 25, 1966, box 8, folder 16, OBP.

53. Lawrence R. Gustin, "North Siders Flock to Urban Renewal Parley," *FJ*, February 11, 1969.

54. Thomas J. Sugrue, *The Origins of the Urban Crisis: Race and Inequality in Postwar Detroit* (Princeton, NJ: Princeton University Press, 1996), 33–88; David Schuyler, *A City Transformed:*

Redevelopment, Race, and Suburbanization in Lancaster, Pennsylvania (University Park: Pennsylvania State University Press, 2002); Simon, *Boardwalk of Dreams*, 132–70; Samuel Zipp, *Manhattan Projects: The Rise and Fall of Urban Renewal in Cold War New York* (New York: Oxford University Press, 2010).

55. "Testimony regarding Public Hearing on Equal Housing Opportunities," box 8, folder 21, OBP; Lewis Morrissey, "Enforcement of Housing Code, Realtors Hit at Rights Hearing," *FJ*, December 1, 1966; *FJ*, December 11, 1966; DCD, *Annual Reports* (Flint, MI: City of Flint, 1970–72); Craig Hall, "The Origin, Rise, and Fall of the St. John Community: Memoir of an American Tragedy" (master's thesis, University of Michigan–Flint, 2001), 21–30.

56. Memorandum, Leo Wilensky to Thomas Kay, November 16, 1965, CMF; *PCC*, November 1, 1971, F-GC/FPL.

57. ULF, "Position Paper on Martin-Jefferson."

58. *PCC*, August 30, 1971, F-GC/FPL; Newsletter, DCD, September 1971, CDN; Peter F. Yost, "Renewal Area Residents Air Insurance Gripes," *FJ*, n.d., ca. October 1971.

59. Ailene Butler, "Testimony regarding Public Hearing on Equal Housing Opportunities in Flint, Michigan," November 1966, box 8, folder 21, OBP; *FJ*, May 29, 1971; Allan R. Wilhelm, "It's 300 Acres of Uncertainty," *FJ*, June 30, 1967; Janice K. Wilberg, "An Analysis of an Urban Neighborhood: St. John Street," n.d., UM-F/HDSP.

60. Newsletter, DCD, "Save Our City Program Announced," September 1973, CDN.

61. Alvin Goldberg, "Relocation Problems Tied to Changes in Neighborhood," *FJ*, December 21, 1966; *FJ*, June 30, 1967; Olive Beasley to Donald Riegle Jr., May 14, 1973, box 1, OBP.

62. Newsletter, DCD, January 1971, CDN.

63. "Meet the St. John Citizens District Council," *Sound Off*, June 1974, CDN.

64. Kenneth B. Moore, "Charges of Discrimination in Buying of Homes for Urban Renewal," *FJ*, May 24, 1966.

65. DCD, *Annual Reports*; USBC, *1970 Census of Population and Housing, Census Tracts, Flint, Michigan, Standard Metropolitan Statistical Area* (Washington, DC: USGPO, 1972).

66. Donald Riegle Jr. to Henrik Staffroth, December 5, 1967, box 7, folder 18, DWRP.

67. Olive Beasley to Arthur L. Johnson, November 18, 1965, box 8, folder 15, OBP.

68. Memorandum, GCDC to Olive Beasley, December 1, 1966, box 11, folder 25, OBP.

69. Box 1, 77-7.9-6, CSMP.

70. Memorandum, James L. Rose to Burton I. Gordin, January 25, 1965, box 11, folder 23, OBP.

71. Burton I. Gordin, "Weekly Report," November 5, 1965, box 92, Civil Rights, Weekly Reports, 1964–1965, GRP.

72. Memorandum, GCDC to MCRC, December 1, 1966, box 11, folder 25, OBP.

73. Flint Human Relations Commission, *1965 Annual Report* (Flint, MI: City of Flint, 1966).

74. Chantland Wysor to Olive Beasley, n.d., box 12, folder 37, OBP.

75. *FJ*, March 20, 22, 1963, May 6, 8, 12, July 10, 28, November 12, 1964, January 27, 1965.

76. CPA, *General Neighborhood Renewal Plan*; Ferris, "Staff Report and Recommendations," 3.

77. Ferris, "Staff Report and Recommendations," 17. Also, see Memorandum, Olive Beasley to James L. Rose, April 10, 1967, box 22, folder 17, OBP.

78. Ferris, "Staff Report and Recommendations"; Memorandum, Beasley to Rose; "Statement of Claim," February 23, 1966, box 10, folder 45, OBP; Memorandum, Olive Beasley to Arthur L. Johnson, May 24, 1966, box 10, folder 45, OBP; Beasley, "Weekly Report from Flint," April 27, 1968, box 8, folder 18, OBP.

79. Arnold R. Hirsch, *Making the Second Ghetto: Race and Housing in Chicago, 1940–1960* (Cambridge: Cambridge University Press, 1983), 212–58; Wendell Pritchett, *Brownsville, Brooklyn: Blacks, Jews, and the Changing Face of the Ghetto* (Chicago: University of Chicago Press, 2002), 97–100, 114–28.

80. Olive Beasley to Arthur L. Johnson, August 26, 1965, box 8, folder 15, OBP; Burton I. Gordin, "Weekly Report, December 18, 1964," box 92, Civil Rights, Weekly Reports, 1964–1965, GRP.

81. Memorandum, Archie LeFlore to Olive Beasley, November 21, 1977, box 22, folder 19, OBP.

82. Memorandum, Robert L. Adams to Olive Beasley, November 5, 1965, box 10, folder 47, OBP; Memorandum, Beasley to Rose, 4; *FJ*, January 11, 25, 27, February 3, May 2, 1966. On color-blind ideology, see Matthew D. Lassiter, "The Suburban Origins of 'Color-Blind' Conservatism: Middle-Class Consciousness in the Charlotte Busing Crisis," *Journal of Urban History* 30, no. 5 (May 2004): 549–82. On the metaphor of containment, see Arnold R. Hirsch, "Containment on the Home Front: Race and Federal Housing Policy from the New Deal to the Cold War," *Journal of Urban History* 26, no. 2 (January 2000): 158–89.

83. "St. John Unit Opposes Plan for Freeway," *FJ*, May 22, 1974.

84. SPA, *Diagnostic Survey*, 49; League of Women Voters of Flint, *Human Aspects of Relocation* (Flint, MI: League of Women Voters of Flint, 1972); G. Dale Bishop and Janice K. Wilberg, "Occupied Housing Units by Tenure and Race," in *A Demographic Analysis of Neighborhood Development Project Areas: Flint, Michigan* (Flint: University of Michigan–Flint, 1972).

85. Woody Etherly Jr., interview by Wanda Howard, April 1, 1994, BPOHPF.

86. Myron Orfield and Thomas Luce, *Genesee County Metropatterns: A Regional Agenda for Community and Stability in Genesee County, Michigan* (Minneapolis: Metropolitan Area Research Council, 2003); Patricia A. Baird, *Analysis of Impediments to Fair Housing in Genesee County* (Flint, MI: GCMPC, 2007).

87. Memorandum, Olive Beasley to Helen Harris, October 27, 1976, box 28, folder 13, OBP.

88. Jon C. Teaford, *The Rough Road to Renaissance: Urban Revitalization in America, 1940–1985* (Baltimore: Johns Hopkins University Press, 1990); Jackson, *Model City Blues*.

89. Amanda I. Seligman, *Block by Block: Neighborhoods and Public Policy on Chicago's West Side* (Chicago: University of Chicago Press, 2005); Kevin M. Kruse, *White Flight: Atlanta and the Making of Modern Conservatism* (Princeton, NJ: Princeton University Press, 2005).

90. "Spatial Distribution of Non-White Families in Flint, Michigan, 1960–1968," in *Diagnostic Survey*, SPA; Genesee County Health Department, *Michigan Health Survey: Fourth Baseline Report, Flint City* (Flint: Flint City Health Department and Michigan Department of Public Health, 1970); Henthorn, "Catholic Dilemma," 1–42.

91. Olive Beasley, "Weekly Report," August 19, 1965, box 8, folder 15, OBP; Olive Beasley, "Weekly Report," August 4, September 29, 1966, box 8, folder 16, OBP; Memorandum, Olive Beasley to Burton Levy, May 19, 1967, box 10, folder 41, OBP; Olive Beasley to Executive Committee, Flint Urban Coalition, July 29, 1971, box 10, folder 24, OBP.

92. Michael J. Riha, "Charge Is Made of Blockbusting," *FJ*, March 27, 1973; Michael J. Riha, "Realtor Finds No 'Blockbusting' Activity Here," *FJ*, April 5, 1973; Henthorn, "Catholic Dilemma," 1–42.

93. Michael J. Riha, "Hundreds in Anderson Area Want It Stable, Integrated," *FJ*, November 21, 1976.

94. *FJ*, June 19–20, 22, 1977; "Looks, Safety Are Priorities in Flint," *FJ*, December 27, 1981.

95. Kenneth B. Moore, "Flint Block Clubs Are Waging War," *FJ*, June 19, 1977.

96. Carol D. Rugg, "On Guard," *FJ*, June 18, 1981.

97. Moore, "Flint Block Clubs"; Kenneth B. Moore, "North Cook Efforts Are Paying Off," *FJ*, June 22, 1977.

98. Kruse, *White Flight*, 79–97.

99. Michael J. Riha, "Black Mother Says 'White Flight' Hurts," *FJ*, May 11, 1975.

100. Bruce J. Schulman, *The Seventies: The Great Shift in American Culture, Society, and Politics* (New York: Free Press, 2001), 56–58.

101. Michael J. Riha, "Fear Drives Family from Flint Home," *FJ*, May 15, 1975.

102. USBC, *1970 Census of Population and Housing*, tables p-1, p-2; ULF, *How Wide the Gap: A Report on the Conditions of Blacks in the City of Flint* (Flint, MI: ULF, n.d., ca. 1976); Gregory L. McKenzie et al., *Flint Update: 1978* (Flint, MI: DCD, 1979), 1–13; DCD, *Flint: 1983 Profile*, 8–26.

103. Kruse, *White Flight*, esp. 234–66.

104. ULF, *How Wide the Gap*.

105. R. L. Polk and Company, *Profiles of Change* (Flint, MI: R. L. Polk, 1978).

106. *FJ*, December 9, 1980.

107. Box 3, folder 35, EBHP.

108. *PCC*, April 26, 1976, F-GC/FPL; Michael J. Riha, "Most Council Members Agree That Red-Lining Exists in Flint," *FJ*, April 27, 1976; *FJ*, December 16, 1976, May 5, June 28, 1977; Flint, Michigan, Mayor's Task Force on Redlining and Disinvestment, *Redlining and Disinvestment Study and Recommendations* (Flint, MI: City of Flint, 1977); Amy C. Kantor and John D. Nystuen, "De Facto Redlining: A Geographic View," *Economic Geography* 58, no. 4 (October 1982): 309–28.

Chapter 8

1. Sidney Fine, *Expanding the Frontiers of Civil Rights: Michigan, 1948–1968* (Detroit: Wayne State University Press, 2000), chap. 15; Stephen Grant Meyer, *As Long as They Don't Move Next Door: Segregation and Racial Conflict in American Neighborhoods* (Lanham, MD: Rowman and Littlefield, 2000), 172–96; Thomas J. Sugrue, *Sweet Land of Liberty: The Forgotten Struggle for Civil Rights in the North* (New York: Random House, 2008), 356–448.

2. Michael J. Riha, "Council Begins an Inquiry into Lending Practices," *FJ*, April 15, 1976.

3. Mayor's Task Force on Redlining and Disinvestment, *Redlining and Disinvestment Study and Recommendations* (Flint, MI: City of Flint, 1977); *FJ*, December 16, 1976, May 5, June 28, 1977; Amy C. Kantor and John D. Nystuen, "De Facto Redlining: A Geographic View," *Economic Geography* 58, no. 4 (October 1982): 309–28.

4. For the text of the HUD law, see HUD Act of 1968, Pub. L. no. 90-448, 82 Stat. 476. On the Fair Housing Act of 1968, see 42 U.S.C. 3601–3619.

5. Brian D. Boyer, *Cities Destroyed for Cash: The FHA Scandal at HUD* (Chicago: Follett, 1973); R. Allen Hays, *The Federal Government and Urban Housing: Ideology and Change in Public Policy* (Albany: SUNY Press, 1985), 107–36; Jill Quadagno, *The Color of Welfare: How Racism Undermined the War on Poverty* (New York: Oxford University Press, 1994), 100–115; Kevin Fox Gotham, "Separate and Unequal: The Housing Act of 1968 and the Section 235 Program," *Sociological Forum* 15, no. 1 (2000): 13–37, and *Race, Real Estate, and Uneven Development: The Kansas City Experience, 1900–2000* (Albany: SUNY Press, 2002), 127–42.

6. HUD and FHA, *Analysis of the Flint, Michigan, Housing Market as of December 1, 1970*, n.d., box 10, folder 40, OBP.

7. Richard J. Meister, "Beecher: Coldwater Settlement to Community in Crisis," n.d., V-4, Suburbs-Beecher File, PA.

8. Tom Dinell, *The Influences of Federal, State, and Local Legislation on Residential Building in the Flint Metropolitan Area* (Ann Arbor: IHA, 1951), 22–24; Peirce Lewis, "Geography in the Politics of Flint" (PhD diss., University of Michigan, 1958), 20. On self-built housing and working-class suburbanization, see Richard Harris, *Unplanned Suburbs: Toronto's American Tragedy, 1900–1950* (Baltimore: Johns Hopkins University Press, 1996); Becky M. Nicolaides, *My Blue Heaven: Life and Politics in the Working-Class Suburbs of Los Angeles, 1920–1965* (Chicago: University of Chicago Press, 2002), 9–119; Andrew Wiese, *Places of Their Own: African American Suburbanization in the Twentieth Century* (Chicago: University of Chicago Press, 2004), 11–93.

9. On federal housing policies, redlining, and racial restrictions, see David M. P. Freund, *Colored Property: State Policy and White Racial Politics in Suburban America* (Chicago: University of Chicago Press, 2007).

10. "Report on the Impact of Section 235/236 Housing on the Beecher School District, Genesee County (Flint), Michigan," n.d., 1–5, box 2, Beecher School District, AAP.

11. Andrew R. Highsmith, "Demolition Means Progress: Race, Class, and the Deconstruction of the American Dream in Flint, Michigan" (PhD diss., University of Michigan, 2009), 386–410.

12. "Report on the Impact of Section 235/236 Housing," 6.

13. Gene Merzejewski, "Beecher Woes Now Minor Compared with Strife of '70s," *FJ*, October 5, 1984.

14. Kenneth T. Jackson, *Crabgrass Frontier: The Suburbanization of the United States* (New York: Oxford University Press, 1985), 190–218; Freund, *Colored Property*.

15. Nationwide, the median annual income for families participating in the Section 235 program was $5,760. See Anthony Valanzano to Glenn R. Davis, December 31, 1970, Subject Correspondence, 1966–1973, box 25, PRO 6-2 HOA, December 22–31, 1970, RG 207.

16. Mark L. Ladenson, "Race and Sex Discrimination in Housing: The Evidence from Probabilities of Homeownership," *Southern Economic Journal* 45, no. 2 (October 1978): 559–75; Chris Bonastia, "Hedging His Bets: Why Nixon Killed HUD's Desegregation Efforts," *Social Science History* 28, no. 1 (Spring 2004): 19–52. Specifically, Section 235 required participants to pay all real estate costs up to 20 percent of their income or all mortgage-related expenses at a special 1 percent interest rate, whichever amount was lower. The federal government covered all of the borrowers' mortgage-related expenses beyond these limits. Because participants in the program paid these costs on a sliding scale that varied depending on income and the value of the mortgage, there was no standard amount that borrowers owed.

17. John W. Finney, "$5-Billion Plan on Housing Voted by Senate, 67 to 4," *NYT*, May 29, 1968; P. N. Brownstein, "New Aids Outlined for Home Owning," *WP*, September 14, 1968; Saul Klaman, "Federal Support Threatens Private Role in Housing," *NYT*, September 20, 1970; US Civil Rights Commission, *Home Ownership for Lower Income Families* (Washington, DC: USGPO, 1971), 174; John C. Welcher, "The Paradox of Housing Costs," *WSJ*, October 26, 1977; Gotham, *Race, Real Estate, and Uneven Development*, 121–42; Marshall H. Medoff, "Race and Sex Discrimination in Housing: Comment," *Southern Economic Journal* 46, no. 3 (January 1980): 946–49; Edward G. Goetz, "Housing Dispersal Programs," *Journal of Planning Literature* 18, no. 1 (August 2003): 3–16.

18. "'Credit Crunch' Is Linked to Housing Shortage," *FJ*, November 23, 1969; Medoff, "Race and Sex Discrimination in Housing," 946–49.

19. "New Bill Stimulates Housing Construction," *Chicago Tribune*, August 13, 1968; Dave

Felton, "First in U.S. under Low-Income Plan: Watts 'Home of Hope' Built in Only 96 Hours," *Los Angeles Times*, November 13, 1968; H. Erich Heinemann, "Savings Bank Group Seeking Sponsors for Housing Projects," *NYT*, May 1, 1969; "Sivart to Hold 7th Mortgage Seminar Nov. 12," *CD*, November 8, 1969; "Realty Seminar Tonight," *CD*, January 29, 1970; "Low-Cost Seminar Set-Up," *WP*, March 14, 1970; Glenn Fowler, "Assurance Given to Homebuilders," *NYT*, January 21, 1970.

20. "Report on the Impact of Section 235/236 Housing," 11.

21. HUD and FHA, *Analysis of the Flint, Michigan, Housing Market as of December 1, 1970.*

22. Welcher, "The Paradox of Housing Costs"; Jack Rosenthal, "Romney, in Shift, Freezes Disputed Home Aid to Poor," *NYT*, January 15, 1971; John Herbers, "U.S. Now Big Landlord in Decaying Inner City," *NYT*, January 2, 1972.

23. John Herbers, "Lag in Housing Spurs a Drive to Alter Law," *NYT*, March 22, 1971.

24. FHC, *Annual Report, 1971* (Flint, MI: FHC, 1972), 8; Leon V. Whitfield, *A Study to Determine the Characteristics of Homeowners and Multifamily Developments in Default on Federally Subsidized Mortgages* (Flint, MI: GCMPC, 1974).

25. Eugene A. Gulledge to Frederick Deming, February 18, 1971, Housing Production and Mortgage Credit, Correspondence, box 2, RG 31; Peter F. Yost, "Federal Guidelines Promote Integrated Housing," *FJ*, October 17, 1971; Robert E. Mitchell and Richard A. Smith, "Race and Housing: A Review and Comments on the Content and Effects of Federal Policy," *Annals of the American Academy of Political and Social Science* 441 (January 1979): 177; Quadagno, *Color of Welfare*, 106–9; Chris Bonastia, "Why Did Affirmative Action Fail during the Nixon Era? Exploring the 'Institutional Homes' of Social Policies," *Social Problems* 47, no. 4 (November 2000): 524; Bonastia, "Hedging His Bets," 27.

26. "Civil Rights Unit Says Housing Bias Study Shows HUD Has Failed to Change Pattern," *WSJ*, June 11, 1971; Fowler, "Assurance Given to Homebuilders."

27. Whitfield, *Study to Determine.*

28. Peter Braestrup, "HUD's Biggest Housing Effort Runs into Trouble in Michigan," *WP*, February 16, 1971.

29. "Report on the Impact of Section 235/236 Housing," 7–9.

30. Harold Ford, "Beecher: Ten Years after Paul Cabell's Suicide," *FV*, June 3–16, 1982.

31. Braestrup, "HUD's Biggest Housing Effort."

32. "Report to Joint Economic Committee on Beecher School District," May 25, 1971, 1–9, box 10, folder 12, OBP.

33. "Report on the Impact of Section 235/236 Housing," 11.

34. Braestrup, "HUD's Biggest Housing Effort."

35. "Report to Joint Economic Committee," 7.

36. Gerald A. Ziegler, "Some 235 Housing Criticism Based on Bigotry, Jean Says," *FJ*, n.d., ca. 1971.

37. "Report to Joint Economic Committee," 6; "M. M. Township and City Could Get Fed. Monies," *GCH*, October 25, 1972.

38. Herbers, "Lag in Housing"; Subject Correspondence, 1966–1973, box 25, PRO 6-2 HOA, December 22–31, 1970, RG 207.

39. Mrs. Virginia Simpson to Senator Alan Cranston, August 7, 1970, Subject Correspondence, 1966–1973, box 86, PRO 6-2 HOA, November 10–17, 1970, RG 207.

40. See Richard Karp, "Painting over the Cracks: Section 235, 'The National Housing Scandal,' Is Back in Business," *Barron's National Business and Financial Weekly*, December 15, 1975.

41. "Report on the Impact of Section 235/236 Housing," 23.

42. "Audit Review of Section 235 Single Family Housing," December 10, 1971, Office of Undersecretary, Van Dusen, box 32, RG 207.

43. *PFCC*, July 23, 1970, 469, F-GC/FPL; Angela Sawyer, interview by Wanda Howard, n.d., ca. 1994, BPOHPF.

44. Meister, "Beecher," VIII–6.

45. "Riegle Claims Housing Will Take a Community Effort," *GCH*, September 27, 1972.

46. "Delinquency Rate Up on Mortgages," *NYT*, September 2, 1971; Herbers, "U.S. Now Big Landlord"; John Herbers, "Housing Reform Bill Lags as Nation's Crisis Grows," *NYT*, July 24, 1972; "Delinquency Rising on Mortgage Loans," *NYT*, November 25, 1972; Michael H. Schill and Susan M. Wachter, "The Spatial Bias of Federal Housing Law and Policy: Concentrated Poverty in Urban America," *University of Pennsylvania Law Review* 143, no. 5 (May 1995): 1313.

47. "Report to Joint Economic Committee," 8.

48. Meister, "Beecher," VIII-6.

49. US Commission on Civil Rights, *Report on the Racial and Ethnic Impact of the Section 235 Program* (Washington, DC: US Commission on Civil Rights, 1971); John Herbers, "Rights Panel Says U.S. Housing Plan Aids Segregation," *NYT*, June 11, 1971.

50. "Audit Review of Section 235 Single Family Housing."

51. Ibid., 6.

52. "Report on Audit of Section 236 Multifamily Housing Program," January 29, 1972, 68–72, box 2, 235–36 Audit Reports—January–February 1972, AAP.

53. Thomas R. Hammer, *Evaluation of Development Potentials for Metropolitan Flint, Michigan* (Evanston, IL: Northwestern University, 1986), 2.

54. Whitfield, *Study to Determine*, 1–12.

55. Howard L. Gay to Donald Riegle Jr., March 23, 1970, Subject Correspondence, 1966–1973, box 86, PRO 6-2 HOA, May 15–20, 1970, RG 207.

56. Fowler, "Assurance Given to Homebuilders."

57. Gotham, "Separate and Unequal," 22.

58. Ronald Kessler, "Patman: FHA Aids Profiteers," *WP*, July 31, 1970; "Audit Review of Section 235 Single Family Housing," 12–21; Fred Ferretti, "U.S. Looks into Profits on Homes," *NYT*, February 20, 1972; H. Erich Heinemann, "F.H.A.—from Suburb to Ghetto," *NYT*, May 7, 1972; Charles Haynes to George Romney, June 29, 1973, Subject Correspondence, box 140, PRO 6-2 HOA, 1973, RG 207; Quadagno, *Color of Welfare*, 105–6.

59. "Audit Review of Section 235 Single Family Housing," 18–21; B. E. Birkle to George Romney, November 2, 1972, Subject Correspondence, box 129, PRO 6-4 HOA, July–December 1972, RG 207; Hays, *Federal Government and Urban Housing*, 114.

60. "Report to Joint Economic Committee," 6; "Report on the Impact of Section 235/236 Housing," 17; Braestrup, "HUD's Biggest Housing Effort."

61. "Report to Joint Economic Committee," 4; "Report on the Impact of Section 235/236 Housing," 13.

62. "Report on the Impact of Section 235/236 Housing," 16; David V. Gregg to President Richard Nixon, July 6, 21, 1970, Subject Correspondence, 1966–1973, box 86, PRO 6-2 HOA, September 17–28, 1970, RG 207; Randall Coates to George Romney, January 22, 1971, box 2, Beecher School District, AAP.

63. "Report on the Impact of Section 235/236 Housing," 16.

64. Braestrup, "HUD's Biggest Housing Effort."

65. Merzejewski, "Beecher Woes Now Minor."

66. Henthorn, "Catholic Dilemma," 27.

67. Box 10, folder 24, OBP.

68. John Harvey to Randall Coates, n.d., box 4, folder 39, EBHP.

69. "Field Representative's Notes," box 10, folder 12, OBP.

70. "Court Rules Beecher Building Moratorium Illegal," *FJ*, July 14, 1970; "Township Declines Appeal of Beecher Case," *FJ*, July 30, 1970.

71. "Building Moratorium Ends, Beecher Plans for Enrollment Jump," *FJ*, n.d., ca. July 18, 1970.

72. Robert Lewis, "New '235' Housing Starts Delayed for Area Probe," *FJ*, February 9, 1971; Gerald A. Ziegler, "Beecher '235' Probe May Trigger Investigations in Other Areas," *FJ*, February 18, 1971; Gerald A. Ziegler, "Krapohl Wants FHA Housing Stalled," *FJ*, March 31, 1971.

73. Gerald A. Ziegler, "HUD Beecher Report Called Surprisingly Candid," *FJ*, May 31, 1971.

74. Fowler, "Assurance Given to Homebuilders."

75. William C. Whitbeck to Edgar B. Holt, October 7, 1971, box 4, folder 39, EBHP.

76. George Romney to Randall Coates, May 25, 1971, box 10, folder 12, OBP; Ziegler, "HUD Beecher Report Called Surprisingly Candid."

77. Romney to Coates, 1. Something similar happened in 1970 in Chicago's south suburbs. See "HUD Halts Suburban Housing; Blames Segregation," *NYT*, July 4, 1970.

78. Harold Ford, "Paul Cabell: A Man in the Middle," *FV*, June 3, 1982.

79. "Beecher: Ten Years after Paul Cabell's Suicide," *FV*, June 3, 1982; "Death of the Middleman," *Time*, March 20, 1972.

80. Loudon Wainwright, "The Man in the Middle," *Life*, July 21, 1972, 55–66; Tom Perry, "Tragedy of Paul Cavell [*sic*] Death Is Indictment of Community . . . If It Forgets," *FS*, March 11, 1972; Timothy Penn, "Beecher Remembers the Goals of Paul Cabell," *FJ*, May 5, 1972; "Frankness on the Race Issue," *CSM*, March 10, 1972.

81. "Black Majority Elected in Beecher," *FTB*, August 18, 1977.

82. Leon V. Whitfield, *Housing Market Analysis and Feasibility Study* (Flint, MI: Genesee County Model Cities Development Corporation, 1972), IV-4; Ford, "Beecher: Ten Years after Paul Cabell's Suicide." For a more thorough articulation of this historiographical point, see Andrew R. Highsmith, "Prelude to the Subprime Crash: Beecher, Michigan, and the Origins of the Suburban Crisis," *Journal of Policy History* 24, no. 4 (October 2012): 572–611.

83. Braestrup, "HUD's Biggest Housing Effort."

84. Memorandum, Thomas M. Hutchinson to William C. Whitbeck, March 3, 1971, box 2, Beecher School District, AAP.

85. Meister, "Beecher," VII-2; "Report on the Impact of Section 235/236 Housing," 30; Betty Layton to Jack Brooks, November 6, 1970, Subject Correspondence, 1966–1973, box 86, PRO 6-2 HOA, November 6–9, 1970, RG 207.

86. "Report on the Impact of Section 235/236 Housing," 35.

87. "Report to Joint Economic Committee," 6.

88. "M. M. Township and City Could Get Fed. Monies."

89. Braestrup, "HUD's Biggest Housing Effort"; Gene Merzejewski, "School Chief Says Beecher's Woes Today Financial, Not Racial," *FJ*, October 5, 1984.

90. This was not unique to Flint. See Bonastia, "Hedging His Bets," 19–52.

91. *FI*, April 15, 1971.

92. *DI*, May 20, 1970.

93. *FJ*, September 29–30, 1971; *DI*, October 6, 1971.

94. "Moore, Alger Win Seats on School Board," *DI*, June 14, 1972.

95. *DI*, November 14, December 12, 1973, March 27, September 18, 1974; Ann C. Emmons, "A 'Sore Thumb' Reminisces," *FJ*, July 18, 1976; Michael Moore, *Here Comes Trouble: Stories from My Life* (New York: Grand Central, 2011), 247–302.

96. Howard Gillette Jr., *Camden after the Fall: Decline and Renewal in a Post-Industrial City* (Philadelphia: University of Pennsylvania Press, 2006), 169–90; Freund, *Colored Property*.

97. *GBN*, April 20, 1972; Bryan L. Steffens, "Vienna Township Raises Minimum House Size," *FJ*, January 19, 1973; *SCN*, March 1, 1973.

98. *FJ*, December 6, 1970.

99. Whitfield, *Study to Determine*.

100. Harold Black and Aaron Blumberg, "Housing Plan: Refinement of Assignments and Initial Portion of Implementation Methodology," December 1972, box 15, folder 22, OBP; Harold Black and Aaron Blumberg, "Implementation Methodology: P-330, Working Paper No. 1," January 1973, box 8, folder 12, EBHP; Harold Black and Aaron Blumberg, "Working Paper No. 2, n.d., ca. 1973, box 8, folder 11, EBHP.

101. Dennis F. Herrick, "Zaccaria against County Housing Plan," *FJ*, September 20, 1973.

102. "Subsidized Housing Is Recommended," *GBN*, March 15, 1973.

103. Paul Siegel, "Township Passes Housing Code," *GBN*, May 16, 1974; Olive Beasley, "Weekly Report," October 6, 1966, box 8, folder 16, OBP; Chris Craig, "The Fight for Open Housing in Flint, Michigan," 10–12, UM-F/HDSP.

104. "Gaines, Argentine, Form Housing Authorities," *SCN*, July 11, 1974; "Mundy Twp. Approves Housing Ordinance," *SCN*, July 25, 1974.

105. "Davison Twp. Forms Housing Commission," *FJ*, October 20, 1974; Ed Backus, "Official Support Is Lacking on County Public Housing," *FJ*, November 30, 1972; "Will Townships Form Housing Commissions?" *FJ*, May 14, 1974.

106. RVPDC, *Subsidized Rental and Housing Rehabilitation Programs Available in Region V* (Flint, MI: RVPDC, 1980); Whitfield, *Study to Determine*.

107. Joe T. Darden and Joshua G. Bagakas, *Analysis of Impediments to Fair Housing in Genesee County including the City of Flint: A Report Submitted to Genesee County and City of Flint* (Flint, MI: Genesee County and City of Flint, 1997), 24; Patricia A. Baird, *Analysis of Impediments to Fair Housing in Genesee County* (Flint, MI: GCMPC, 2007). For evidence from elsewhere, see Michael N. Danielson, *The Politics of Exclusion* (New York: Columbia University Press, 1976); Jackson, *Crabgrass Frontier*, 219–30; David L. Kirp, John P. Dwyer, and Larry A. Rosenthal, *Our Town: Race, Housing, and the Soul of Suburbia* (New Brunswick, NJ: Rutgers University Press, 1997); Gillette, *Camden after the Fall*, 169–90; Freund, *Colored Property*, 243–381.

Chapter 9

1. Kenneth A. Mines to Peter L. Clancy, August 29, 1975, box 16, folder 9, OBP.

2. Ibid.

3. Thomas J. Sugrue, *Sweet Land of Liberty: The Forgotten Struggle for Civil Rights in the North* (New York: Random House, 2008), 163–99.

4. *Taylor v. New Rochelle Board of Education*, 288 F.2d 600 (1961); *Bell v. School Board, City of Gary*, 324 F.2d 209 (1963); *Webb v. Board of Education of the City of Chicago*, 223 F. Supp. 466 (1963).

5. Olive Beasley to Clarence Wood, March 29, 1976, box 38, folder 8, OBP; Davison M. Douglas, *Jim Crow Moves North: The Battle over Northern School Desegregation, 1865–1954* (New

York: Cambridge University Press, 2005), 167–218; Rita A. Scott, "The Status of Equal Opportunity in Michigan's Public Schools," n.d., 11, box 9, folder 57, EBHP; "Report Criticizes Mott Program, Educational System, Community Ed. Program in Schools," *FS*, March 6, 1971.

6. Flint Citizens Information Committee to NAACP Civil Rights Committee, February 1964, box 2, folder 47, EBHP.

7. *OMFBE*, December 9, 1964, F-GC/FPL.

8. William S. Price III, "Implications in U.S. Supreme Court Ruling on Segregation in Public Schools," *Amplifier*, n.d., ca. 1955.

9. Darcy DeMille, "NAACP National Officer Confers with Flint School Officials," *FM*, October 1964.

10. Matthew D. Lassiter, "De Jure / De Facto Segregation: The Long Shadow of a National Myth," in *The Myth of Southern Exceptionalism*, ed. M. D. Lassiter and Joseph Crespino (New York: Oxford University Press, 2010), 33.

11. Donald Riegle Jr. to Elizabeth Coy, April 7, 1970, box 14, folder 6, DWRP.

12. On the multiple meanings and valences of de facto segregation, see Andrew R. Highsmith and Ansley T. Erickson, "The Strange Career of De Facto Segregation: Race and Region in the Scholarly Imagination," unpublished manuscript, October 1, 2014.

13. *FJ*, March 6, 1969.

14. "Flint NAACP May Force School Busing Issue," *FS*, May 1, 1971; Michael Schwartz, "NAACP Mulls Possibility of Suit against Ed. Board," *FJ*, September 15, 1971.

15. Ernest Holsendolph, "HEW Chief Sued on Segregation," *NYT*, July 4, 1975; "Sirica Gives HEW Order on Schools," *NYT*, July 22, 1976.

16. *Swann v. Charlotte-Mecklenburg Board of Education*, 402 U.S. 1 (1971).

17. *Keyes v. School District No. 1, Denver, Colorado*, 413 U.S. 189 (1973).

18. Jeffrey Mirel, *The Rise and Fall of an Urban School System: Detroit, 1907–1981* (Ann Arbor: University of Michigan Press, 1993), 293–398.

19. *Davis v. School District of the City of Pontiac*, 309 F. Supp. 734 (1970).

20. Sugrue, *Sweet Land of Liberty*, 478–80.

21. Michael J. Riha, "500 Attend Antibusing Meeting of SOS," *FJ*, September 15, 1971.

22. Gary Orfield, *Must We Bus? Segregated Schools and National Policy* (Washington, DC: Brookings Institution, 1978); Jennifer L. Hochschild, *The New American Dilemma: Liberal Democracy and School Desegregation* (New Haven, CT: Yale University Press, 1984). Both Orfield and Hochschild have found that school desegregation is most successful when it occurs on a metropolitan scale. They argue that because metropolitan initiatives maintain black enrollments at lower levels than city-only plans, whites are less likely to flee as the geographic scope of desegregation widens.

23. Sandra Faye Urquhart Brown, "Taxpayer and Student Equity in Twenty-One Selected School Districts in Genesee County, Michigan, 1973–1983" (PhD diss., University of Michigan, 1985).

24. *Bradley v. Milliken*, 338 F. Supp. 582 (1971).

25. *FJ*, September 15, 21, December 8, 1971, January 18, February 16, 21, 1972.

26. *Milliken v. Bradley*, 418 U.S. 717 (1974); Mirel, *Rise and Fall of an Urban School System*, 293–398.

27. *OMFBE*, September 17, 1975, 118a, F-GC/FPL; Jennifer G. Sweet, "Compromising between Equity, Choice, and Quality: Voluntary Magnet School Policy in Flint and Grand Rapids" (master's thesis, University of Michigan, 1993).

28. "An Open Letter to the Flint Community from the Flint Board of Education," *FJ*, November 3, 1975.

29. Olive Beasley, "Salient Community Issues for Commission Visit," November 24, 1975, box 6, OBP.

30. Craig Carter, "Flint Plan for Desegregation 'Not Acceptable,' HEW Says," *FJ*, December 24, 1975.

31. Ed Hayman, "11 School Councils Report at Hearing on Desegregation," *FJ*, January 20, 1976; Sweet, "Compromising between Equity, Choice, and Quality," 27.

32. Craig Carter, "Mott Foundation Backs School Desegregation," *FJ*, January 29, 1976.

33. Box 48, folder 30, OBP.

34. Judy Samelson, "NAACP's Busing Demand Extended to All of County," *FJ*, January 18, 1976; Olive Beasley to Althea Simmons, September 10, 1976, box 40, folder 23, OBP.

35. *OMFBE*, March 10, 1976, 211a–c, F-GC/FPL.

36. Sweet, "Compromising between Equity, Choice, and Quality," 29. For teachers and other employees, the board adopted the so-called Singleton Rule, which mandated that each school's staff should reflect the racial composition of the entire district within five percentage points.

37. Craig Carter, "Brannon: Board Overly Optimistic on 'Option' Plan," *FJ*, March 18, 1976.

38. Craig Carter, "HEW Rejects Flint's Plan for Schools," *FJ*, May 26, 1976.

39. "Memorandum Re: Proposed Additional Input for Consensus Statement," March 22, 1976, box 36, folder 6, OBP.

40. *Holman v. School District of the City of Flint* (1976); Ed Hayman, "Suit Demands Cross-District Busing in 21 Area Districts," *FJ*, July 21, 1976.

41. "Fact Sheet, *Holman et al. v. Flint Community Schools et al.*," n.d., box 29, folder 22, OBP.

42. Daryl Michael Scott, *Contempt and Pity: Social Policy and the Image of the Damaged Black Psyche, 1880–1996* (Chapel Hill: University of North Carolina Press, 1997).

43. "Deposition of Edward R. Hintz, July 13, 1977," *Holman, v. School District of the City of Flint* (1976).

44. *Holman v. School District of the City of Flint* (1976).

45. Ibid.

46. Ernest Holman, interview by Andrew R. Highsmith, March 1, 2006.

47. Betsy Anderson, "Cross-District Busing Lawsuit Inching Along," *FJ*, February 5, 1980.

48. "Flint's Legal Battles Took Years to End," *FJ*, October 14, 1984.

49. Betsy Anderson, "Suit Seeking Flint-Suburban Busing Dismissed," *FJ*, August 27, 1980.

50. CSMF, Board of Trustees, "A Long, Fruitful Relationship: Position Paper on Mott Foundation Grants to Flint Board of Education," April 21, 1977.

51. Ibid., 2–3.

52. *FJ*, July 5–6, 10, 1977.

53. "Model Use of Money," *Time*, April 12, 1968.

54. Charles Stewart Mott to James O. Eastland, February 2, 1956, box 29, 77-7.6-1.6, CSMP.

55. Charles Stewart Mott to Genesee County Taxpayers Association, n.d., ca. 1945, box 18, 77-7.1-60, CSMP.

56. Box 19, 77-7.8-12.4, CSMP. Wallace was a proponent of Mott's community education program.

57. Box 52, 78-8.7-1, FJMP.

58. Robert Lewis, "Flint Told: Submit School Busing Plan," *FJ*, May 9, 1978.

59. Drew S. Days to Olive Beasley, December 8, 1978, box 12, folder 7, EBHP.

60. *FJ*, May 11, 1978.

61. The drop in enrollments actually began in 1966, well before the HEW proceedings commenced. See "School Enrollment Expected to Drop in Flint for 5th Year," *FJ*, August 25, 1972; *PCC*, March 29, 1976, 116, F-GC/FPL.

62. Olive Beasley to Althea Simmons, May 15, 1978, box 40, folder 22, OBP.

63. Craig Carter, "Flint May Repeat Bid for 34 Mills to Run Schools," *FJ*, June 15, 1978.

64. Olive Beasley to William H. Oliver, August 14, 1978, box 12, folder 36, OBP.

65. Craig Carter, "Flint School and Library Taxes OK'd," *FJ*, August 9, 1978.

66. *OMFBE*, April 28, 1980, 291, F-GC/FPL.

67. As part of the consent decree, the board of education also agreed to forbid student transfers that undermined racial balances at magnet schools. See *FJ*, May 1, 5, 1980.

68. Betsy Anderson, "Desegregation Pact: Now, to Implement It," *FJ*, April 29, 1980; Sweet, "Compromising between Equity, Choice, and Quality," 29.

69. *FJ*, May 5, 1980.

70. Betsy Anderson, "Some Protest Settlement of Desegregation Suit," *FJ*, May 1, 1980.

71. David Fenech, "Nine Hired to Promote Racial Plan in Schools," *FJ*, January 8, 1981.

72. Anderson, "Desegregation Pact"; *FJ*, May 1–5, 1980.

73. Betsy Anderson, "2 Mostly Black Schools May Miss Enrollment Goals," *FJ*, September 29, 1981.

74. Betsy Anderson, "Federal Money to Aid End of School Bias Is Cut," *FJ*, August 21, 1981.

75. Fenech, "Nine Hired"; Sweet, "Compromising between Equity, Choice, and Quality," 35–40.

76. On the national trend, see Gary Orfield, Susan E. Eaton, and the Harvard Project on School Desegregation, *Dismantling Desegregation: The Quiet Reversal of Brown v. Board of Education* (New York: New Press, 1997).

77. Marcia McDonald, "Busing Yields Mixed Results," *FJ*, January 12, 1992.

78. Michael J. Riha, "The Mott Foundation Impact: A Tour Helped Shift the Focus in Flint," *FJ*, June 13, 1982.

79. I borrow the "place over people" framework from Bruce Schulman. See Bruce J. Schulman, *From Cotton Belt to Sunbelt: Federal Policy, Economic Development, and the Transformation of the South* (New York: Oxford University Press, 1991).

Chapter 10

1. Ronald Edsforth, *Class Conflict and Cultural Consensus: The Making of a Mass Consumer Society in Flint, Michigan* (New Brunswick, NJ: Rutgers University Press, 1987), 224; John Greenwald, "How GM Broke Down," *Time*, November 9, 1992; Steven P. Dandaneau, *A Town Abandoned: Flint, Michigan, Confronts Deindustrialization* (Albany: SUNY Press, 1996), 8.

2. Judith Stein, *Pivotal Decade: How the United States Traded Factories for Finance in the Seventies* (New Haven, CT: Yale University Press, 2010); Jefferson Cowie, *Stayin' Alive: The 1970s and the Last Days of the Working Class* (New York: New Press, 2010).

3. "Ripples of Strike Spreading Farther," *FJ*, October 30, 1970; William Serrin, *The Company and the Union: The "Civilized Relationship" of the General Motors Corporation and the United Automobile Workers* (New York: Random House, 1973).

4. John A. Larson, Tevfik F. Nas, and Charles T. Weber, *Methodology for the Study of Urban*

Economic Development in Flint (Flint, MI: PURA, 1986), 14; "Flint, the Central Business District: A Report," n.d., box 1, folder 10, DCD Files, GHCC.

5. *FJ*, June 25, 1972, June 18, 1978.

6. Ed Cray, *Chrome Colossus: General Motors and Its Times* (New York: McGraw-Hill, 1980); Bryan D. Jones and Lynn W. Bachelor, *The Sustaining Hand: Community Leadership and Corporate Power* (Lawrence: University Press of Kansas, 1986), 57; Bruce J. Schulman, *The Seventies: The Great Shift in American Culture, Society, and Politics* (New York: Free Press, 2001); Karen R. Merrill, *The Oil Crisis of 1973–1974: A Brief History with Documents* (Boston: Bedford Books of St. Martin's, 2007).

7. *General Motors Corporation Annual Report for 1975* (Detroit: GM, 1976), 4, F-GC/FPL.

8. Barry Bluestone and Bennett Harrison, *The Deindustrialization of America: Plant Closings, Community Abandonment, and the Dismantling of Basic Industry* (New York: Basic Books, 1982).

9. Lawrence R. Gustin, "Elges Predicts Tough Times, Then Products Creating Jobs," *FJ*, January 5, 1974.

10. Lawrence R. Gustin, "GM Layoffs to Affect 15,300 Here," *FJ*, February 15, 1974.

11. Lawrence R. Gustin, "Estes: GM Outlays Cushion Area Layoffs," *FJ*, December 13, 1974; Thomas R. Hammer, *Evaluation of Development Potentials for Metropolitan Flint, Michigan* (Evanston, IL: Northwestern University, 1986), 2.

12. MESC, *Annual Planning Report for Flint SMSA Fiscal Year 1977* (Lansing: MESC, 1978), LM.

13. David Vizard, "Half of Flint's Black Teens Believed Jobless," *FJ*, March 22, 1979.

14. Box 5, folder 9, EBHP; *Flint, Michigan, 1977–78: Jobless Persons—Heads of Household* (Flint, MI: R. L. Polk, 1978); *FJ*, March 18, 22, 1979.

15. Colin Gordon, *Mapping Decline: St. Louis and the Fate of the American City* (Philadelphia: University of Pennsylvania Press, 2008), 181–219.

16. CRCM, *Municipal Government Economic Development Incentive Programs in Michigan* (Detroit: CRCM, 1986); SEMCOG, *Tax Incentives / Tax Abatements in Southeast Michigan* (Detroit: SEMCOG, 1990); Anthony Lupo II, "Local Development and Tax Abatements" (master's thesis, University of Michigan–Flint, 2000); Greg LeRoy, *The Great American Jobs Scam: Corporate Tax Dodging and the Myth of Job Creation* (San Francisco: Berrett-Koehler, 2005).

17. Michael Moore, "GM Demands Their Taxes to Be Cut in Half . . . or Else," *FV*, December 1978; Lupo, "Local Development and Tax Abatements." By 1993 the city of Flint had granted thirty-seven of these tax cuts to General Motors. See Theodore J. Gilman, "Urban Redevelopment in Omuta, Japan, and Flint, Michigan: A Comparison," in *The Japanese City*, ed. P. P. Karan and Kristin Stapleton (Lexington: University Press of Kentucky, 1997).

18. Tim Retzloff, "Moore to Flint? Suburban Omission in *Roger and Me*," unpublished essay, n.d., ca. 2003.

19. *PCC*, March 10, 1986, 62, F-GC/FPL.

20. Barry Wolf, "Beware Sign Should Take Place of Welcome to GM," *FJ*, January 28, 1981.

21. *CERP Technology Letter*, March 1982, box 1, folder 2, CERP Monthly Technology Letter, 1981–1982, MWP.

22. *PCC*, December 18, 1980, 697–700, F-GC/FPL; Michael Moore, "Flint and Me," *Money*, July 7, 1996, 86–87.

23. *PCC*, January 8, 1979, 6, F-GC/FPL.

24. Gordon, *Mapping Decline*, 181–219; Laura A. Reese and Gary Sands, *Money for Nothing: Industrial Tax Abatements and Economic Development* (Lanham, MD: Lexington Books, 2012).

25. *General Motors Annual Report: 1978* (Detroit: GM, 1979), F-GC/FPL.

26. Timothy Penn, "Flint Is Leading GM Innovation, Official Says," *FJ*, October 28, 1977.

27. Mary Jane Bolle, "Overview of Plant Closing and Industrial Migration Issues in the United States," 1980, part 2, box 34, folder 20, Plant Closings, 1975–1980, UAWR; "Recent Research on Plant Closings, Hon. William D. Ford of Michigan," May 8, 1984, part 2, box 33, folder 13, Plant Closings, 1984, UAWR.

28. *General Motors Annual Report: 1980* (Detroit: GM, 1981), F-GC/FPL.

29. *CERP Technology Letter*, March 1982.

30. *General Motors Annual Report: 1982* (Detroit: GM, 1983), F-GC/FPL; Richard C. Noble, "GM Chief Says 'Megajobs' Could Be Lost," *FJ*, November 25, 1981.

31. Albert Lee, *Call Me Roger* (Chicago: Contemporary Books, 1988); Maryann Keller, *Rude Awakening: The Rise, Fall, and Struggle for Recovery of General Motors* (New York: Harper Perennial, 1989). Between 1977 and 1981, GM's use of computerized technology increased by 500 percent. For more on technology and automation at GM, see Harley Shaiken, "Detroit Downsizes U.S. Jobs," *Nation*, October 11, 1980; box 1, folder 1, MWP.

32. Box 53, folder 26, OBP; *FJ*, April 11, July 7, August 21, September 17, October 3, 15, 26, November 9, December 7, 1978, February 11, April 11, May 20, July 29, 1979, August 28, September 17, 24, November 5, 1980, April 1, 26–27, May 21, 1981; "The Jarvis-Headlee-Tisch Con Job," *FV*, September 1978; Peter Schrag, *Paradise Lost: California's Experience, America's Future* (Berkeley: University of California Press, 1999); Lisa McGirr, *Suburban Warriors: The Origins of the New American Right* (Princeton, NJ: Princeton University Press, 2001), 238–39; Robert O. Self, *American Babylon: Race and the Struggle for Postwar Oakland* (Princeton, NJ: Princeton University Press, 2003). The Headlee legislation established a strict cap on state taxes and required local governments to obtain voter approval for new levies. Yet it also forbade state officials from cutting outlays to local governments below 1978 levels.

33. "Referendums: Rising Impatience," *Time*, November 17, 1980; "Proposal A—Michigan Property and Sales Taxes, Adopted March 15, 1994," http://www.lib.umich.edu/govdocs/propa.html; Lawrence W. Reed, "The Headlee Amendment: Serving Michigan for 25 Years," August 4, 2003, http://www.mackinac.org/article.aspx?ID=5574.

34. Julie Greenwalt, "Auto Worker Dean Hazel Leads a Tax Revolt, but the IRS Is Convinced It Has His Number," *People*, April 6, 1981; Bob Sherefkin, "We the People Wane: Tax-Protest Movement Losing Steam since April 15," *FJ*, April 27, 1981.

35. Many chroniclers of the 1970s tax revolt have focused on white suburban homeowners as the leaders of the antitax movement. Within this body of literature, corporate antitax campaigns often receive short shrift. See, for instance, Schrag, *Paradise Lost*; Self, *American Babylon*. On American business leaders' support for tax cuts in the 1970s and 1980s, see Benjamin C. Waterhouse, *Lobbying America: The Politics of Business from Nixon to NAFTA* (Princeton, NJ: Princeton University Press, 2014).

36. "The Westfall Awareness Papers," box 1, folder 10, MWF; Dawson Bell, "GM Begins Drive to Ease Tax Bite," *DFP*, March 22, 1985.

37. Mike Westfall, "Overestimating Wages: Is GM Covering Up Something?" *652 Technogram*, February 1982; Mike Westfall, "Things Changing in Flint—and Not Always for Better," *FJ*, July 7, 1985; "The Westfall Awareness Papers: The Ralph Nader Interview," July 1985, box 1, folder 8, MWP; "Sharp Stands His Ground on Tax Fight," *Eye Opener*, July 28, 1987.

38. *PCC*, May 25, 1988, 167–68, F-GC/FPL.

39. Dawson Bell, "GM's Battle for Lower Taxes Grinds to an End," *DFP*, February 14, 1992; Deborah L. Cherry, "Genesee County's Struggle to Survive" (master's thesis, University of Michigan–Flint, 1993).

40. Mike Davis, *Prisoners of the American Dream: Politics and Economy in the History of the US Working Class* (London: Verso, 1986).

41. *FJ*, March 22, 1979; Michael B. Amspaugh, "Two Economic Organizations Making Plans to Disband," *FJ*, August 12, 1987; Robin Widgery, *Shifts in the Flint Area Economic Structure during the 1980s* (Flint, MI: PURA, 1991).

42. *CERP Technology Letter*, March 1982.

43. USBC, *1990 Census of Population: Social and Economic Characteristics; Michigan* (Washington, DC: USGPO, 1990), tables 1–3, 5, 9, 13, 173, http://www.census.gov/prod/cen1990/cp2/cp-2-24-1.pdf. On the feminization of poverty, see Diana Pearce, "The Feminization of Poverty: Women, Work, and Welfare," *Urban and Social Change Review* 11 (1978): 28–36; Annelise Orleck, *Storming Caesars Palace: How Black Mothers Fought Their Own War on Poverty* (Boston: Beacon, 2005); Gertrude Schaffner Goldberg, "Feminization of Poverty in the United States: Any Surprises?" in *Poor Women in Rich Countries: The Feminization of Poverty over the Life Course*, ed. Goldberg (New York: Oxford University Press, 2009), 230–65.

44. Thomas Sedgewick, *Study for Possible Utilization of Air Rights above Freeways* (Flint, MI: Sedgewick, Sellers, and Associates, 1965), 1–11. On the impact of freeways, see John F. C. DiMento and Cliff Ellis, *Changing Lanes: Visions and Histories of Urban Freeways* (Cambridge, MA: MIT Press, 2013), 143–208.

45. Angela Sawyer, interview by Wanda Howard, n.d., ca. 1994, BPOHPF. In Floral Park the freeway and interchange formed a large barrier between the FHC's Howard Estates project and the downtown and cultural districts. See Memorandum, Olive Beasley to Arthur L. Johnson, August 16, 1966, box 11, folder 21, OBP; box 12, folder 37, OBP; "City Withdrawing Support for Change in I-475 Route," *FJ*, March 26, 1971; *FJ*, October 26, 1971; Timothy Penn, "Hello Freeway, Goodby[e] Customers," *FJ*, September 4, 1977.

46. US Department of Transportation, FHA, and MSHD, *Environmental Section 4[f] Statement Corridor/Adjustment, I-475 and Pierson Road, Genesee County, Michigan*, 1974, part 4, box 21, folder 22, Department of State Highways and Transportation, 1974, GLP.

47. See "Councils Push to Move I-475," *Model Cities News*, July–August 1971, BHL.

48. Boxes 8–9, Thomas W. Stephens Papers, WPRL.

49. Memorandum, Dan McRill to Members of the GCMPC, May 21, 1976, box 15, folder 29, OBP.

50. Michael Moore, "General Motors Pulls Out: In Flint, Tough Times Last," *Nation*, June 6, 1987.

51. "Flint," *CERP Technology Letter*, July 1983, box 1, folder 7, MWP; Jones and Bachelor, *Sustaining Hand*; Theodore J. Gilman, *No Miracles Here: Fighting Urban Decline in Japan and the United States* (Albany: SUNY Press, 2001), 137–67.

52. Japan's "Toyota City" was the model for Flint's Buick City development. See Alan Lenhoff, "Japan Now Teaches the Teacher," *DFP*, October 22, 1980; John Koten, "GM Plans to Rebuild Flint, Mich., Plants As a Car Factory in Style of 'Toyota City,'" *WSJ*, January 20, 1983; *FJ*, February 27, June 14, July 12, 1983, January 25, May 14, 25, July 6, November 22, 1984, August 1, 21, February 12, September 17, 1985, November 6–7, 1986; John Holusha, "New Ways at 2 G.M. Plants," *NYT*, April 10, 1984; Dale Buss, "GM Gears Up Buick City in Its Biggest Effort to Cut Costs, Boost Efficiency at Older Site," *WSJ*, February 21, 1985.

53. Mike Westfall, "Restructuring Strategies in the Auto Industry and Blue Collar Survival," n.d., box 1, folder 5, Letters and Speeches, MWP.

54. Arthur M. Spinella, "Buick City: Hawk Bares Talons," *WAW*, June 1983; Paul Eisenstein, "Aphrodisia for the Old Shop," *Monthly Detroit*, August 1983.

55. Spinella, "Buick City," 22.

56. "Buick City: A Re-awakening," *Inside Buick*, December 1984, 10; Sharon Dunn, "B-O-C 'Heart' Pumps Energy," *Inside Buick*, Spring 1986, 14; "New G.M. Layoffs Announced," *NYT*, November 9, 1986.

57. Moore, "General Motors Pulls Out"; *Genesee County Market Profile* (Flint, MI: *Flint Journal*, 1990).

58. Michael Moore, "The Houston Post—Flint's Fastest Growing Newspaper?" *FV*, March 20, 1981. See also "He's Going to Houston to Look for Open Doors," *FJ*, February 5, 1981; David B. Crary and Carol Hogan, *Willow Run and Related Plant Closings: Causes and Impacts* (Ypsilanti: Institute for Community and Regional Development, Eastern Michigan University, 1992), 3-31; Lynn Ryan Mackenzie, "Capitalism, Power, and Community Well-Being: Developing a Model for Understanding the Effect of Local Economic Configuration on Cities and Their Citizens" (PhD diss., Cornell University, 2005).

59. Alex Kotlowitz, "The Exodus Has Begun," *FV*, February 6-19, 1981.

60. *CERP Technology Letter*, March 1982.

61. Neal R. Peirce, "Will Mott Foundation Save Flint's Ravaged Inner-City?" *Detroit News*, April 28, 1983.

62. Rick Matthews, "What's Good for GM: Deindustrialization and Crime in Four Michigan Cities, 1975-1993" (PhD diss., Western Michigan University, 1997).

63. Alex Kotlowitz, "Life in the Slow Lane: Surviving in the Unemployment Capitol," *MV*, December 1983.

64. Michael Moore, "Dance Band on the *Titanic*: AutoWorld and the Death of Flint," *MV*, September 1984.

65. *FJ*, February 17, 1970, February 2, 14, June 6, 1971; Charles Stewart Mott to Gerald F. Healy, March 11, 1970, box 2, folder 37, GFHP; Lawrence R. Gustin, "Big Plans on the Horizon for Central Flint," *FJ*, June 8, 1971; Harding Mott to Ronald O. Warner, n.d., box 16, 95-1.1-373, ROWP; "Specific Problem Areas: Tippy Dam," n.d., box 16, 95-1.1-373, ROWP; Gilman, "Urban Redevelopment," 208.

66. Gustin, "Big Plans"; 91.136.93, 91.136.95, 91.136.96, 91.136.97, 91.136.117, FACIF.

67. Alison Isenberg, *Downtown America: A History of the Place and the People Who Made It* (Chicago: University of Chicago Press, 2005), 255-317.

68. Joseph A. Anderson to Saul Seigel, February 25, 1972, 91.136.92, FACIF; FACI Press Memorandum, February 28, 1972, 91.136.92, FACIF.

69. *Centric 80: A Revitalization Strategy for Flint; Report to Flint Area Conference, Inc.* (Flint, MI: Lybrand, Ross Brothers, and Montgomery, 1972); 91.136.92, FACIF.

70. 91.136.95, FACIF; Dolan, "Mott Grants Pose Community Challenge," *FJ*, December 24, 1975; *FJ*, January 24, 1974, December 24, 1975.

71. *Site Investigation: Riverfront Site Proposal, University of Michigan–Flint* (Flint, MI: Sasaki, Dawson, DeMay Associates, 1972); *Assembling a Riverfront Site for the University of Michigan–Flint* (Flint, MI: Mayor's Implementation Committee, 1972); William L. Pereira Associates, *University of Michigan–Flint Impact Study* (Flint, MI: William L. Pereira Associates, 1974); Michael J. Riha, "Downtown M: The Future Is Now," *FJ*, July 23, 1974; William L. Pereira Associates, *Flint: The Center City; Plan and Program Due to Development of the University Impact Area* (Flint, MI: Pereira Associates, 1975); Lawrence R. Gustin, "Flint's 'Great Leap Forward' under Way," *FJ*, June 14, 1976; "Riverfront Center, Flint Michigan—Concept," June 12, 1978, box 1, FDF; "Riverfront Center, City of Flint, Michigan," October 27, 1978, box 1, FDF; "Project Description: Urban Development Action Grant, Festival Marketplace," October 31, 1983, box 22, folder 4, OBP;

Flint: New Life (Flint, MI: FACI, n.d., ca. 1983); Jack A. Litzenberg, "Urban Development Action Grant: Application for Assistance," n.d., box 1, folder 18, GFHP; "Relocation of the University of Michigan–Flint Campus: An Unparalleled Opportunity for the Entire Flint Community," n.d., box 46, folder 3, OBP; Robert W. Heywood, *A Work in Progress: Portrait of the University of Michigan–Flint* (Flint: Regents of the University of Michigan, 1996).

72. Katherine Ford Beebe, "An Evaluation of Three Urban Riverfront Parks: Lessons for Designers" (PhD diss., University of Michigan, 1984).

73. *FJ*, May 14, June 14, 1976, March 27, 1977, January 22, 1978, August 18, 1978, February 3, 1979; box 36, folder 29, OBP; Barry Wolf, "In the Shadow of the Hyatt: Flint's Worst Neighborhood," *FV*, November 13–24, 1981; Gilman, "Urban Redevelopment," 204.

74. 91.136.114, 91.136.92, FACIF; Olive Beasley to Carl Bekofske, November 28, 1977, box 21, folder 31, OBP; George F. Lord and Albert Price, "Growth Ideology in a Period of Decline: Deindustrialization and Restructuring, Flint Style," *Social Problems* 39, no. 2 (May 1992): 155–69.

75. Box 12, folder 4, EBHP.

76. Lee Bergquist, "There's a Whole Lot of Developin' Goin' on in Flint," *FJ*, June 19, 1980.

77. Isenberg, *Downtown America*, 255–317.

78. *FJ*, June 16–17, 23, August 29, October 3–4, 1980, March 20, April 10, July 8, December 20, 1981, June 18, July 9–10, August 1, 12, October 1, 1982, July 9, October 2, 11, 15, 1983.

79. Isenberg, *Downtown America*, 255–61.

80. Flint Convention and Visitors Bureau, "AutoWorld," in *A Visitor's Guide to Flint* (Flint, MI: Flint Convention and Visitor's Bureau, n.d.), F-GC/FPL; "AutoWorld: The Amusement Park That Couldn't," *FV*, April 30–May 13, 1982; Daniel Zwerdling, "And Then There's the Disneyland Solution," *FV*, July 2–15, 1982; Moore, "Dance Band on the *Titanic*"; *FJ*, February 3, 1985; "Flint, MI (Circa Now and Whenever)," unpublished essay, n.d.; 91.136.1-1, AutoWorld Files, PA. See also AWCF. For the final quotation in this paragraph, see www.toysaregoodfood .com. On nostalgia and urban development, see Isenberg, *Downtown America*, 255–317.

81. *DFP*, September 19, 1984.

82. Joseph A. Anderson, "Tourism in Flint," n.d., JAANP.

83. The former employee's quotation is available at www.toysaregoodfood.com. See also *FJ*, January 21, 1985.

84. *Bay City Times*, December 21, 1984; "AutoWorld Closing Tied to Dispute over Losses," *FJ*, January 20, 1985.

85. See AWCF; www.toysaregoodfood.com; Gary Flinn, "A Tale of Two Failed Tourist Attractions," http://home.comcast.net/~steelbeard1/flinn071907.htm. For more on AutoWorld, see Stephen C. Wisniewski, "'Dirty Factory Town or 'A Good City'? Neoliberalism and the Cultural Politics of Rust Belt Urban Revitalization" (PhD diss., University of Michigan, 2013).

86. Lord and Price, "Growth Ideology in a Period of Decline," 155–69; Dandaneau, *Town Abandoned*, 159–72; Gilman, *No Miracles Here*, 77–183.

87. Oscar Newman, *Defensible Space: Crime Prevention through Urban Design* (New York: Macmillan, 1972); Mike Davis, *City of Quartz: Excavating the Future in Los Angeles* (London: Verso, 1990); Michael Sorkin, ed., *Variations on a Theme Park: The New American City and the End of Public Space* (New York: Hill and Wang, 1992); Neil Smith, *The New Urban Frontier: Gentrification and the Revanchist City* (New York: Routledge, 1996).

88. Hammer, *Evaluation of Development Potentials*, 160.

89. Moore, "Flint and Me"; Moore, "General Motors Pulls Out."

90. See, for example, Ben Hamper, "I, Shop Rat," *FV*, March 5–18, 1982; Ben Hamper, *Rivethead: Tales from the Assembly Line* (New York: Warner Books, 1991); Kotlowitz, *There Are No*

Children Here: The Story of Two Boys Growing Up in the Other America (New York: Doubleday, 1991); Alex Kotlowitz, *The Other Side of the River: A Story of Two Towns, a Death, and America's Dilemma* (New York: Nan A. Talese, 1998).

91. Box 1, folder 26, MWF. See also Michael Moore, dir., *Roger and Me*, DVD (orig. 1989; Burbank, CA: Warner Home Video, 2003).

92. Box 1, folder 26, MWF.

93. Box 1, folder 26, MWP.

94. Hugh F. Semple Sr., "Film Maker Seeks Investigation of Mayor's Office," *FE*, September 21, 1992.

95. *General Motors Plant Closings: Hearing before the Subcommittee on Labor of the Committee on Labor and Human Resources* (Washington, DC: USGPO, 1987); Ralph Nader, "Robots, Robots Everywhere," *Tri-City News*, November 18, 1981; Mike Westfall, "Man vs. Machine: Automation Comes to the Work Place," *Alicia Patterson Foundation Reporter*, Spring 1984, box 1, folder 3, MWF.

96. *FE*, March 31, 1992; Donald Mosher, ed., *We Make Our Own History: The History of UAW Local 659* (Flint, MI: UAW Local 659, 1993), 106–7.

97. Billy Durant Automotive Commission, *A Living Agreement: A Presentation to the General Motors Corporation by the Greater Flint Community* (Flint, MI: Billy Durant Automotive Commission, 1997), F-GC/FPL.

98. Lisa Eddy, "General Manager of Buick Offers Encouraging News, Cites Longevity," *Flint Editorial*, August 15, 1991.

99. Alan K. Binder and Deebe Ferris, *General Motors in the Twentieth Century* (Southfield, MI: Ward's Communications, 2000), 191.

100. Greg Gardner, "Buick City's Demise," *WAW*, June 1997, 23–25; Dale Buss, "GM's Company Town," *Automotive News*, June 28, 1999; Gilman, *No Miracles Here*, 137–83.

101. Bill Vlasic and Brett Canton, "The Fall of Flint," *DFP*, December 11, 2005.

102. Flint Chamber of Commerce, *Flint Community Salute to Buick* (Flint, MI: Flint Chamber of Commerce, 1953).

103. Warren Cohen, "The End of an Era for Autos," *USNWR*, July 12, 1999.

Epilogue

1. Dale Buss, "GM's Company Town," *Automotive News*, June 28, 1999.

2. Old 97's, "Buick City Complex," *Satellite Rides* (Elektra Records, 2001), http://itunes.com.

3. "Mad Mac," *Uncommon Sense*, June 2007.

4. Nathan Geisler et al., "Adversity to Advantage: New Vacant Land Uses in Flint," August 2009, http://www.mlive.com/fljournal/pdfs/Adversity_to_Advantage.pdf.

5. Michael Moore, "Flint and Me," *Money*, July 7, 1996, 86–87.

6. On the persistence of urban renewers, see Andrew R. Highsmith, "Decline and Renewal in North American Cities," *Journal of Urban History* 37, no. 4 (July 2011): 619–26.

7. Thomas R. Hammer, *The Once and Future Economy of Metropolitan Flint, Michigan* (Flint, MI: CSMF, 1997), 15.

8. James C. Cobb, *The Selling of the South: The Southern Crusade for Industrial Development, 1936–1980* (Baton Rouge: Louisiana State University Press, 1982); Bruce J. Schulman, *From Cotton Belt to Sunbelt: Federal Policy, Economic Development, and the Transformation of the South* (New York: Oxford University Press, 1991).

9. "Wage Advantage: Low-Cost Work Force Now a Selling Point for County," *FJ*, December 7, 2004.

10. Kevin P. Balfe, *Analysis of the Metropolitan Flint Labor Market* (Evanston, IL: Institute for Urban Economic Development Studies, Northwestern University, 1988), 39–59; "Genesee County Joins Automation Alley as Foundation Member," *PR Newswire*, October 21, 2003, 1.

11. Flint-Genesee Economic Growth Alliance, "It All Starts Here," n.d., ca. 2000, http:// ref.michigan.org/cm/attach/48A9CD56-A9EF-40F7-87AA-4E2ABC76C6B7/GrowthAllian ceIncentivePackage.pdf; City of Flint Department of Community and Economic Development, "Renaissance Zone Information," n.d., http://www.cityofflint.com/economic/zoning.html; City of Flint, "Doing Business with the City," n.d., http://www.cityofflint.com/About_Flint/ business_ctr.asp.

12. Todd Seibt, "GM Money Makes Political Honey," *FJ*, December 17, 2004; Melissa Burden, "Volt to Power GM's, Flint's Future," *FJ*, September 16, 2008; "GM News—United States— Company Information," n.d., http://media.gm.com/media/us/en/gm/company_info.html.

13. *FJ*, December 1, 8, 16, 2004.

14. *FJ*, October 21, 2004. On the economic impact of tax incentives, see Louise Story, "As Companies Seek Tax Deals, Governments Pay High Price," *NYT*, December 1, 2012.

15. *FJ*, November 20, 2004.

16. Rick Haglund and Todd Seibt, "GM Ax Falling on Local Plant," *FJ*, November 21, 2005.

17. James M. Miller, "Flint's Spark: Spark Plug Production a Part of City for Generations," *FJ*, April 9, 2006; Ron Fonger, "Wrecking Ball to Fell another Flint Giant," *FJ*, April 1, 2008. The string of factory shutdowns extended all the way through 2013, when officials from GM announced the permanent closure of two additional plants: the Flint East facility and the Grand Blanc Weld Tool Center. See Sarah Schuch, "Fast Facts: Economic Impact, Status of GM Plants in Genesee County," *FJ*, February 5, 2013.

18. Bob Wheaton, "On the Grow," *FJ*, January 30, 2005; Bernie Hillman, "Housing Boom Putting School Space at a Premium," *Fenton Press*, February 27, 2005; Rhonda Sanders, "Burton's Boom," *FJ*, May 1, 2005; Bob Wheaton, "Boom Town," *FJ*, September 25, 2005; *GBN*, January 27, 2008; Tom Henderson, "Tech. Industry Remains a Bright Spot in the Region, Report Says," *Crain's Detroit Business*, November 10, 2008.

19. Don L. Boroughs, Robert F. Black, and Sara Collins, "Counties in Crisis," *USNWR*, November 25, 1991, 52; Myron Orfield and Thomas Luce, *Genesee County Metropatterns: A Regional Agenda for Community and Stability in Genesee County, Michigan* (Minneapolis: Metropolitan Area Research Council, 2003); *FJ*, April 2, December 12, 2006, April 1, 10, 15, August 12, October 7, 2007, January 22–23, 27, May 16, 2008.

20. *FJ*, July 11, 2008.

21. *NYT*, November 1–31, 2008; *FJ*, November 1–31, 2008.

22. "GM Collapses into Government's Arms," *WSJ*, June 2, 2009.

23. See *NYT*, July 1–15, 2009.

24. Bill Vlasic and Nick Bunkley, "When Auto Plants Close, Only White Elephants Remain," *NYT*, July 30, 2009; Tom Krisher, "GM Layoffs: Thousands of Factory Jobs Likely to Be Cut," *Huffington Post*, August 3, 2009, http://www.huffingtonpost.com/2009/08/03/gm-layoffs -thousands-of-f_n_250414.html; Nick Bunkley, "After 6,000 Take Buyouts, GM to Lay Off Thousands," *NYT*, August 3, 2009.

25. "GM News—United States—Company Information."

26. Christopher Swope, "The Man Who Owns Flint," *Governing*, January 2008, 52–57;

Adam Geller, "Abandoning Flint, Michigan: As Homeowners Move Out, Fires Move In," *Insurance Journal*, June 18, 2009, http://www.insurancejournal.com/news/midwest/2009/06/18/101510.htm; Laura Angus, "Rash of Fires in Flint Thought to Be for 'Perverted Political Purpose,'" *FJ*, March 25, 2010; Gordon Young, "The Incredible Shrinking American City," *Slate*, July 16, 2010, http://www.slate.com/id/2260473/; Teresa Gillotti and Daniel Kildee, "Land Banks as Revitalization Tools: The Example of Genesee County and the City of Flint, Michigan," n.d., http://www.geneseeinstitute.org/downloads/Revitalization_Tools.pdf. By the beginning of 2010, an additional 6,600 area property owners were heading into tax foreclosure. See Ron Fonger, "6,600 Properties in Genesee County Have Overdue Taxes, Headed to Foreclosure," *FJ*, January 14, 2010.

27. Khalil AlHajal, "Foreclosure Fight Drags on in Genesee County," *FJ*, November 12, 2010.

28. On previous dislocations due to urban renewal, see Andrew R. Highsmith, "Demolition Means Progress: Urban Renewal, Local Politics, and State-Sanctioned Ghetto Formation in Flint, Michigan," *Journal of Urban History* 35, no. 3 (March 2009): 348–68; Andrew R. Highsmith, "Demolition Means Progress: Race, Class, and the Deconstruction of the American Dream in Flint, Michigan" (PhD diss., University of Michigan, 2009), 403–81.

29. On prices, see http://abclocal.go.com/wjrt/story?section=news/local&id=7823748.

30. See USBC, "Flint (City) Quick Facts from the U.S. Census Bureau," http://quickfacts.census.gov/qfd/states/26/2629000.html.

31. Ibid.

32. USBC, "2010 Census Interactive Population Map," http://2010.census.gov/2010census/popmap/.

33. USBC, "Owosso (City) Quick Facts from the U.S. Census Bureau," http://quickfacts.census.gov/qfd/states/26/2661940.html.

34. USBC, "American FactFinder—Results for Goodrich (Village), Michigan," http://factfinder2.census.gov/faces/tableservices/jsf/pages/productview.xhtml?pid=DEC_10_DP_DPDP1; USBC, "American FactFinder—Results for Montrose (City), Michigan," http://factfinder2.census.gov/faces/tableservices/jsf/pages/productview.xhtml?pid=DEC_10_DP_DPDP1; USBC, "American FactFinder—Results for Beecher (CDP), Michigan," http://factfinder2.census.gov/faces/nav/jsf/pages/searchresults.xhtml; USBC, "American FactFinder—Results for Swartz Creek (City), Michigan," http://factfinder2.census.gov/faces/nav/jsf/pages/searchresults.xhtml.

35. Myron Orfield, *American Metropolitics: The New Suburban Reality* (Washington, DC: Brookings Institution Press, 2002), 28; Orfield and Luce, *Genesee County Metropatterns*, 3; Eyal Press, "The New Suburban Poverty," *Nation*, April 23, 2007; Elizabeth Kneebone and Alan Berube, *Confronting Suburban Poverty in America* (Washington, DC: Brookings Institution Press, 2013).

36. Orfield and Luce, *Genesee County Metropatterns*, 1–7; Blake Thorne, "Flint among First to Get Free School Lunches for All Courtesy of U.S. Government," *FJ*, September 4, 2011.

37. Orfield, *American Metropolitics*, 49–64; USBC, *Profile of Selected Economic Characteristics for Flint City, Michigan*; USBC, *Race and Hispanic or Latino Summary File for County Subdivisions in Genesee County, Michigan* (Washington, DC: USGPO, 2003); Keith Schneider, "Address Segregation in Land Use Plans," *DFP*, April 14, 2003; Keith Schneider, "Michigan Apartheid: Reforming Land Use Policy Can Help Most Segregated State," April 17, 2003, http://www.mlui.org/growthmanagement/fullarticle.asp?fileid=16480; Bettie Landauer-Menchik, "How Segregated Are Michigan's Schools? Changes in Enrollment from 1992–93 to 2004–05," in *Policy Report 27* (East Lansing: Education Policy Center at Michigan State University, 2006); Civil Rights Proj-

ect / Proyecto de Derechos Civiles, "Michigan Fact Sheet," in *A Divisive Choice: The Segregation of Charter School Students* (Los Angeles: Civil Rights Project / Proyecto de Derechos Civiles, 2010); Dominic Adams, "Sixty Years after Landmark Court Ruling, Four Flint Schools 'Intensely Segregated,'" *FJ*, May 16, 2014, http://www.mlive.com /news/flint /index.ssf/2014/05/60_years _later_four_flint_scho.html.

38. Erica Frankenberg, Chungmei Lee, and Gary Orfield, *A Multiracial Society with Segregated Schools: Are We Losing the Dream?* (Los Angeles: Civil Rights Project / Proyecto de Derechos Civiles, 2003); Jonathan Kozol, *The Shame of the Nation: The Restoration of Apartheid Schooling in America* (New York: Three Rivers, 2005); Gary Orfield and Erica Frankenberg with Jongyeon Ee and John Kuscera, "Brown at 60: Great Progress, a Long Retreat, and an Uncertain Future," May 15, 2014, http://civilrightsproject.ucla.edu /research / k-12-education /integration -and-diversity/brown-at-60-great-progress-a-long-retreat-and-an-uncertain-future/ Brown-at -50– 051614-repost.pdf.

39. On Uptown Developments and the new building and demolition projects downtown, see Matt Bach, "Redevelopment Brings New Energy for Downtown Flint," *Michigan Municipal Review*, March–April 2010, http://www.mml.org/resources/publications/mmr/issue/march -april2010/flint.html; Daniel Duggan, "Detroit, Flint May Be Diamonds in the Rough, Real Estate Investors Say," *Crain's Detroit Business*, November 10, 2010; Bowdeya Tweh, "Leaders in Former Auto Powerhouse Work through Recovery Process," *Times of Northwest Indiana*, July 10, 2011; Blake Thorne, "Genesee Towers Implosion, by the Numbers," *FJ*, December 18, 2013, http://www.mlive.com /news/flint /index.ssf/2013/12/genesee_towers_implosion_by_th .html. For more on Uptown Developments, see http://www.uptowndevelopments.com /#.

40. "Flint and Genesee County," http://www.michigan.org /hot-spots/flint-genesee-county/; "Flint, Michigan, Attractions," http://www.flint.travel /members/attractions/Attractions Tour; "Event Highlights," http://www.flint.travel /pages/ EventHighlights.

41. Bach, "Redevelopment Brings New Energy for Downtown Flint."

42. "Carriage Town Group Discusses Berridge Hotel Renovations," *East Village Magazine*, January 31, 2007; Matt Bach, "Berridge Hotel Renovations Complete; Hotel Set to Open as Loft Apartment Complex," *FJ*, December 17, 2008; Julie Morrison, "Berridge Hotel Reopens as Renovated Berridge Place Loft Apartments in Flint," *FJ*, December 18, 2008. On gentrification in American cities, see Neil Smith, *The New Urban Frontier: Gentrification and the Revanchist City* (New York: Routledge, 1996); Alison Isenberg, *Downtown America: A History of the Place and the People Who Made It* (Chicago: University of Chicago Press, 2005), 255–311; Suleiman Osman, *The Invention of Brownstone Brooklyn: Gentrification and the Search for Authenticity in Postwar New York* (New York: Oxford University Press, 2011).

43. William Crockerham, "The 'Vehicle City' Has Become 'University Town,'" February 22, 2009, http://www.examiner.com /article/the-vehicle-city-has-become-university-town; Duggan, "Detroit, Flint May Be Diamonds in the Rough"; Kristin Longley, "Kiplinger: Flint One of 11 'Comeback Cities' for 2011," *FJ*, April 11, 2011; Chrissie Thompson and Aaron Kessler, "GM to Add 4,200 in 8-State Hiring Blitz," *DFP*, May 9, 2011; Cathy Shafran, "GM Invests in $7.5 Million Office Space at Flint Engine Operation Facility," *FJ*, March 22, 2012; Cathy Shafran, "Latest Flint Jobs Numbers Tell Story of Growth—Stronger than Nationally, but with a Long Way to Recovery," *FJ*, May 9, 2012; Tweh, "Leaders in Former Auto Powerhouse"; Jeremy Allen, "Flint's Diplomat Pharmacy Named One of America's Fastest Growing Businesses by *Inc.* Magazine," *FJ*, August 26, 2012.

44. Joe Lawlor, "Despite Proposed Tax Increase, Flint's Budget Deficit Still Cast a Shadow over Finance Talks," *FJ*, July 22, 2008; Matthew Dolan, "Michigan Sizes up Taking over Flint,"

WSJ, November 10, 2011; Beata Mostafavi, "What Happened Last Time? A Look Back at Flint's 2002 State Takeover," *FJ*, November 10, 2011; Shafran, "Latest Flint Jobs Numbers."

45. Kristin Longley, "Flint Mayor Dayne Walling Warns of 'Tough Cuts' Ahead, Remains Optimistic in State of the City Speech," *FJ*, February 25, 2010; Carter Dougherty, "Budget Cuts Take Their Toll on Essential City Services," *Fiscal Times*, July 19, 2010, http://www.thefiscaltimes .com/Articles/2010/07/19/Budget-Cuts-Take-Their-Toll-on-Essential-City-Services.aspx #page1; Kristin Longley, "Flint Mayor Dayne Walling Talks Public Safety, Budget Cuts on CNN," *FJ*, August 5, 2011.

46. Geller, "Abandoning Flint, Michigan"; Angus, "Rash of Fires in Flint"; David Harris, "Flint Holds on to Dubious Title of Arson Capital of the Nation despite Downturn in Fires," *FJ*, June 13, 2012; Dominic Adams, "Suspicious Fires Burn Bright in Flint," *FJ*, March 25, 2013.

47. Charlie LeDuff, "Riding Along with the Cops in Murdertown, U.S.A.," *NYT*, April 15, 2011; David Harris, "Flint Ranks as Nation's Most Violent, FBI Statistics Show," *FJ*, Mary 23, 2011.

48. Paul Abowd, "Michigan's Hostile Takeover," *Mother Jones*, February 15, 2012; Kristin Longley, "Flint Emergency: Timeline of State Takeover," *FJ*, December 1, 2012.

49. Ryan Holeywell, "Emergency Financial Managers: Michigan's Unwelcome Savior," *Governing*, May 2012; Mark Niquette and Chris Christoff, "Half of Michigan Blacks Lose Local Power in Detroit Takeover," *Bloomberg News*, March 15, 2013, http://www.bloomberg.com/news/2013 -03-15/half-of-michigan-s-blacks-lose-local-control-in-detroit-takeover.html. On the specific cities and school districts under state control, see http://www.michigan.gov/treasury/0,1607,7 -121-1751_51556–201116-,00.html.

50. Bill Vlasic and Brett Canton, "The Fall of Flint," *DFP*, December 11, 2005.

51. For journalistic perspectives on urban decay and life in Flint after deindustrialization, see Gordon Young, *Teardown: Memoir of a Vanishing City* (Berkeley: University of California Press, 2013); Edward McClelland, *Nothin' but Blue Skies: The Heyday, Hard Times, and Hopes of America's Industrial Heartland* (New York: Bloomsbury, 2013), chaps. 5, 15. On the fascination with "disaster porn" and "ruin porn," see David Sirota, "Our Addiction to Disaster Porn," *In These Times*, January 30, 2010, http://www.inthesetimes.com/article/5487/; Noreen Malone, "The Case against Economic Disaster Porn," *New Republic*, January 22, 2011, http://www.tnr.com/ article/metro-policy/81954/Detroit-economic-disaster-porn; Joann Greco, "The Psychology of Ruin Porn," *Atlantic Cities*, http://www.theatlanticcities.com/design/2012/01/psychology-ruin -porn/886/.

52. For examples of works that emphasize urban decline and abandonment, see Barry Bluestone and Bennett Harrison, *The Deindustrialization of America: Plant Closings, Community Abandonment, and the Dismantling of Basic Industry* (New York: Basic Books, 1982); Kenneth T. Jackson, *Crabgrass Frontier: The Suburbanization of the United States* (New York: Oxford University Press, 1985), esp. 246–82; Robert A. Beauregard, *Voices of Decline: The Postwar Fate of US Cities* (Oxford: Blackwell, 1993); Steven P. Dandaneau, *A Town Abandoned: Flint, Michigan, Confronts Deindustrialization* (Albany: SUNY Press, 1996); Thomas J. Sugrue, *The Origins of the Urban Crisis: Race and Inequality in Postwar Detroit* (Princeton, NJ: Princeton University Press, 1996); Jefferson Cowie, *Capital Moves: RCA's Seventy-Year Quest for Cheap Labor* (New York: New Press, 1999); Kevin M. Kruse, *White Flight: Atlanta and the Making of Modern Conservatism* (Princeton, NJ: Princeton University Press, 2005); Colin Gordon, *Mapping Decline: St. Louis and the Fate of the American City* (Philadelphia: University of Pennsylvania Press, 2008).

53. Ron Fonger, "General Motors Invests $600 Million in Flint Plant as Executive Tells Workers, 'You Earned This,'" *FJ*, December 16, 2013, http://www.mlive.com/news/flint/index .ssf/2013/12/general_motors_commits_600_mil.html.

54. See, for example, David E. Cole, *The Automotive Industry, General Motors, and Genesee County* (Ann Arbor: Transportation Research Institute, University of Michigan, 1987); Hammer, *The Once and Future Economy of Metropolitan Flint, Michigan*, 37; Geisler et al., "Adversity to Advantage"; Dana Hedgpeth and Neil Irwin, "Out like Old Flint," *WP*, March 24, 2006.

55. On shrinking cities, see Philipp Oswalt, ed., *Shrinking Cities*, vol. 1, *International Research* (Ostfildern, Germany: Hatje Cantz, 2005), and *Shrinking Cities*, vol. 2, *Interventions* (Ostfildern, Germany: Hatje Cantz, 2006); Brent D. Ryan, *Design after Decline: How America Rebuilds Shrinking Cities* (Philadelphia: University of Pennsylvania Press, 2012). On Kildee and the Genesee County Land Bank, see Swope, "The Man Who Owns Flint"; David Streitfeld, "An Effort to Save Flint, Mich., by Shrinking It," *NYT*, April 22, 2009; Geller, "Abandoning Flint, Michigan"; Angus, "Rash of Fires in Flint"; Gillotti and Kildee, "Land Banks as Revitalization Tools"; Young, *Teardown*.

56. On the competing ideas about the Genesee County Land Bank and shrinking cities, see Andrew Heller, "Is the Genesee County Land Bank a Savior or Slumlord?" *FJ*, March 3, 2009; Young, *Teardown*.

57. On gays and lesbians and urban revitalization, see Bryant Simon, *Boardwalk of Dreams: Atlantic City and the Fate of Urban America* (New York: Oxford University Press, 2004), 160–70.

58. Gordon Young, "Faded Glory: Polishing Flint's Jewels," *NYT*, August 20, 2009. On gentrification, see Smith, *The New Urban Frontier*; Lance Freeman, *There Goes the 'Hood: Views of Gentrification from the Ground Up* (Philadelphia: Temple University Press, 2006); Loretta Lees, Tom Slater, and Elvin Wyly, *Gentrification* (New York: Routledge, 2008); Japonica Brown-Saracino, ed., *The Gentrification Debates: A Reader* (New York: Routledge, 2010); Osman, *Invention of Brownstone Brooklyn*.

59. Dan Barry, "Amid Ruin of Flint, Seeing Hope in a Garden," *NYT*, October 18, 2009; Sara Schuch, "The Growth of the Urban Farming Movement," *FJ*, June 10, 2011. On urban farming, see Jennifer Cockrall-King, *Food and the City: Urban Agriculture and the New Food Revolution* (Amherst, NY: Prometheus Books, 2012).

60. See, for instance, Vlasic and Canton, "Fall of Flint"; Matt Taibbi, "Apocalypse, New Jersey: A Dispatch from America's Most Desperate Town," *Rolling Stone*, December 11, 2013, http://www.rollingstone.com/culture/news/apocalypse-new-jersey-a-dispatch-from-americas-most-desperate-town-20131211; Laura Dimon, "This Is America's Most Apocalyptic, Violent City—and You've Probably Never Heard of It," *Policymic*, December 26, 2013, http://www.policymic.com/articles/77225/this-is-one-of-america-s-most-violent-cities-and-it-deserves-more-attention.

61. Khalil AlHajal, "Roof Repairs Too Costly, So Giant Car Ad Stems Leaks at Flint Home," *FJ*, April 7, 2011.

62. For a more complete articulation of this point, see Highsmith, "Decline and Renewal in North American Cities," 619–26.

63. Carl Crow, *The City of Flint Grows Up: The Success Story of an American Community* (New York: Harper and Brothers, 1945).

Index

Page numbers in italics refer to illustrations and tables.

Made in the USA
Lexington, KY
03 May 2018